Performance of Induction Machines

Performance of Induction Machines

Editor

Ryszard Palka

MDPI • Basel • Beijing • Wuhan • Barcelona • Belgrade • Manchester • Tokyo • Cluj • Tianjin

Editor
Ryszard Palka
West Pomeranian University
of Technology
Poland

Editorial Office
MDPI
St. Alban-Anlage 66
4052 Basel, Switzerland

This is a reprint of articles from the Special Issue published online in the open access journal *Energies* (ISSN 1996-1073) (available at: https://www.mdpi.com/journal/energies/special_issues/Performance_Induction_Machines).

For citation purposes, cite each article independently as indicated on the article page online and as indicated below:

LastName, A.A.; LastName, B.B.; LastName, C.C. Article Title. *Journal Name* **Year**, *Volume Number*, Page Range.

ISBN 978-3-0365-4785-5 (Hbk)
ISBN 978-3-0365-4786-2 (PDF)

Contents

About the Editor

Ryszard Palka

Ryszard Palka, D.Sc. Ph.D. Eng.; Head of Department of Electrical Machines and Drives, West Pomeranian University of Technology, Szczecin, Poland. In 1987–2005, he was with the Institute of Electrical Machines, Traction and Drives, TU Braunschweig, Germany. Areas of research: electromagnetic field theory, numerical field calculations, optimization of electromagnetic fields, electrical machines, high-temperature superconductivity. He is the author of about 320 refereed journal articles, conference papers, technical reports, and co-author of 4 books. He is a member of the IEEE, International Compumag Society, Polish Society of Theoretical and Applied Electrical Engineering, International Maglev Board, and Committee on Electrical Engineering, Polish Academy of Sciences.

Preface to "Performance of Induction Machines"

Induction machines are one of the most important technical applications for both the industrial world and private use. Since their invention (along with the achievements of Galileo Ferraris, Nikola Tesla, and Michal Doliwo-Dobrowolski), they have been widely used in different electrical drives and as generators, thanks to their features such as reliability, durability, low price, high efficiency, and resistance to failure. The main objective of this book is to contribute to the development of induction machines in all areas of their applications.

The methods for designing and using induction machines and drives are, in many aspects, very similar to the methods used in other electric machines but have their own specificity. Many issues discussed here are based on the fundamental achievements of authors such as Nasar, Boldea, Yamamura, Tegopoulos, and Kriezis, who laid the foundations for the development of induction machines, which are still relevant today. The control algorithms presented here are based on the achievements of Blaschke (field vector-oriented control) and Depenbrock or Takahashi (direct torque control), who created standards for the control of induction machines.

Today's induction machines must meet very stringent requirements of reliability, availability, high efficiency, and performance. Thanks to the application of highly efficient numerical algorithms, it is possible to design induction machines and entire very complex drive (or generator) systems faster and at a lower cost. At the same time, progress in the field of materials science and technology enables the development of ever more complex machine topologies.

The Guest Editor hopes that the publication of this collection of scientific papers, dedicated to the topic of induction machines, will contribute to the dissemination of modern knowledge about these machines among specialists in this subject.

Ryszard Palka
Editor

energies

Editorial

The Performance of Induction Machines

Ryszard Palka

Faculty of Electrical Engineering, West Pomeranian University of Technology, Sikorskiego 37, 70-313 Szczecin, Poland; ryszard.palka@zut.edu.pl

Citation: Palka, R. The Performance of Induction Machines. *Energies* **2022**, *15*, 3291. https://doi.org/10.3390/en15093291

Received: 3 April 2022
Accepted: 27 April 2022
Published: 30 April 2022

1. Introduction

Induction machines are one of the most important technical applications for both the industrial world and private use. Since they were invented (the achievements of Galileo Ferraris, Nikola Tesla, and Michal Doliwo-Dobrowolski), they have been widely used thanks to such features as reliability, durability, low price, high efficiency, and resistance to failure. Induction machines are used in different electrical drives and as generators.

The main objective of this Special Issue on "The Performance of Induction Machines" is to contribute to the development of induction machines in all areas of their applications.

2. A Short Review of the Contributions in This Special Issue

Eleven scientific papers were collected in this Special Issue [1–11]. These papers concern many aspects of the theory, design, control, optimization, supervision, and use of induction machines (IMs).

In [1], the authors present a review of IMs over the last 75 years. This paper cites many important articles describing the recent development of IM technology (e.g., [12]) and gives many valuable tips on the proper use and operation of IMs. The paper also contains many tips on the selection of appropriate materials for IM structures, windings, and insulation.

In [2], the thermal conductivity of soft magnetic materials in electric traction machines was studied and presented. Within this study, eight different soft magnetic materials were analyzed. An analytical approach was introduced to calculate the thermal conductivity of these materials. Temperature-dependent measurements of the electric resistivity were performed to obtain sufficient data for the analytical algorithm. Finally, an experimental approach was performed, and the thermal diffusivity, density, and specific heat capacity were determined. The accuracy study of all measurements shows good agreement for all materials. This is of great importance in all types of electrical machines; the selection of the appropriate soft magnetic material is a significant influencing factor on the overall efficiency of all drives.

In [3], a procedure for the accurate modeling of ring induction motors (RIMs) is proposed. This modeling was carried out based on the measured data for the torque-slip characteristic and using the equivalent circuit of the RIM. The use of the Monte Carlo method allowed for significant improvement in the modeling results in terms of both the torque-slip characteristic and extended Kloss equation of RIMs.

Different methods of IMs' optimization were summarized in [4]. The main purpose of this paper was to develop methods to reduce the computational effort in the design and optimization of IMs using the finite element method or analytical methods. For this, indirect machine models, such as the Response Surface Model, Kriging Model, or Artificial Neural Networks, have been proposed. With the help of the above algorithms, it is possible to optimally select the geometrical sizes of machines with a given structure. By means of appropriate analysis, the Response Surface Model seeks to relate a response of input variables that influence the output of the system. Kriging is able to exploit the spatial correlation of data in order to predict the shape of the objective function based only on limited information. These surrogate models replace the machine model and estimate the output parameters of the machine based on the input parameters. The optimization environment using the model and parameter selection procedures was applied to the design of a traction IM.

In [5], a procedure for the methodical selection of the most suitable model for design optimization is presented. The model selection based on the electromagnetic field calculation is presented. For this purpose, models of different value ranges and levels of detail were considered. The model selection approach was explained in detail and applied based on a coupled electromagnetic–thermal simulation of an exemplary IM. The results show that this model selection can be used to methodically determine the most suitable model in terms of its value range, level of detail, and computational effort for a given multiphysical problem.

The work in [6] deals with fault detection in IMs. This is a fundamental element of the electric power industry, manufacturing enterprise, and services; hence, considerable efforts have been carried out on developing reliable, low-cost procedures for fault diagnosis in IMs since the early detection of any failure may prevent the machine from suffering catastrophic damage. In this paper, a straightforward procedure was introduced for identifying and classifying faults in IMs working as motors. The proposed approach is based on the analysis of the startup transient current signal through the current signal homogeneity and fourth central moment (kurtosis) analysis. It was applied for training a feed-forward, backpropagation artificial neural network used as a classifier. From experimentally obtained results, it was demonstrated that the brought-in scheme attained high certainty in recognizing and discriminating among different IM conditions, i.e., a machine in good physical condition, a machine with one broken rotor bar, a machine with two broken rotor bars, a machine with damage on the bearing outer race, and a machine with an unbalanced mechanical load.

The next contribution [7] presents a predictive rotor field-oriented angle compensation approach for IM drives. The algorithm proposed here makes it possible to improve the traditionally used algorithms for controlling IMs, which are very sensitive to changes in the resistance and impedance of the machine circuit. Therefore, the d-axis and q-axis currents in the rotation reference frame are predicted based on the model and compared with the feedback current to correct the rotor field-oriented angle. A very similar predictive algorithm was used to control linear induction motors (LIMs) in urban transit applications, where the motors are usually required to operate at peak thrust and the main parameters responsible for the precise peak tracking (the rotor resistance and the mutual inductance) vary in a very wide range [13].

The field vector-oriented control is one of the most advanced and widely accepted methods used for the rotary machine torque control. It was first conceptualized by Blaschke [14]. Direct Torque Control is yet another vector control technique. It was introduced by Depenbrock [15] and Takahashi [16]. The control algorithms presented here, as well as other algorithms used to control IMs, are based on the above fundamental achievements, which belong to the canon of literature on IM control.

Garbiec and Jagiela [8] have presented a validated computational algorithm that enables the inclusion of the nonsinusoidal or asymmetrical voltage supply in the multi-harmonic field-circuit model of IM. The development of the strongly coupled multi-harmonic field model concept effectively accounted for the nonlinearity and asymmetry of the voltage supply in the calculation of the operating characteristics of a high-speed IM with a solid rotor in a steady-state complex-valued finite element modeling framework. The multi-harmonic field-circuit model may become an effective tool in the process of designing IMs, in particular for reducing losses due to higher harmonics of the magnetic field of various origins.

The next paper [9] presents an adaptable simulation of an IM with a downstream protection scheme. For this purpose, a special algorithm was proposed to implement both static and dynamic modeling of a three-phase IM due to possible faults and high-performance requirements. This algorithm has been tested against several conventional methods. It was observed that during the stable condition of the machinery, it prevented the occurrence of many serious faults. To simulate and examine the behavior of a three-phase IM, the Matlab-Simulink software was used. Many simulations were carried out to obtain realistic characteristics of the analyzed IM, such as torque-speed, efficiency-torque, etc.

Park, S.-U. et al. [10] have investigated the efficiency improvement of the slip frequency in LIMs. In their study, mathematical analysis was conducted for each factor that mutually affects the control of the train. On this basis, the magnitude of the normal force related to the safety of the train is limited. Operating efficiency was improved by varying the slip frequency according to the operating conditions of the train. This algorithm takes advantage of the fact that the generated normal force is a factor that destabilizes the levitation system of the train and is a potential safety problem due to train levitation failure. It also induces additional energy consumption in both the propulsion system and the levitation system, thereby reducing efficiency. The verification of the proposed method was proved through a comparative experiment for the Maglev Train running at Incheon International Airport.

The following paper [11] presents the theory and classification of LIMs. Fundamental achievements on LIMs and rotating IMs were studied here, and specific LIM problems are discussed. Many methods of LIMs' calculation, optimization, and control are identical (or very similar) to the methods applicable to rotating IMs. However, because of the differences between the LIM and the rotary machine, some unconventional analysis techniques and modeling methods have been developed. The electromagnetic calculations of the rotary motor are reasonably simple because of the motor's "infinite" character and the possibility of applying many simplifications, thus limiting the solution region and speeding up the calculations even further. This paper provides an overview of linear transportation systems—levitated, non-levitated, with synchronous motors, with induction motors, and with superconducting induction motors—and focuses on the application of a LIM as a major constituent of such systems. Thus, solutions to the following problems are presented there: the development of new analytical solutions and finite element methods for LIM evaluation [17]; self-developed LIM adaptive control methods [13]; LIM performance under voltage supply (non-symmetrical phase current values); method for the power loss evaluation in the LIM reaction rail [13,18]; the temperature rise prediction method of a traction LIM; and the discussion of the performance of the superconducting LIM (superconducting propulsion and levitation [19]). The addressed research topics have been chosen for their practical impact on the advancement of a LIM as the preferred urban transport propulsion motor [20].

The problems presented in the last two papers are based on the foundational achievements of authors such as Nasar and Boldea [21], Yamamura [22], and Tegopoulos and Kriezis [23], who laid the foundations for the development of LIMs, which are still relevant today.

3. Conclusions

The Special Issue, "The Performance of Induction Machines", highlights the variety of problems faced by designers and users of induction machines.

Some of the presented approaches, e.g., design, control, optimization, and fault detection in induction machines, may also be adapted and applied to other related applications of electrical machines. Therefore, the Guest Editor hopes that the collected papers may be inspiring for the readers, leading to the further development of new methods of designing and using modern, high-efficient electrical machines.

Funding: This research received no external funding.

Acknowledgments: The Guest Editor is thankful for the invaluable contributions from the authors, reviewers.

Conflicts of Interest: The author declares no conflict of interest.

References

1. De Souza, D.F.; Salotti, F.A.M.; Sauer, I.L.; Tatizawa, H.; de Almeida, A.T.; Kanashiro, A.G. A Performance Evaluation of Three-Phase Induction Electric Motors between 1945 and 2020. *Energies* **2022**, *15*, 2002. [CrossRef]
2. Groschup, B.; Rosca, A.; Leuning, N.; Hameyer, K. Study of the Thermal Conductivity of Soft Magnetic Materials in Electric Traction Machines. *Energies* **2021**, *14*, 5310. [CrossRef]
3. Tomczyk, K.; Makowski, T.; Kowalczyk, M.; Ostrowska, K.; Beňko, P. Procedure for the Accurate Modelling of Ring Induction Motors. *Energies* **2021**, *14*, 5469. [CrossRef]
4. Nell, M.; Kubin, A.; Hameyer, K. Multi-Stage Optimization of Induction Machines Using Methods for Model and Parameter Selection. *Energies* **2021**, *14*, 5537. [CrossRef]
5. Nell, M.; Kubin, A.; Hameyer, K. Approach for the Model and Parameter Selection for the Calculation of Induction Machines. *Energies* **2021**, *14*, 5623. [CrossRef]
6. Martinez-Herrera, A.L.; Ferrucho-Alvarez, E.R.; Ledesma-Carrillo, L.M.; Mata-Chavez, R.I.; Lopez-Ramirez, M.; Cabal-Yepez, E. Multiple Fault Detection in Induction Motors through Homogeneity and Kurtosis Computation. *Energies* **2022**, *15*, 1541. [CrossRef]
7. Liu, Y.; Zhao, J.; Yin, Q. Model-Based Predictive Rotor Field-Oriented Angle Compensation for Induction Machine Drives. *Energies* **2021**, *14*, 2049. [CrossRef]
8. Garbiec, T.; Jagiela, M. Accounting for Slot Harmonics and Nonsinusoidal Unbalanced Voltage Supply in High-Speed Solid-Rotor Induction Motor Using Complex Multi-Harmonic Finite Element Analysis. *Energies* **2021**, *14*, 5404. [CrossRef]
9. Le Roux, P.F.; Ngwenyama, M.K. Static and Dynamic Simulation of an Induction Motor using Matlab/Simulink. *Energies* 2022; *in press*.
10. Park, S.-U.; Mok, H.-S.; Lim, J.-W.; Seo, H.-U.; Oh, S.-H. Efficiency Improvement by Deriving the Optimal Operating Slip Frequency of a Linear-Induction-Style Maglev Train. *Energies* **2020**, *13*, 6544. [CrossRef]
11. Palka, R.; Woronowicz, K. Linear Induction Motors in Transportation Systems. *Energies* **2021**, *14*, 2549. [CrossRef]
12. Barnett, R.D. Induction motors: Early development [History]. *IEEE Power Energy Mag.* **2022**, *20*, 90–98. [CrossRef]
13. Woronowicz, K.; Palka, R. Optimised Thrust Control of Linear Induction Motors by a Compensation Approach. *Int. J. Appl. Electromagn. Mech.* **2004**, *19*, 533–536. [CrossRef]
14. Blaschke, F. The Principle of Field Orientation as Applied to the New Transvector Closed-Loop Control System for Rotating Machines. *Siemens Rev.* **1972**, *34*, 217–220.
15. Depenbrock, M. Direct Self-Control of the Flux and Rotary Moment of a Rotary-Field Machine. U.S. Patent 4,678,248, 7 July 1987.
16. Takahashi, I.; Noguchi, T. A new quick response and high efficiency control strategy of an induction motor. *IEEE Trans. Ind. Appl.* **1986**, *IA-22*, 820–827. [CrossRef]
17. Woronowicz, K.; Abdelqader, M.; Palka, R.; Morelli, J. 2-D quasi-static Fourier series solution for a linear induction motor. *Int. J. Comput. Math. Electr. Electron. Eng.* **2018**, *37*, 1099–1109. [CrossRef]
18. Yan, W.; Chen, H.; Liu, X.; Ma, X.; Lv, Z.; Wang, X.; Palka, R.; Chen, L.; Wang, K. Design and multi-objective optimisation of switched reluctance machine with iron loss. *IET Electr. Power Appl.* **2019**, *13*, 435–444. [CrossRef]
19. Patel, A.; Hopkins, S.; Giunchi, G.; Figini Albisetti, A.; Shi, Y.; Palka, R.; Cardwell, D.; Glowacki, B. The Use of an MgB_2 Hollow Cylinder and Pulse Magnetized (RE)BCO Bulk for Magnetic Levitation Applications. *IEEE Trans. Appl. Supercond.* **2013**, *23*, 6800604. [CrossRef]
20. Kircher, R.; Klühspies, J.; Palka, R.; Fritz, E.; Eiler, K.; Witt, M. Electromagnetic Fields Related to High Speed Transportation Systems. *Transp. Syst. Technol.* **2018**, *4*, 152–166. [CrossRef]
21. Nasar, S.A.; Boldea, I. *Linear Motion Electric Machines*; John Wiley & Sons: New York, NY, USA, 1976.
22. Yamamura, S. *Theory of Linear Induction Motors*, 2nd ed.; Halster Press: New York, NY, USA, 1979.
23. Tegopoulos, J.A.; Kriezis, E.E. *Eddy Currents in Linear Conducting Media*; Elsevier Science: New York, NY, USA, 1985.

energies

MDPI

Article

A Performance Evaluation of Three-Phase Induction Electric Motors between 1945 and 2020

Danilo Ferreira de Souza [1,2,*], Francisco Antônio Marino Salotti [2], Ildo Luís Sauer [2], Hédio Tatizawa [2], Aníbal Traça de Almeida [3] and Arnaldo Gakiya Kanashiro [2]

1 Faculty of Architecture, Engineering and Technology—FAET, Campus Cuiabá, Federal University of Mato Grosso—UFMT, Cuiaba 78060-900, Brazil
2 Institute of Energy and Environment—IEE, University of São Paulo—USP, São Paulo 05508-010, Brazil; fsalotti@iee.usp.br (F.A.M.S.); illsauer@iee.usp.br (I.L.S.); hedio@iee.usp.br (H.T.); arnaldo@iee.usp.br (A.G.K.)
3 Department of Electrical and Computer Engineering, Institute of Systems and Robotics, University of Coimbra, 3030-290 Coimbra, Portugal; adealmeida@isr.uc.pt
* Correspondence: danilo.souza@ufmt.br; Tel.: +55-65-3615-8780

Abstract: In the late 19th century, the three-phase induction motor was the central element of productivity increase in the second industrial revolution in Europe and the United States. Currently, it is the main load on electrical systems in global terms, reaching approximately 70% of electrical energy consumption in the industrial sector worldwide. During the 20th century, electric motors underwent intense technological innovations that enabled significant performance gains. Thus, this work analyses the performance changes in squirrel-cage rotor three-phase induction electric motors (SCIMs) with mechanical powers of 3.7 kW, 37 kW, and 150 kW and speed ranges corresponding to two poles and eight poles, connected to a low voltage at a frequency of 60 Hz and tested between 1945 and 2020. The study confirms accumulated performance gains of above 10% in some cases. Insulating materials for electrical conductors have gone through several generations (cotton, silk, and currently, varnish). Improvements to the housing for cooling, the bearings, the quality of active materials, and the design were the elements that enabled the high gains in performance. The first commercial two-pole SCIM with a shaft power of 4.4 kW was marketed in 1891, with a weight/power ratio of 86 kg/kW, and until the 2000s, this value gradually decreased, eventually reaching 4.8 kg/kW. Between 2000 and 2020, this ratio showed a reversed trend based on improvements in the performance of SCIMs. More active materials were used, causing the weight/power ratio to reach 8.6 kg/kW. The MEPS (minimum energy performance standards) of SCIMs had an essential role in the performance gain over the last three decades. Data collection was via tests at the Electrical Machines Laboratory of the Institute of Energy and Environment of the University of São Paulo. The laboratory has a history of tests on electrical equipment dating from 1911.

Keywords: three-phase induction motor; squirrel-cage rotor; energy efficiency; motor performance

Citation: de Souza, D.F.; Salotti, F.A.M.; Sauer, I.L.; Tatizawa, H.; de Almeida, A.T.; Kanashiro, A.G. A Performance Evaluation of Three-Phase Induction Electric Motors between 1945 and 2020. *Energies* **2022**, *15*, 2002. https://doi.org/10.3390/en15062002

Academic Editor: Ryszard Palka

Received: 10 February 2022
Accepted: 7 March 2022
Published: 9 March 2022

1. Introduction

The production of mechanical force was one of the fundamental human demands in the transformation process that took homo sapiens from an animal in nature like the others to the construction of megacities and technological mastery [1].

The process of producing mechanical force went through several phases. The domestication of animals represented an essential step in automation and the increase in labour productivity, necessary for changing the way of life from hunter and gatherer to farmer/shepherd [2]. With the use of other domesticated animals, homo sapiens could perform an activity without the need to use muscular strength directly [3].

The production of mechanical force was primarily responsible for the first two great industrial revolutions. The first industrial revolution began in England around 1750–1760, lasting until somewhere between 1820 and 1840, and was marked by the development and application of the steam engine in industrial manufacturing processes [4]. The second industrial revolution replaced steam engines or gas engines with electric motors [5].

In the late 19th century, new electric motors were more economical. They required less maintenance, took up less space, ran at a more uniform speed, and allowed a cleaner environment [6]. Within just one generation after its introduction in the 1880s, the electric motor drive had replaced steam as the preferred means of providing motive power (Figure 1) [7].

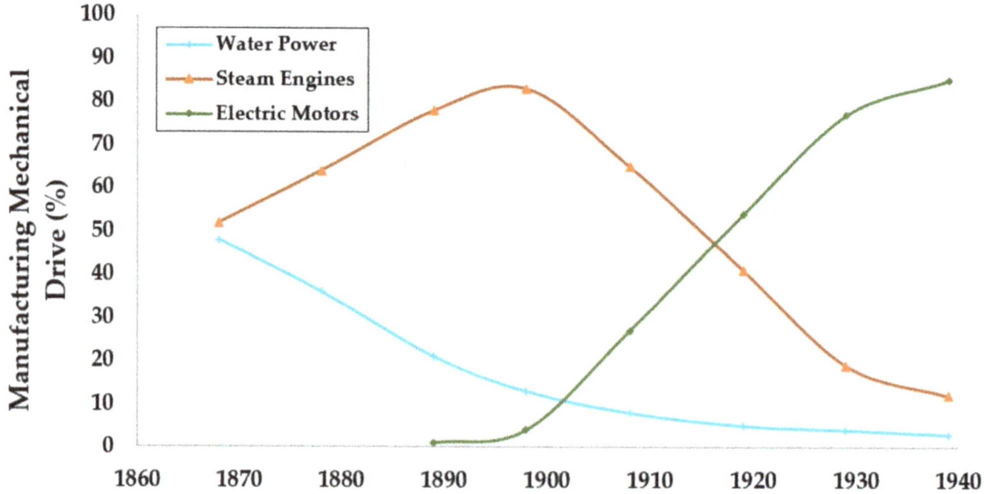

Figure 1. Percentage of mechanical drive manufacturing from hydraulic power, steam engines, and electric motors per year. Source: adapted from [7,8].

With large-scale electric motors electrifying industrial plants, the industrial plants gained flexibility. It was no longer necessary to be close to a stream in order to use mechanical energy from water or a coal mine for direct use of coal in a steam engine [9].

The mechanical force arising from water (Figure 2a) or steam engines (Figure 2b) was generally available from a single central axis and later, when subdivided, ran the entire length of the factory, with high losses in the gears and the emission of noise and vibrations throughout the industrial plant. In some cases, the engines served different industrial buildings. The connections made by belts and gears could drive hammers, presses, looms, and other machines, transferring mechanical energy horizontally between walls and vertically through industrial floors [10]. Due to the large distances and inevitable friction in these units, 60% to 80% of the transmitted energy was lost [11]. Everything required continuous lubrication by thousands of drip lubricators, with workers having direct access to the rotating parts, thus remaining exposed to high possibilities of work accidents [12].

Electric motors proved to be more efficient and more economical, and they reduced reliance on the complex mechanical shaft, pulley, and belt systems to distribute the mechanical drive from the central plant throughout the plant. The drive was located close to the load, and the energy was transferred by small electrical conductors [17] (Figure 2c).

Figure 2. An industrial organization based on: (**a**) mechanical drive from hydraulic energy; (**b**) steam engines; and (**c**) electric motors. Source: [13–16].

The motors could reliably be fractionally coupled to the mechanical load with the electric drive, making it possible to establish an industrial flow in the manufacturing process. By splitting the mechanical drives, flexibility in maintenance was also gained, and islands with independent operation were possible. During a breakdown, it was not necessary to stop the entire plant. This freedom revolutionized industrial design and layout and provided the possibility of optimization in process control and better working conditions, leading to significant advances in productivity [8,18,19].

Currently (21st century), electric motors are the driving force of modern industrial society. Electric motors drive domestic refrigerators, pump water for heating, and drive ventilation, enabling the distribution of compressed air and the movement of loads on conveyor belts, in addition to keeping cities' water supplies flowing [20].

Several electric motor technologies have been developed. However, only three have become mainstream in industrial stationary electric drives. They are:

1. Direct-current motors, which were the first to be developed. Their applications are limited to situations in which speed control is essential, because by controlling the voltages applied to the rotor windings (armature) and the stator windings (field), it is possible to fine-tune the speed over a wide range. However, they are expensive, have high sparking caused by switching currents, and cannot be connected directly to the electrical grid, requiring a converter. They also have a high need for maintenance compared to other technologies.
2. Alternating-current synchronous motors, which are mainly reserved for high-load drives that require a constant speed. They are also expensive and require additional starting devices and electronic converters to feed the rotor winding (field) separately from the stator winding (armature). Synchronous motors are high-efficiency motors, as they have low rotor losses.
3. Cage rotor induction electric motors, which since their development have been the predominant choice in residential (single-phase), commercial, and industrial (three-phase SCIMs) environments. SCIMs correspond to about 87% of the total alternating-current electric motors used in the industry [21]. Several factors make SCIMs suitable for the broadest range of applications. Some of these are highlighted in Table 1.

Table 1. Characteristics of SCIMs.

Advantages	References
Low acquisition and maintenance cost compared to competing technologies;	[22–24]
The simple constructive characteristics make its manufacture simple compared to competing technologies;	[25,26]
Simple replacement due to a high degree of standardization of housings and connections;	[27,28]
Long service life;	[29,30]
A high degree of speed control using the variable speed drive (VSD), also enabling the saving of electrical energy;	[31,32]
Small dimensions can be used in compact places;	[33]
Does not produce sparking, making it easier to apply in classified areas (Ex areas);	[34–36]
High starting torque compared to other competing technologies;	[25,26]
Quiet compared to competing technologies;	[37–39]
There is no electrical contact between the rotor and the stator; the connection is made only by the bearings, thus giving high operating safety;	[25,26]
They can be powered directly by alternating current without the need for electronic converters;	[25,26]
Easy detection of faults of various natures (electrical, mechanical, thermal, and environmental)	[40–43]
Known production chain and easy access to the mineral resources necessary for the construction of SCIMs. They do not depend on high-volatility materials in the supply chain, such as, for example, the rare earth magnets present in permanent magnet synchronous electric motors (PMSMs).	[44–46]

SCIMs are seen as having undergone little change from their development to the present day, especially when compared to the obvious advances in electronics, communication, and information technologies. Hence, this research sought answers to the following questions:

I. What are the most significant changes that SCIMs have undergone throughout their history?
II. Has the performance of SCIMs changed since their development?
III. Has the volume of SCIMs changed over time?

2. Materials and Methods

Section 3.1 is a review of the literature that shows the historical development of SCIMs, focusing on the main technological innovations, material improvements, and various projects. The sources of information are technical documents from SCIM manufacturers or scientific articles that described the processes and the main events that caused the changes in the mass/power ratio between 1890 and 1990, contributing to answering question I.

In Section 3.2, a literature review is presented discussing the variations in performance between 1935 and 2012, contributing to answering questions I and II.

In Section 3.3, the primary data collected in the Technical Test Reports of the Laboratory of Electrical Machines of the Institute of Energy and Environment (IEE) of the University of São Paulo (USP) are presented and discussed.

Between 1945 and 1996, the Technical Test Reports were only available in printed form. Thus, it was necessary to digitize the data and collect them into a spreadsheet. Between 1997 and 2020, the Technical Test Reports were already available in digital format for processing and analysis.

The Laboratory of Electrical Machines of the IEE-USP has a technical collection of approximately 21,000 technical reports. For this analysis, reports with the following characteristics were considered:

(a) Three-phase induction electric motors with squirrel-cage rotor—SCIMs;
(b) Technical reports of new SCIMs;

(c) SCIMs tested according to current regulations, with the availability of test data at full load;
(d) SCIMs in which the nameplate data were made available by the manufacturer;
(e) SCIMs powered at low voltage (up to 600 volts);
(f) SCIMs for power supply at the industrial frequencies of 60 Hz or 50 Hz, tested at 60 Hz;
(g) SCIMs produced for continuous operation.

Using the conditions expressed in a–g, 359 technical reports of tested SCIMs with speeds corresponding to 2, 4, 6, or 8 poles, with a motor rated output power of 3.7, 37, or 150 kW, were collected for the evaluation of the change in performance between 1945 and 2020. The assessment seeks to answer the questions (I and II) that motivated this research, based on the data collected.

The results are organized into three different output power (kW) categories. The chosen groups include low power (3.7 kW), medium power (37 kW), and high power (150 kW). As the groups chosen to represent SCIMs are of significantly different dimensions, the production processes used in the manufacturing process and the standards of precision/quality of the materials are also different, even when dealing with the same equipment.

The number of poles of the electric motor determines the rotation speed, due to the arrangement and distribution of the electrical conductors of the windings located in the stator slots. In the SCIM market, historically, four speeds have been the most used. Between 80% and 90% of all the SCIMs sold have between 2 and 8 poles; therefore, this research evaluates them in this speed range. In fact, 4-pole SCIMs are dominant, representing between 45 and 70% of SCIMs [21,47].

Using the conditions expressed in a–g, 28 SCIMs with speeds corresponding to 2 poles and motor rated output powers of 3.7, and 4.4 kW were used to evaluate the change in the mass/power ratio between 2000 and 2020, seeking to answer the questions (I and III) that motivated this research, based on the data collected.

The National Institute of Metrology, Quality and Technology (INMETRO) accredits the Laboratory of Electrical Machines at IEE-USP, following ABNT NBR ISO/IEC 17025:2017 [48] under No. CRL 0011. INMETRO periodically carries out audits in accredited laboratories, aiming to guarantee the quality of the measurement results. INMETRO is a signatory to the mutual recognition agreements of the International Laboratory Accreditation Cooperation (ILAC) and the Inter-American Accreditation Cooperation (IAAC), thus following a world standard of quality and reliability.

This research, therefore, used data from standardized performance tests. This is because there may be differences between values measured in neutral laboratories and values reported by manufacturers [49], and when using the measured data, errors and uncertainties are reduced.

3. Results and Discussion

3.1. The Improvements in SCIMs

All the technological and theoretical bases for electric motor development were already advanced by the end of the 19th century. Direct-current motors were on the market, and alternating-current motors were in the full developmental stage, with research ongoing in Europe and the United States. The first patent for the electric motor with asynchronous technology was filed by the engineer Nikola Tesla [50] in 1888 and accepted in 1889 [51] in New York. The asynchronous motor became known as an induction motor, based on its working principle. However, Tesla's proposal was similar to the current single-phase auxiliary winding motors, operating with a wound rotor. The text that explained the working principle of the new electric induction motor was published by Nikola Tesla in 1988 with the title "A new system of alternate current motors and transformers" [52].

Parallel to Nikola Tesla's experiments in the USA were those of Galileo Ferraris in Italy. In 1885, Ferraris developed the idea that two out-of-phase currents could be used to produce two magnetic fields that could be combined to produce a rotating field, without the need for switching or moving parts, opening the door to AC electric motors [53–55].

The three-phase squirrel-cage rotor induction motor (SCIM) closest to the type we have today was developed by a German company AEG (Allgemeine Elektricitäts-Gesellschaft), headed by the Russian engineer Mikhail Dolivo-Dobrovolsky between 1888 and 1890 [56]. The electric motor developed by the Dobrovolsky team had very favourable characteristics such as high starting torque, more straightforward construction features, robustness in construction, and low maintenance needs. However, it also had the inconvenience of needing to be powered by a three-phase alternating-current system, which was not yet commercial. Until then, the available electrical systems were single-phase and two-phase systems. This type of supply does not provide efficient starting of the Tesla-mounted motor (starting torque practically non-existent), in addition to imposing some degree of vibration during operation. The SCIM has a high starting torque and does not need auxiliary windings and accessories such as a capacitor and a centrifugal starter, in addition to having a lower operating current compared to a single-phase motor. However, three-phase electric power generation, transmission, and distribution systems were quickly implemented with the objective of feeding the attractive SCIMs [57–59].

Dobrovolsky and the AEG company gained fame for the great invention. The artist Irene Ahrens created the illustration in Figure 3, which was exhibited in Berlin. The engineer appears in the sky, entering the Hall of Fame with his SCIM shown near his feet.

Figure 3. Mikhail Dolivo-Dobrovolsky entering the Hall of Fame with his SCIM. Source: [57,60].

In 1891, at AEG, Dobrovolsky coordinated the first serial production of SCIMs with shaft powers between 0.4 and 7.5 kW. The first SCIMs assembled had a performance of approximately 80% for the power range produced and very high mass by today's standards. The first commercial two-pole SCIM with a shaft power of 4.4 kW was marketed in 1891. These SCIMs had a mass/power ratio of 86 kg/kW, as shown in Figure 4.

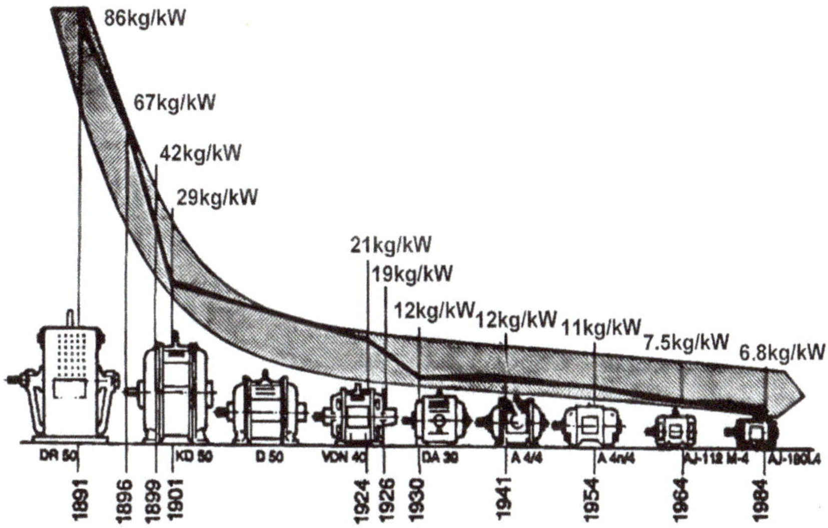

Figure 4. Improvements in SCIM mass/power ratio between 1891 and 1984. Source: [61–63].

The company AEG published the famous image represented in Figure 4, which shows the mass/power ratio from the first SCIMs manufactured by the company in 1891 to the SCIMs manufactured in 1984. The optimization of materials for electrical, magnetic, and mechanical purposes, combined with solid technological innovations, made it possible to reach a ratio of 6.8 kg/kW in 1984, representing only 8% of the total mass of the two-pole SCIMs with an axle power of 4.4 kW produced in 1891, as AEG's first commercial units.

The concept of the SCIM has not changed since the beginning of its commercialization; however, the volume has changed considerably (Figure 4).

The technological progress of SCIMs has been remarkable, stimulated by strong competition and by processes, technological innovations, and improvements in materials. According to Browning (1997) [64], the changes in the mass/power ratio resulted in better operational characteristics, even more excellent reliability, versatility, and longer life.

Browning (1997) [64] identified the following improvements in SCIMs:

- The change from open housing to closed housing;
- The change from plain bearings to anti-friction bearings. (In 1945, 35–40% of SCIMs used plain bearings);
- The change from cotton-insulated wires to varnished wires in the stator windings;
- Construction of the squirrel-cage rotor using copper or cast aluminium bars.

The adoption of industry standards has played a significant role in the progress of SCIMs [28,64]. An example is the thermal classification of insulating materials, which first appeared in 1898. In 1911, standardization by the AIEE Standards (now IEEE—Institute of Electrical and Electronics Engineers) established temperature limits for SCIMs. The 1915 edition of the AIEE Standards included definitions of insulation classes A, B, and C and the materials assigned to those classes. In 1929, the first SCIMs built to NEMA standards were made available on the market, setting standard dimensions and operating characteristics for specific ratings for the first time. Users were given the ability to directly replace SCIMs via the concept of stock electric motors for quick replacement in case of failure [64].

It is possible to observe in Figure 5 the tremendous technological innovations that were decisive in reducing the mass and volume of SCIMs.

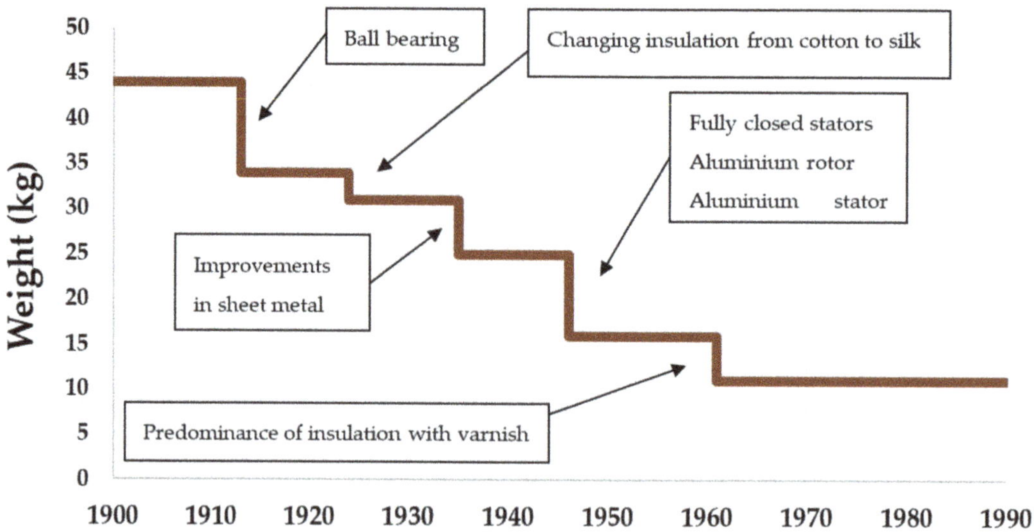

Figure 5. Chronology of 0.75 kW SCIM mass reduction between 1900 and 1990. Source: [65–68].

At the beginning of the 20th century, the first major technological innovation was the development of ball bearings, replacing the traditional plain bearings that were bulky, heavy, and required lubrication with oil. With the new bearings and the reduction in friction losses, the mass and volume of the SCIMs decreased considerably.

Between 1913 and 1940, there were gains in the quality of materials, improving compaction and making it possible to reduce the volume of copper and iron used in SCIMs and to reduce losses. In the 1940s, rotors previously built using iron sheets began to be developed using cast aluminium, adding more mass reduction, as shown in Figure 5. In addition, in the 1940s, with successive advances in metallurgy, SCIM housings could be built in an increasingly closed way and could maintain the cooling of the windings located in the stator.

In the early 1960s, a series of advances in insulation systems were instrumental in reducing the volume of SCIMs. Between 1960 and 1970, SCIMs went through five generations of materials used to construct insulation for electrical conductors. In the first SCIMs, the insulation was composed of paper, and later cotton. Then, insulation with varnish predominated until the present day. Figure 6 shows in white the area necessary to accommodate electrical conductors of the same metallic volume inside the stator magnetic package slot for different insulation technologies [63].

The first significant innovation in SCIM insulation systems was the replacement of the double layers of cotton between the conductors and the sheets with two layers of silk, allowing a reduction of approximately 59% of the groove area in the metal sheets (ferromagnetic material) of the stator. The second major innovation was the introduction of varnish used in conjunction with silk, giving an area reduction of over 2%, as shown in Figure 6. Subsequently, improvements in the quality of silk and varnish allowed an area equivalent to be reached of only 22% of the space required for the same electrical conductor using cotton as an insulator.

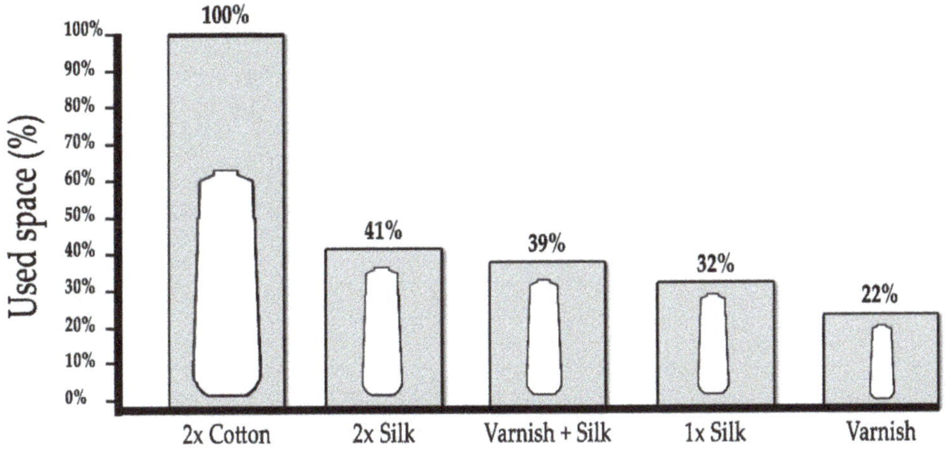

Figure 6. Space used by different insulation technologies for the same SCIM output power. Source: adapted from [63].

Successive technological innovations and improvements in electrical, magnetic, and mechanical materials achieved significant volume compaction in SCIMs between 1903 and 1974 [28], as illustrated in Figure 7a. Figure 7b shows the changes in appearance and frame dimensions of SCIMs of different powers from the open construction of 1904 to those used in the 1970s, similar to today's drip-proof and fully fan-cooled SCIMs.

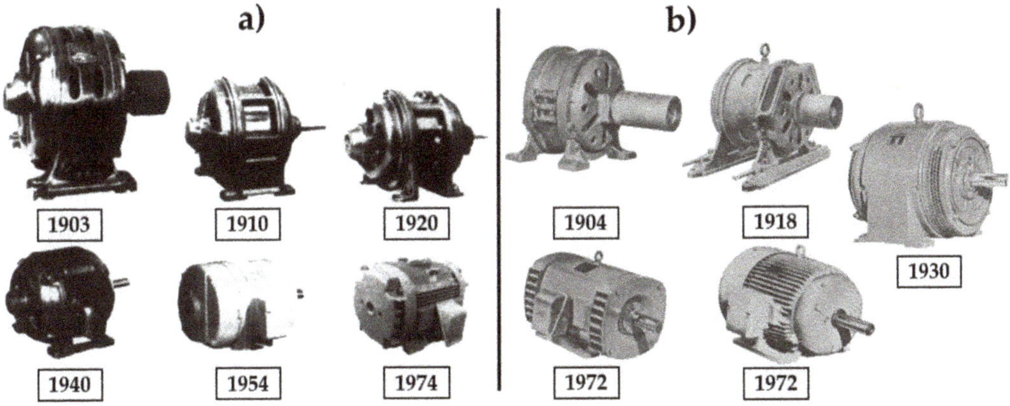

Figure 7. Dimension trends and housing changes of 11 kW 4-pole SCIMs between 1903 and 1974. (**a**,**b**): the changes in appearance and frame dimensions of SCIMs of different powers from the open construction of 1904 to those used in the 1970s. Source adapted from: [28,53].

Figure 7a presents SCIMs designed for operation at 220 volts and 11 kW, built by General Electric (GE). In Figure 7a, it is possible to observe the changes in the NEMA 404 housing over the years, and two significant innovations are evident in the images of SCIMs between the years 1920 and 1954: axial extension of the rotor at the rear and the closed housing, seen from 1954 and made possible by the improvement in insulation systems, enabling the transfer of heat from the windings to the outside.

Figure 7b shows a small SCIM (1904) built without a fan and considered to be "self-ventilated" by the semi-open housing. As early as 1918, SCIMs used a fan attached to the shaft for cooling. In 1930, the 15 kW SCIM already had a more efficient fan and could adopt a more closed design. In 1972, the engines were already drip-proof. Figure 7b shows 18.5 kW and 45 kW SCIMs. They could be fully enclosed (45 kW), allowing a reduction in SCIM dimensions. As a result of the improved insulation between the conductors and between the conductors and the ferromagnetic material of the sheets, the temperature of the winding wires and the groove walls became more homogeneous, as they were closer together with a thinner insulating layer. The temperature of the set decreased, and for this reason it was possible to increase the power considerably for the same housing. The stator slot was significantly reduced for the same power, and the magnetic section between the slots could be increased. There was also an improvement in the ferromagnetic material, an increase in the magnetic flux, and a consequent decrease in the number of turns per stator coil for the same electrical voltage.

According to Alger and Arnold (1979) [28], to avoid hot spots in the centres of long cores, radial ducts were introduced in the stator and impellers in the rotor operating as fans, creating the airflow through the stator channels. Therefore, the rating given to the NEMA 404 frame with an axle height and length of 25.4 cm and 31.1 cm, respectively, was increased with respect to mechanical power from 5.5 kW in 1897 to 75 kW in 1974, as shown in Table 2.

Table 2. Mechanical power increments in the same frame from 1898 to 1974. Source: [28].

Years	Motor Rated Output Power (kW)	Operating Temperature
1898–1903	5.5	40 °C Thermometer
1903–1905	7.5	40 °C Thermometer
1905–1914	11	40 °C Thermometer
1914–1924	15	40 °C Thermometer
1924–1929	18.5	40 °C Thermometer
1929–1940	22	40 °C Thermometer
1940–1956	30	50 °C Resistance
1956–1961	37	50 °C Resistance
1961–1966	45	50 °C Resistance
1966–1974	75	80 °C Resistance

The reduction in the volume of SCIMs also made it possible to reduce their costs, intensifying the electrification of industrial plants. For example, in 1890, a 3.7 kW SCIM weighed approximately 450 kg and cost about USD 900, and in 1957, a SCIM of the same power weighed around 50 kg and cost USD 110 [64]. Thus, the relationship between value and mass remained practically the same. However, as mass reduced significantly, the price of the SCIM reduced considerably, since the cost of an SCIM is fundamentally a function of the quantity and quality of materials used.

The company Hitachi produced three SCIMs of 3.7 kW in 1910, and in 2010 the total production was already 40 million SCIMs. The company recorded the significant advances that SCIMs have made over more than 100 years in this period. Hitachi divides advances in electric motors into three distinct periods. Between 1830 and 1890 is the period of inventions, from 1930 to approximately 1950 is the period of scientific initiatives, and between the 1950s and the present day is the period of industrial initiatives [69].

Various technical and technological developments have made Hitachi SCIMs smaller and lighter over the 100 years from 1910 to 2010. Figure 8 presents the leading technologies used by Hitachi that made it possible to reduce the mass of the first SCIM, with a power of 3.7 kW (four poles) manufactured by the company in 1910 with a mass of 150 kg, to approximately 20% of the mass in 2010 (30 kg).

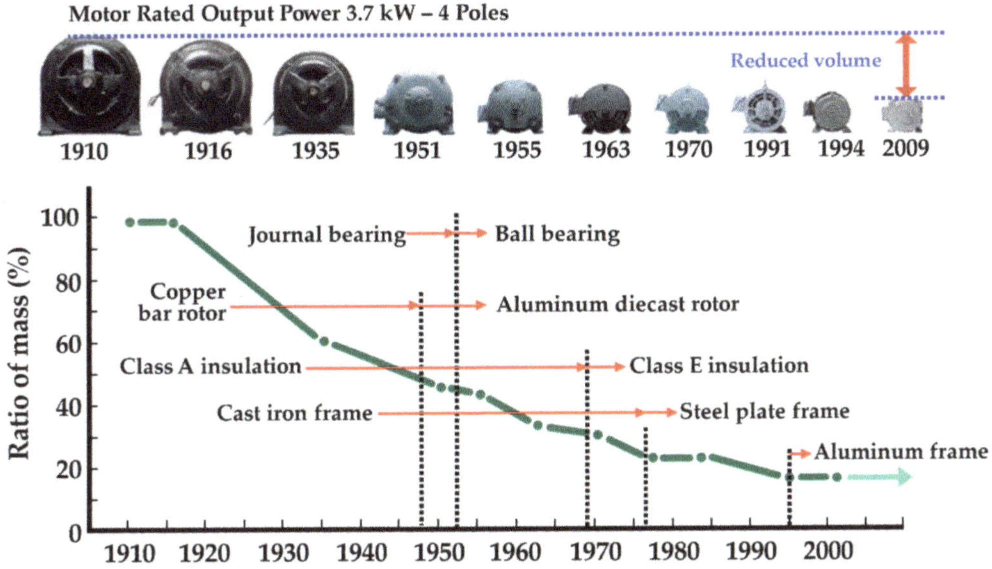

Figure 8. Hitachi SCIM mass changes for 3.7 kW (4 poles) SCIMs. Adapted from [69].

The main changes recorded were the use of aluminium in the rotor in the late 1940s. Later, in the 1950s, bearings improved, moving from sliding technology to ball bearings. In the late 1960s, improvements were made with the application of new insulation classes of varnishes on the wires. In the mid-1970s, cast iron frames gave way to lighter sheet steel frames. In the 1990s, aluminium structures closed the cycle of major technological innovations in Hitachi's first 100 years (1910–2010).

Reduction of Volume and Losses of Ferromagnetic Materials in SCIMs

The first electrical devices to use ferromagnetic materials were developed in the second half of the 19th century. Knowledge about these materials, such as their structure, was absent; as a result, the development of the projects was based on trial and error [70].

For SCIMs to thrive, they needed to advance in generating, transmitting, and distributing electrical energy via alternating current (AC) [71]. Charles Proteus Steinmetz was hired by General Electric (GE) (by Thomas Alva Edison) to improve the AC distribution system. He developed the complex representation of variables sinusoidally in time, which is still in use today [72]. Steinmetz deepened his studies of ferromagnetic materials to better compete with Westinghouse, which manufactured the induction motors invented by Tesla.

The first concepts regarding losses in ferromagnetic materials, traditionally known as iron losses, were developed by Steinmetz [73]. Via understanding how the losses behaved with changes in the intensity of the magnetic field, the General Electric induction motors became competitive, due to the reduction in the volume of material used [74].

Steinmetz's secret was to use increasingly thin sheets. The eddy current losses depend on the square of the sheet thickness, the hysteresis losses, and the square of the magnetic field strength [75]. Steinmetz's discoveries led to more efficient rolling mills that produced thinner and thinner sheets.

Understanding the ferromagnetic losses (hysteresis and eddy current) was decisive for selecting increasingly thin sheets to assemble the stator and rotor magnetic package. Thus, it was possible to impose a greater magnetic flux density in the package of sheets, approaching the limit of the magnetic saturation of the plate. This knowledge contributed to reducing the volume of SCIMs to approximately one third of the initial volume between 1891 and 1901 (Figure 4).

Subsequently, the development of ferromagnetic materials focused on reducing iron losses through heat treatment of the materials and the "doping" of silicon to increase the resistivity of the composite [76], thus enabling the intensification of the magnetic field and consequently reducing the volume of the SCIMs for a defined power.

To better understand the reason for the volume reduction of SCIMs over time, as shown in Figure 4, regarding the reduction provided by the improvement in ferromagnetic materials, it is possible to model the volume of SCIMs from the increase in the intensity of the magnetic field in their structures. This imposition of increasingly intense magnetic fields was one of the main reasons for the reduction in the volume of electric motors since their development.

A mathematical expression that translates the volume/power ratio as a function of the imposed magnetic flux density can be deduced from the electromechanical energy conversion equation, where the phase-induced electromotive force is given by Cardoso et al. [70]:

$$E = 4.44 f N_{eff} \varnothing \tag{1}$$

where f is the frequency (Hz), N_{eff} is the number of adequate turns in series per phase, and $\varnothing$ is the magnetic flux per pole (Wb).

Furthermore:

$$\varnothing = 2\frac{BLR}{p} \tag{2}$$

where B is the density of the magnetic flux in the air gap (T), L is the packet length (m), R is the radius of the air gap (m), and p is the number of pole pairs.

The electric current expressed as a function of the magnetic field in the motor air gap was expanded from the classical magnetomotive force equation FMM = NI = $\Re\phi$, and can be expressed as [70]:

$$I = \frac{\pi p l_g B}{3\sqrt{2}\mu_0 N_{eff}} \tag{3}$$

where μ_0 is the magnetic air permeability (H/m), p is the number of pole pairs, and l_g is the thickness of the air gap (m).

Ignoring any type of losses, the motor power will be given by $P = mEI$, where m is the number of motor phases.

Substituting E and I by their values expressed in Equations (1) and (3) results in:

$$P = \frac{\pi m p^2 l_g n}{3\mu_0} B^2 Vol \tag{4}$$

where $n = \frac{f}{p}$ represents the synchronous motor rotation in rps and $Vol = \pi R^2 L$ represents the motor volume.

Reorganizing Equation (4), it is possible to mathematically verify the volume/power ratio of SCIMs and other equipment that uses ferromagnetic materials, in proportion to the intensity of the internal magnetic field in Equation (5).

$$\frac{Vol}{P} = \frac{1}{K_m B^2} \tag{5}$$

with:

$$K_m = \frac{\pi m p^2 l_g n}{3\mu_0} \tag{6}$$

Equation (5) is inversely proportional to the magnetic flux density characteristic square, expressing a curve similar to that of Figure 4. It suggests that one of the primary explanations for the reduction in the volume (or mass) of SCIMs over the years was the more significant imposition of the magnetic field on its magnetic structure, as shown in Figure 9.

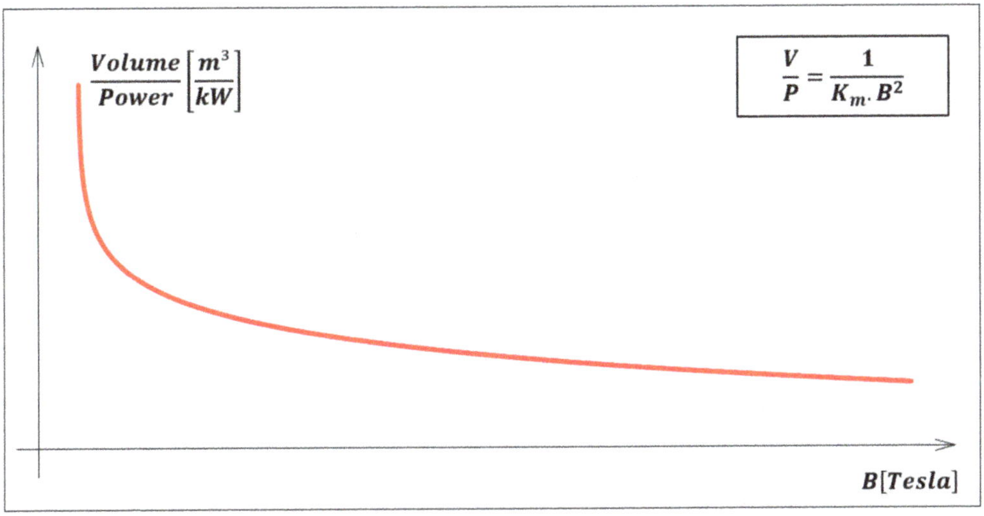

Figure 9. Density of the magnetic flux in the air gap and the volume of SCIMs.

Figure 9 represents the relationship between the improvement in the quality of the ferromagnetic material and the reduction in SCIM volume, considering the same output power. As a result of Equation (5), the curve is theoretical since it is impossible to design an electric machine with unlimited magnetic flux density (B) or a value of B very close to zero. Hence, the curve represents one of the essential reasons for the reduction in the volume of SCIMs from the first units to the present day.

It is estimated that for 110 kW electric motors, the losses in ferromagnetic materials represent, on average, 59% of the total losses [77]. The losses increase with increasing frequency of the electric voltage. These materials have also undergone improvements from the first electric motors to the current ones.

Since Michael Faraday demonstrated electromagnetic induction in 1831 [78], soft magnetic (ferromagnetic) materials have continued to evolve. When iron was the only soft magnetic material available, metallurgists and materials scientists experimented by introducing other elements to improve the efficiency of iron.

The main known losses in ferromagnetic materials are hysteresis and eddy current losses. Hysteresis losses occur through the coercivity of a magnetic material. Each time a material with magnetic characteristics completes an entire cycle of its magnetization curve, the area within this curve measures the energy lost in the magnetization process.

The second primary loss mechanism in soft magnetic materials is eddy currents. Eddy currents are closed electric current paths generated in a conductor whose source is a time-varying magnetic field. These current loops create a magnetic field in opposition to the change in magnetic flux (according to Faraday's law of induction). The energy losses caused by eddy currents scale approximately with the square of the operating frequency and are thus a significant cause of losses in alternating-current machines.

The development of silicon (electrical) steel in about 1900 was a notable event in the advances of soft magnetic materials [79]. Silicon steel still dominates the global soft magnet market and is the material of choice for large-scale transformers and electrical machines such as SCIMs. In 1900, Robert Hadfield, a metallurgist from England, and his team developed unoriented silicon steel by adding up to 3% of silicon to iron and increasing its electrical resistivity (ρ) [80].

The team led by the American metallurgist Norman Goss developed grain-oriented silicon steel in 1933, promoting grain growth along a crystalline direction. The most common applications for silicon steel are large-scale transformers (grain-oriented silicon steel) and electrical machines (unoriented isotropic silicon steel is preferred for rotating machines), for which the economical price is a great benefit [80].

Improvements in magnetic properties were also achieved, from the treatment of iron to minimize chemical impurities to the techniques of slicing the iron into thin sheets. Subsequently, silicon was used to increase the electrical resistance of iron and control the crystal orientation. Figure 10 presents the reduction in losses in the core of electrical machines in watts for each kilogram of ferromagnetic material, highlighting the predominant technological advances of each period.

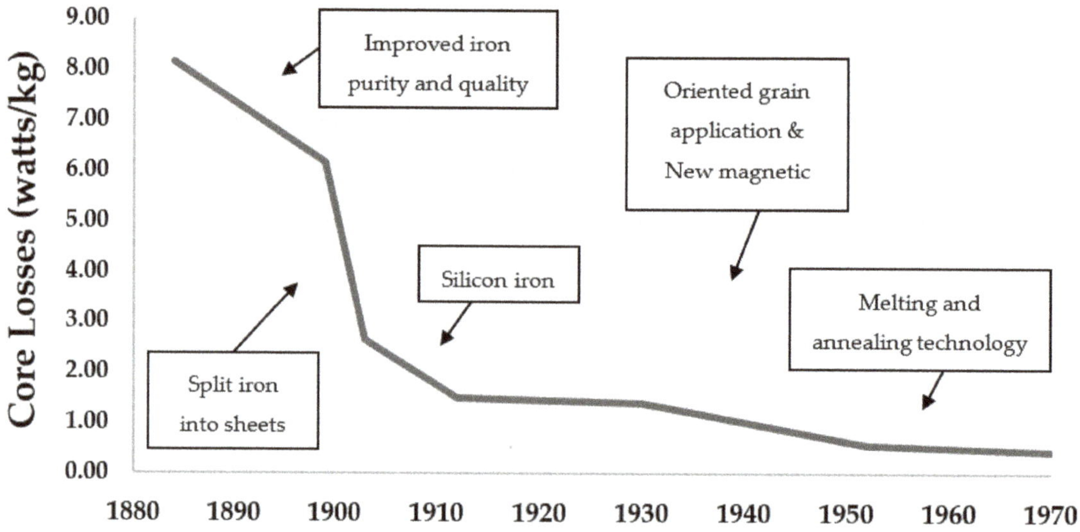

Figure 10. Changes in losses in the core of electrical machines (ferromagnetic material). Source: adapted from [7,76,81].

It is possible to observe in Figure 10 that between 1884 and 1970, the losses in the core of alternating-current electrical equipment reduced from 8.16 W/kg to 0.44 W/kg, which represents an approximately 95% reduction.

Figure 10 shows low frequencies (50 or 60 Hz) and a constant B (T) value, as both directly influence losses in ferromagnetic materials.

Today's primary soft ferromagnetic materials in electric motors are iron and ferrosilicon alloys (2022). However, materials with lower eddy current and hysteresis losses have been developed since the 1970s.

After the energy crisis of the 1970s, the first attempts to use amorphous materials for electric motors were recorded (1981). Mischler et al., demonstrated the low-loss potential of the amorphous stator in a laboratory environment [82].

In 1967, a new class of materials, amorphous alloys, was introduced [83]. In the mid-1970s, interest in amorphous alloys based on iron and cobalt increased, and these materials began to find applications [84]. However, only in 1988 did Hitachi researchers investigate Nb and Cu additives. They added an annealing step to amorphous alloys to produce small-spaced crystallites of iron or cobalt within an amorphous matrix material. The formation of isolated crystallites of transition metals reduced the eddy current losses of these materials compared to traditional amorphous alloys. Despite a higher initial cost than silicon steel, these advanced alloys can reduce the total lifetime costs of electric motors due to reduced losses.

Currently (2022), unique treatments involving thermal manipulation, laser bombardment, and other technologies continue to produce high-performance magnetic materials.

3.2. Changes in the Performance of SCIMS

It was not just the masses and volumes that changed. Successive changes in the performance of SCIMs occurred from the first commercially available versions to the mass manufacturing versions of today (2022).

Several technological advances explained the changes in the performance of SCIMs, from the technological innovations already mentioned to improvements in production processes and the purification of active materials. Sven Sjöberg [66] presented the reasons for the performance gains of SCIMs manufactured by the company ABB Motors after the great cycle of innovations that closed in the 1970s:

(a) Cutting tooling: improving mechanical precision and enabling the elimination of burrs;
(b) Laminated package: lamination mixing, lamination control before pressing, welding, or stator clasping. Quality assessment of raw material sampling before casting (rotor);
(c) Machining the outer surface of the stator core (generally not necessary): reduces surface roughness and improves tolerances;
(d) Stator winding: length of coils, type of winding, filling factor, insulation system, loops, and connections;
(e) Impregnation: good filling results and improvements in thermal exchanges;
(f) The casting of the rotor cage: filling of the slots and the closing rings of the cage, purity of the casting material, and balancing of the rotor;
(g) Alignment of the rotor shaft and machining of the outer surface of the rotor.

Sven Sjöberg (1997) presented the performance changes to SCIMs manufactured by ABB Motors between 1935 and 1996, as shown in Figure 11. According to Sven Sjöberg, performance changes did not result from any performance regulation but occurred due to materials improvements, technological innovations, and improvements in production processes [66].

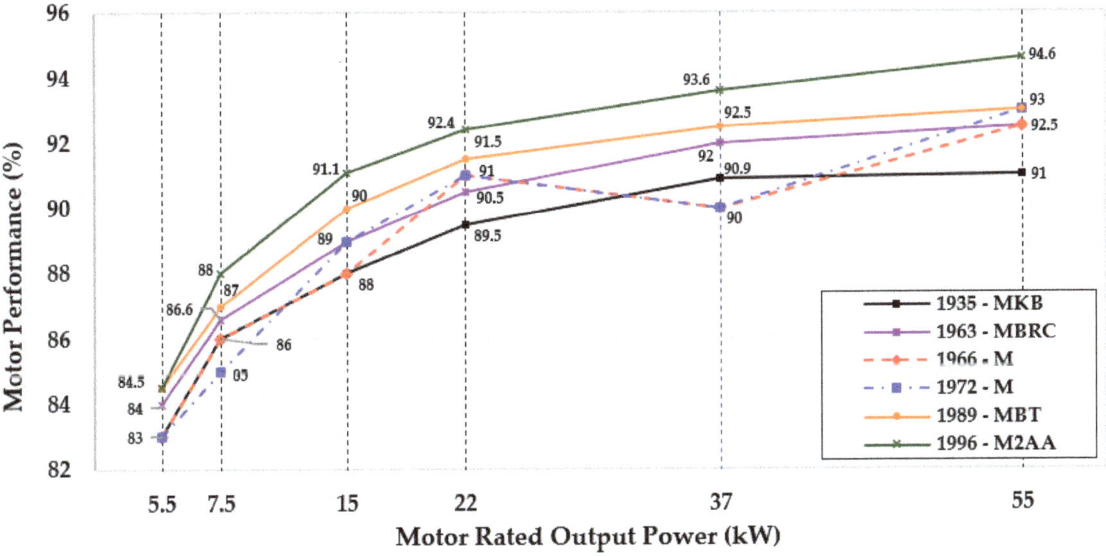

Figure 11. Changes to performance of 4-pole SCIMs between 1935 and 1996. Source: adapted from [65,66].

It is observable in Figure 10 that in the 1960s and 1970s there was a reduction in the average performance of SCIMs, considering a wide range of power. In some periods, the performances were inferior to those obtained by the industry in 1935. The researcher Sven Sjöberg, in his text, does not identify the elements that led to this temporary drop in performance between the 1970s and 1980s.

For the United States Department of Energy (DOE), the 1960s and 1970s were periods of global economic crisis, where SCIM manufacturers built lower-cost equipment compared to previous years. These SCIMs were less efficient, as shown in Figure 12, as they minimized the use of materials such as copper, aluminium, and steel. According to the DOE, these SCIMs had lower initial costs than previous projects. However, they consumed more electrical energy due to their inefficiency, so their use throughout the life cycle was more expensive [85].

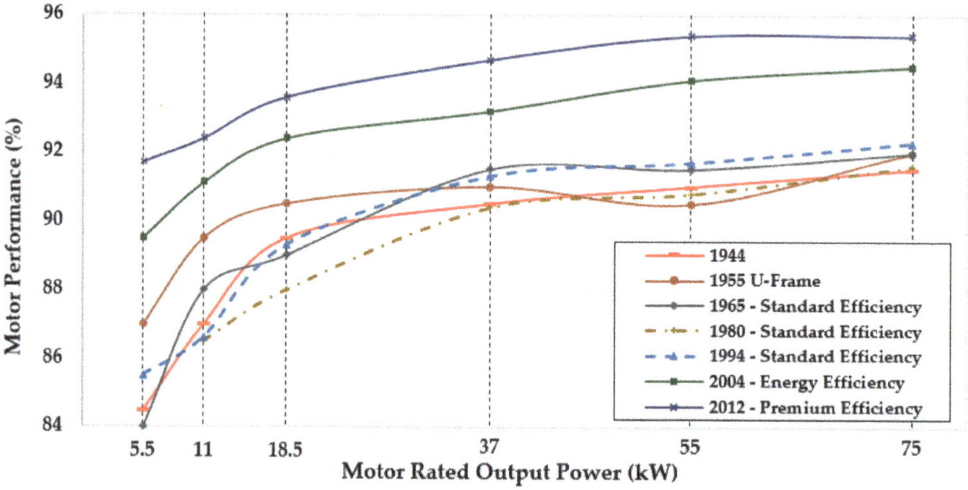

Figure 12. Changes to performance of 4-pole SCIMs between 1944 and 2012. Source: adapted from [85].

Figure 12 shows that the four-pole SCIMs manufactured and marketed in North America in the 1980s had even lower performance than SCIMs manufactured in 1944, which were the first officially registered by the DOE.

According to the DOE, less-efficient and more-compact SCIMs became possible with insulating materials that could withstand high temperatures. These SCIMs were designed to admit higher losses due to the increase in temperature in the coils located in the stator, making it possible to accommodate the winding wires in smaller frames without damaging the insulation [85].

Figure 13 shows the performance changes to four-pole SCIMs with motor rated output powers of 37 and 45 kW, operating at 50 or 60 Hz, at low voltage.

Figure 13 shows the average performance presented by Sjöberg and the DOE. The SCIMs showed a performance reduction between the 1960s and 1980s, and only the data provided by WEG (2015) showed a continuous increase in performance. Figure 13 illustrates the performance data available in the company's publications, beginning in 1960, which was the year the company was established. The data show performances below those obtained in the international market, with high performances recorded for 2010.

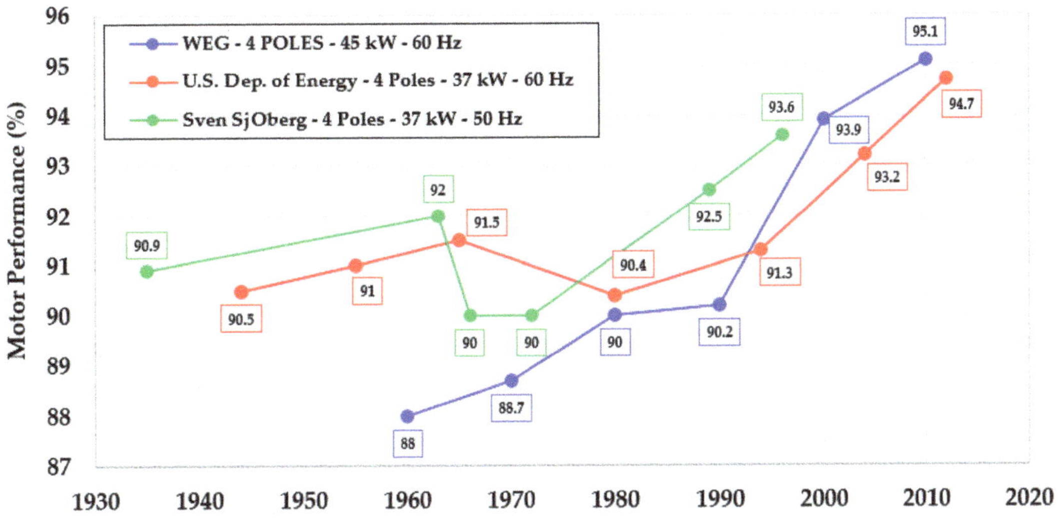

Figure 13. Changes to performance of 4-pole SCIM performance between 1935 and 2012. Source: adapted from [65,66,85].

3.3. Changes in the Performance of SCIMs between 1945 and 2020

SCIMs and most of the electromechanical equipment developed in the 20th century underwent a series of improvements and refinements, from conception through the technological advances in construction processes, mainly in the improvement in the quality of the materials used.

Test results based on data from 1945 and 2020 were used to analyse the change in the performance of 359 SCIMs, with speeds corresponding to two, four, six, or eight poles, at a motor rated output power of 3.7, 37, 150 kW, in order to aid in answering the questions (I and II) that motivated this research.

Figure 14 shows the trends in performance of two-pole SCIMs over time, tested from 1945 to 2020.

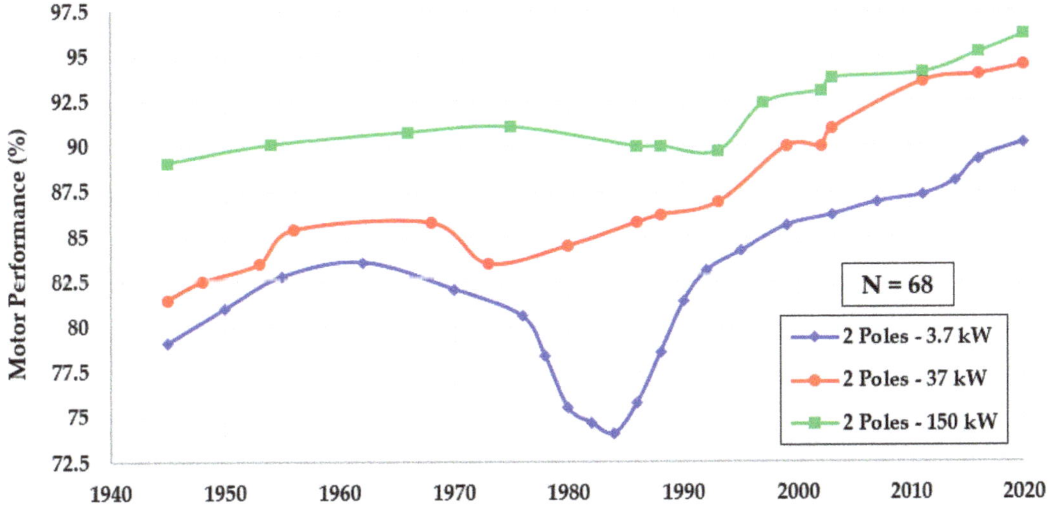

Figure 14. The average performance of 2-pole SCIMs between 1945 and 2020.

Figure 14 shows test results from 68 SCIMs organized into three output power categories and arranged over time. In the years in which results were obtained from more than one SCIM of the same speed and mechanical power, the average performance was calculated for the construction of the figure. In addition, in the years in which there were no SCIMs tested at the output power used in the analysis, the linear regression method was used between the adjacent years in which data were available, in order to construct the figure. The same considerations were applied to Figure 15 (four-pole SCIMs), Figure 16 (six-pole SCIMs), and Figure 17 (eight-pole SCIMs).

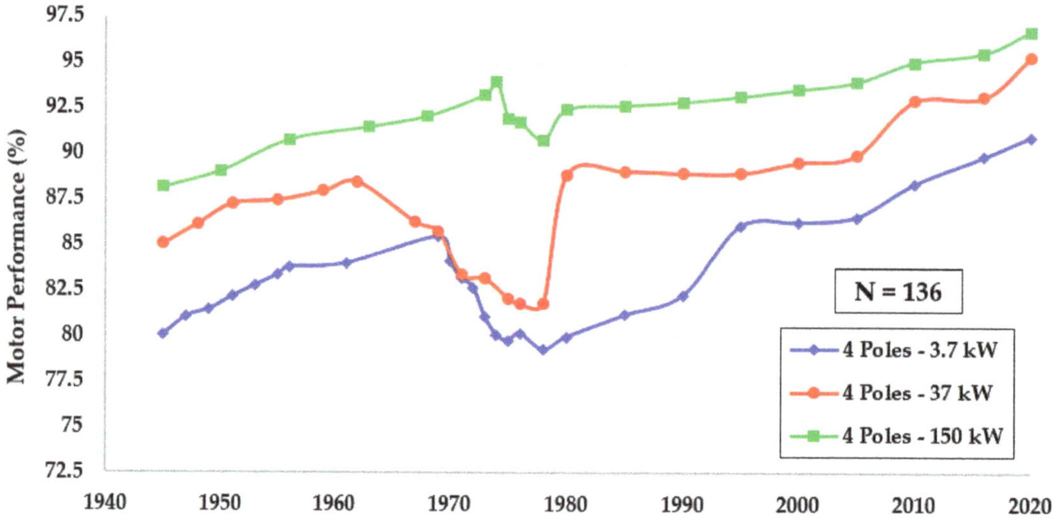

Figure 15. The average performance of 4-pole SCIMs between 1945 and 2020.

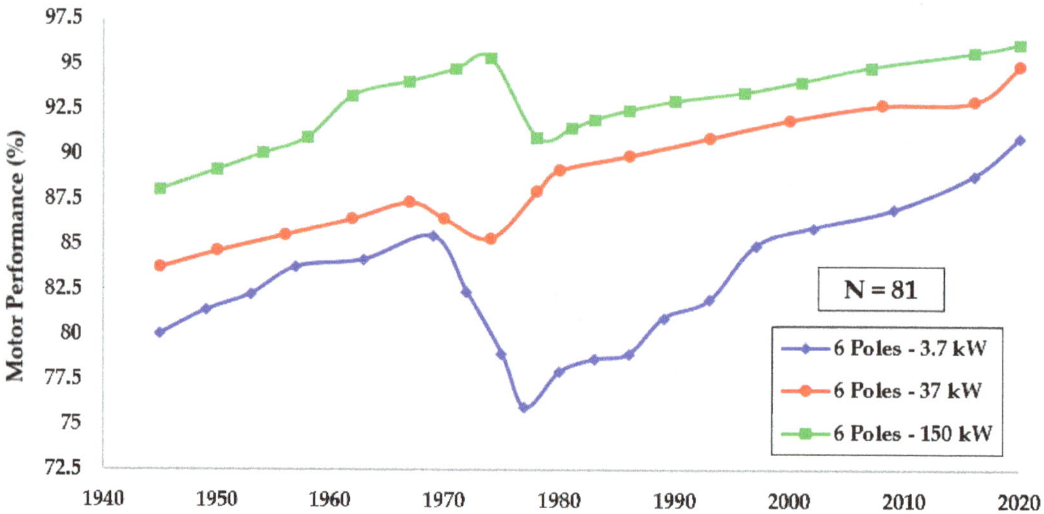

Figure 16. The average performance of 6-pole SCIMs between 1945 and 2020.

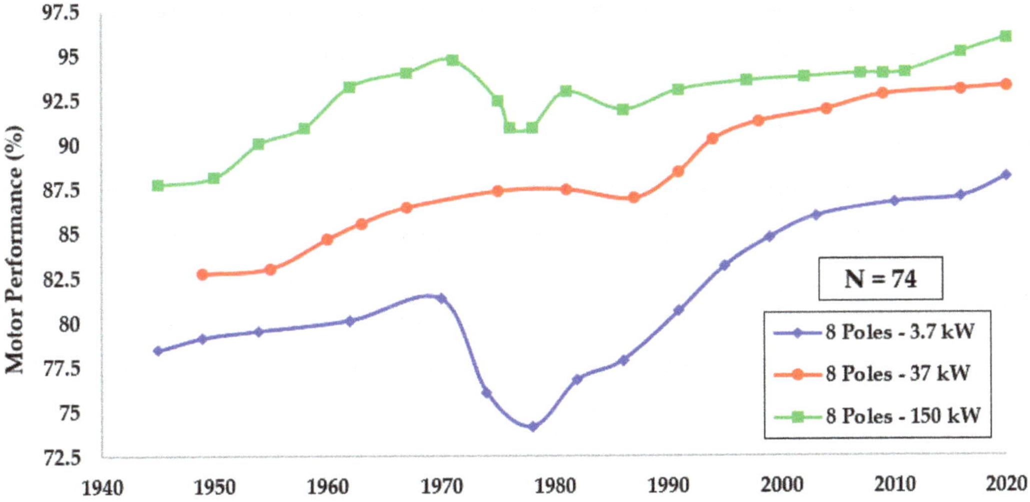

Figure 17. The average performance of 8-pole SCIMs between 1945 and 2020.

Table 3 presents the cumulative performance gain between 1945 and 2020 for the three analysed power values.

Table 3. The average performance of 2-pole SCIMs between 1945 and 2020.

Motor Rated Output Power (kW)	3.7	37	150
Performance (%) 1945	79.1	81.5	89.1
Performance (%) 2020	90.2	94.5	96.2
Accumulated gain (%)	11.1	13	7.1
Loss reduction (%)	53.1	70.3	65.1

Generally, high-power SCIMs are always associated with high performances. They are often subjected to more rigorous quality control routines by the manufacturers and users, who are concerned about the losses in this equipment because they are primarily the predominant industrial electrical loads. This fact results in SCIMs of higher power such as 150 kW having smaller performance gains over that time interval. Medium-power (37 kW) and low-power (3.7 kW) SCIMs are associated with high performance gains, with accumulated values of 13% and 11.1%, respectively, based on the analysed period. In other words, the reduction in losses in two-pole SCIMs between 1945 and 2020 was 53.1% for 3.7 kW, 70.3% for 37 kW, and 65.1% for 150 kW. The trends shown in Figure 14 and Table 3 for two-pole SCIMs are similar to those in Figure 15 and Table 4 for four-pole SCIMs, in Figure 16 and Table 5 for six-pole SCIMs, and in Figure 17 and Table 6 for eight-pole SCIMs.

Table 4. The average performance of 4-pole SCIMs between 1945 and 2020.

Motor Rated Output Power (kW)	3.7	37	150
Performance (%) 1945	80.1	85.1	88.2
Performance (%) 2020	91	95.4	96.8
Accumulated gain (%)	10.9	10.3	8.6
Loss reduction (%)	54.8	69.1	72.9

Table 5. The average performance of 6-pole SCIMs between 1945 and 2020.

Motor Rated Output Power (kW)	3.7	37	150
Performance (%) 1945	80.1	83.8	88.1
Performance (%) 2020	91	95	96.2
Accumulated gain (%)	10.9	11.2	8.1
Loss reduction (%)	54.8	69.1	68.1

Table 6. The average performance of 8-pole SCIMs between 1945 and 2020.

Motor Rated Output Power (kW)	3.7	37	150
Performance (%) 1945	78.5	82.8	87.8
Performance (%) 2020	88.2	94	96
Accumulated gain (%)	9.7	11.2	8.2
Loss reduction (%)	45.1	65.1	67.2

According to Table 4, the loss reduction for four-pole SCIMs was 54.8% for 3.7 kW power, 69.1% for 37 kW, and 72.9% for 150 kW between 1945 and 2020.

According to Table 5, the loss reduction for six-pole SCIMs was 54.8% for 3.7 kW power, 69.1% for 37 kW, and 68.1% for 150 kW between 1945 and 2020.

According to Table 6, the loss reduction for eight-pole SCIMs was 45.1% for 3.7 kW power, 65.1% for 37 kW, and 67.2% for 150 kW between 1945 and 2020.

The three curves (3.7 kW, 37 kW, and 150 kW) showed similar trends in the four figures presented (Figures 14–17), making it possible to separate three periods:

1. Between 1945 and the mid-1960s, SCIMs presented a curve indicating continuous increasing performance gains;
2. Between the 1960s and 1980s, SCIMs showed significant performance drops, in some cases reaching lower levels than the SCIMs marketed in 1945;
3. Between the 1980s and 2020, performance improvement dominated the scenario. It resulted in high levels of performance in the last years of the analysis, presenting a net result, from 1945 to 2020, of gains above 10% in average performance, corresponding to a worst-case reduction of losses of approximately 45%.

Several elements influenced these trends for each of the three periods described above. At first, between 1945 and the mid-1960s, an intensive process of technological innovation was identified, highlighting the following elements that directly influenced the performance gains of SCIMs:

a. Many SCIMs tested in the 1940s still had plain bearings. Sleeve bearings, compared to ball bearings, produce more noise, are larger and heavier, and generally provide greater friction, requiring oil lubrication;
b. In the 1940s, there was a transition from rotors made of iron bars to rotors made of cast aluminium bars. Aluminium has lower electrical resistivity and lower density, and is therefore lighter for the same power;
c. Advances in metallurgy allowed SCIM housings to be built more compactly, improving the safety of operation and maintenance workers, maintaining winding ventilation, and reducing masses and volumes;
d. The insulation system in that period underwent substantial advances, moving from the use of cotton as an insulator to silk, where a significant reduction in the size of the grooves was possible, reducing the size and volume of the SCIMs;
e. Due to the use of silk, it was also possible to insert more copper into the same slot, reducing the most significant losses in SCIMs (Joule losses in the stator winding wires);
f. Improvements in the manufacturing processes of SCIMs were remarkable in that period, whether due to advances in cutting tools or better machining of the active ferromagnetic materials of SCIMs;
g. Between 1884 and 1970, the core losses of AC SCIMS dropped from 8.16 watts/kg to 0.44 watts/kg, which represents an approximately 95% reduction.

In the second period, between the 1960s and 1980s, SCIMs showed significant drops in performance, making it possible to identify the influence of the following elements. In this period, insulation from varnish was developed. The varnish made it possible to withstand high temperatures without compromising the insulation. For this reason, SCIM designs emerged that admitted more significant losses in the stator winding wires due to increased temperature in the coils. Temperatures up to 180 °C, already standardized in the 1970s (Table 7), were observable in some SCIMs.

Table 7. Thermal class of insulation of electrical conductors. Source: [86].

Thermal Class (°C)	Designation Letter
90	Y
105	A
120	E
130	B
155	F
180	H
200	N
220	R
250	-

Cotton and silk operated only as electrical insulators. In contrast, the varnish used, in addition to being an electrical insulator, is a thermal conductor. This factor made it possible to accommodate the winding wires in even more miniature housings without damaging the insulation and to improve cooling with an increased transfer of heat produced mainly in the stator winding wires to the external surface, via the design of the fins on the housing.

When varnish is used to insulate the winding wire, it conducts the temperature rise resulting from the losses in the stator winding wires to the housing (Figure 18). In the process, the fins are designed to increase the contact area with air, thus improving the heat dissipation process and changing the geometry of the SCIM housing.

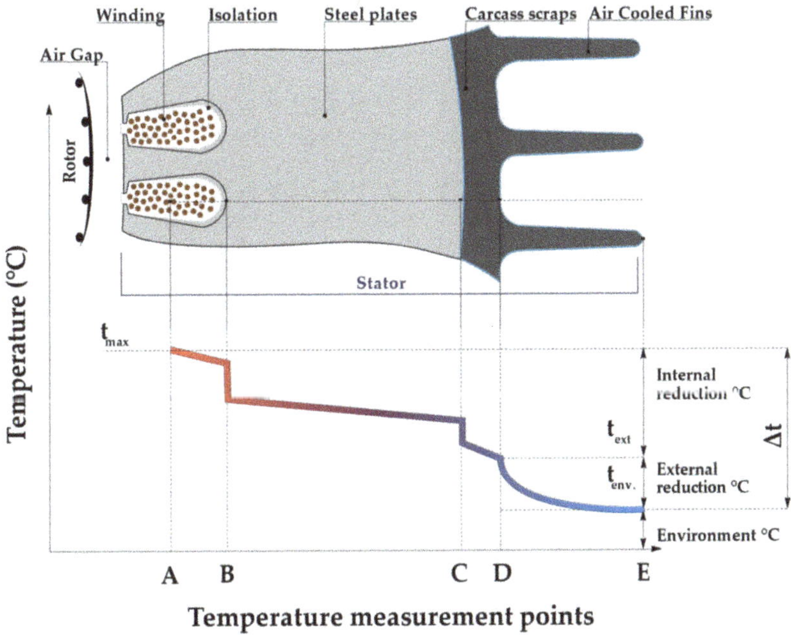

Figure 18. Stator temperature measurement points (A, B, C, D and E). Adapted from [87].

The temperature reduction in SCIMs between points A and E, expressed in Figure 18, can be described as follows:

A—the hottest point of the SCIM, inside the slot that generates the heat from the Joule losses of the stator winding wires;

AB—the temperature reduction resulting from heat transfer from the hottest point to the outer wires of the coil. As air is not a good conductor of heat, there must be no "voids" inside the groove. Therefore, the windings must be compacted and impregnated with varnish, filling the voids as much as possible;

B—the temperature reduction caused by the insulator inserted between the winding wires and the metal plates. It is common to use special paper or synthetic insulating foil to line the groove;

BC—the temperature reduction by thermal conduction in the SCIM core plates;

C—the temperature reduction in the contact between the core and the housing. Precision machining of the housing to reduce surface irregularities is essential in heat conduction;

CD—the temperature reduction by thermal conduction through the shell thickness;

DE—the temperature reduction due to the increase in the SCIM surface exposure caused by the fins.

The reduction in copper mass meant that SCIM manufacturers were able to reduce the final cost of the equipment, since copper is the highest cost input in the construction of SCIMs. This trend was verified in the test reports of the analysed period. An increase in Joule losses (I^2R) in the stator winding wires was mainly observed in relation to previous decades. When the section of the copper conductors reduces, the total mass of the SCIM also reduces. The reduction in copper increased the Joule losses and consequently increased the operating temperature of the SCIMs. The heat generated internally could be more easily dissipated in the housing with varnish.

In the third period, between the 1980s and 2020, improvements in the average performance of SCIMs were evidenced mainly by the following observations.

Minimum performance level policies were applied in the world's largest economies between the 1990s and 2020. The policies that indicate the minimum energy performance of equipment are entitled "minimum energy performance standards (MEPS)," which specify minimum levels of energy performance for commercial purposes. The main objective of MEPS is to guide the performance of the equipment for the consumer and establish a minimum legal requirement for commercialization.

Government bodies usually institute MEPS policies. In the case of SCIMs, MEPS are divided into performance classes, allowing different levels that increase the requirement of a specific minimum performance value according to technological advances and market acceptance. Performance classes for SCIMs internationally are harmonized with the IE code in IEC 60034-30-1 [88], which is widely accepted as the global standard, making performance classes comparable across the various regional energy policy documents for SCIMs. The standard defines efficiency classes from IE1 to IE4 (Figure 19), where IE1 is the lowest, and IE4 is the highest. Similarly, in the United States, performance classes IE1 to IE4 are called Standard, High efficiency, Premium efficiency and Super-Premium efficiency, according to NEMA [89]. The new IE5 class has not been defined in detail; however, it is foreseeable in a future edition of the standard. For IE5 SCIMs, the goal is to reduce losses by about 20% compared to the IE4 class [88,90].

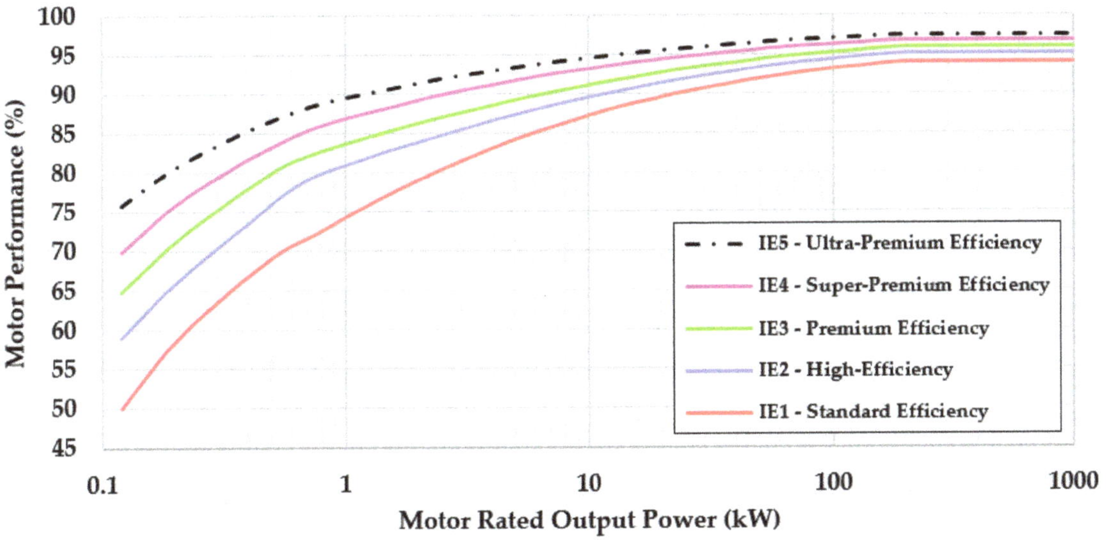

Figure 19. Efficiency levels in the IEC 60034-30-1 (2014) classification standard curves for 50 Hz, 4-pole SCIMs. Source: [88,90].

The SCIMs tested in 2020 were already IE3. Therefore, in the next few years, it should be possible to make another short jump in the performance gain of SCIMs.

The implementation of MEPS for SCIMs took place in the USA and Canada in 1997 and was later gradually applied in other countries, with modifications implemented by each energy agency of the various countries, but maintaining the harmonization as shown in Figure 20.

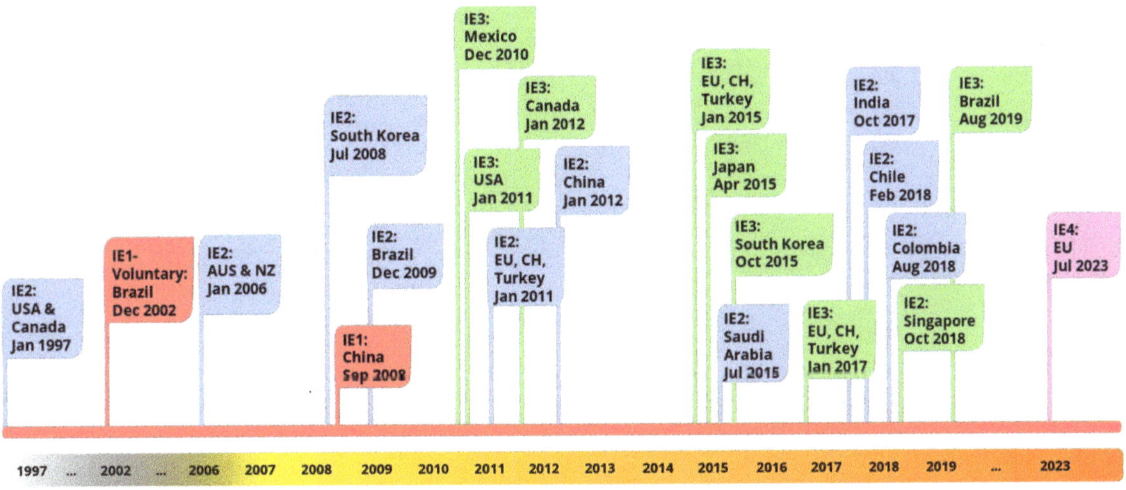

Figure 20. Timeline of global minimum performance standards for SCIMs. Source: [91–97].

To comply with the new legislation, which imposes higher performance indices, the central intervention of the manufacturers, verified in the test reports of the analysed period, was the reduction of Joule losses in the stator, because stator windings started to be built with more copper mass compared to previous decades. This movement also meant that the mass of SCIMs, which until then had decreased with time, began to increase, returning to the levels verified in the 1960s.

During this period, other secondary elements were observed that also influenced the improvement of the performance of SCIMs:

1. Advances in the design of SCIMs through the use of modelling software, enabling structural improvements in the coupling and a reduction in vibrations and noise;
2. Three-dimensional computational modelling of electromagnetic fields, enabling project optimization;
3. Advances in the processes of the casting of steel-silicon sheets;
4. Use of more efficient cooling systems (ventilation).

The three periods described led to profound changes in the mass/power ratio of SCIMs. The analysis presented in Figure 4 demonstrates the falling mass/power ratio and points to the lower levels in the following years needing to be updated. For this reason, Figure 21 was created to answer question III, which was one of the questions motivating this research.

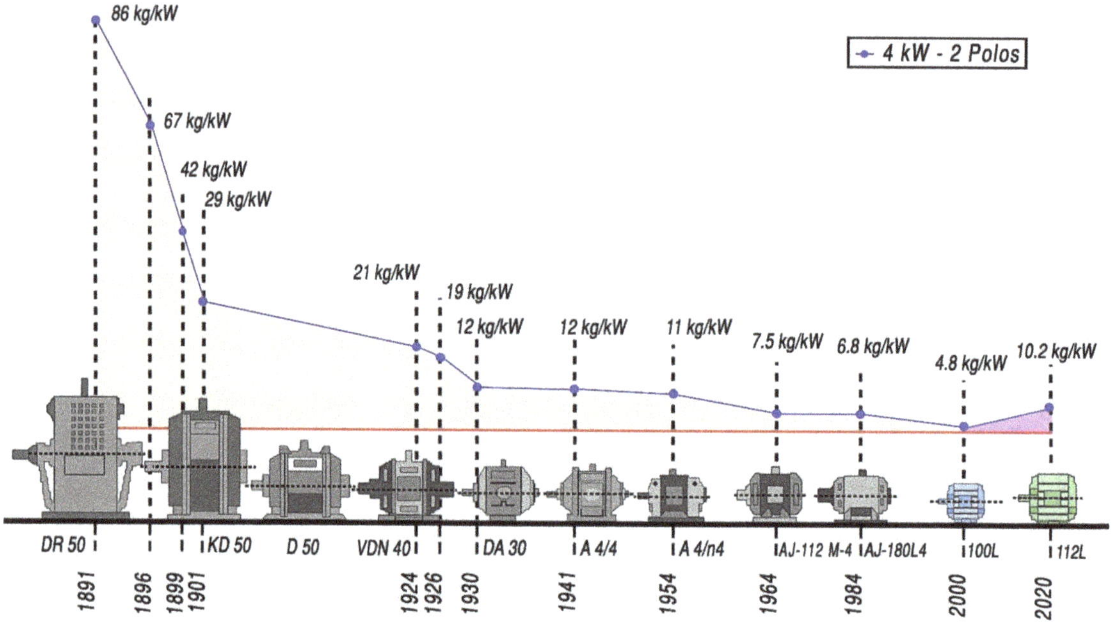

Figure 21. Changes to SCIMs in the mass/power ratio between 1891 and 2020.

The research relied on SCIM test data from 1945 to 2020. However, SCIM mass data was only available in technical reports from 1997 onwards. Before this date, few reports presented a record of the mass of the SCIM under test. Between 1945 and 2020, records of seven SCIMs with power and speeds compatible with Figure 4 and with mass records were discovered. The mass/power ratio found in these seven SCIMs was compatible with the data published by AEG. Thus, Figure 4, containing results between 1891 and 1984, was updated with data obtained in this research (Figure 21).

To create Figure 21, in 2000, 12 two-pole SCIMs with power between 3.7 kW and 4.4 kW were used, and in 2020, 16 SCIMs were used in the same power range and for the same speed. After calculating the mass/power ratio for each SCIM, the arithmetic mean was calculated for each of the two years under analysis.

A significant result verified in Figure 21 was the increase in the mass of SCIMs from the 2000s onwards, reaching the level of 10.2 kg/kW for the same power and speed, returning to levels verified in the 1950s.

The increase in mass was produced mainly by using conductors of a larger section, to reduce the block where the most significant losses in SCIMs are found, that is, the losses from the Joule effect in the wires of the stator windings.

The reduction in volume of an electrical machine can also result in challenges in keeping components cool. In the case of high heating, deterioration of the properties of most materials (such as insulators, coils, and sheets of ferromagnetic material) can occur, causing a reduction in the useful life of the equipment. This is one of the reasons that justify the average increase in the carcass of SCIMs in the last two decades.

There was also an increase in the lengthening of the rotor package, and consequently of the stator windings, significantly increasing the amount of material used in the construction of the high-efficiency motor, as seen in Figure 22.

Figure 22. The difference in the material quantity between Standard SCIM and High-Efficiency SCIM. Source: [98].

In Figure 22, the most significant change made to increase the performance of a 5 HP (3.7 kW) electric motor from 84% to 90.2% was an increase in mass of 27 kg or approximately 33%, while maintaining the same carcass.

For performance gains superior to those shown in Figure 22, increasing the carcass to accommodate the new stator and rotor dimensions was necessary. Figure 21 shows that SCIMs went from a 100L housing in 2000 to a 112L housing twenty years later (2020).

The mass/power ratio depends on the power range and speed, so Figure 21 cannot be directly generalized to other power values without proper adjustments. However, the shape of the curve presents a similar trend for the other power ranges and speeds.

There is no forecast of a continuous increase in the mass/power ratio of SCIMs, as this has been optimized in recent years through technological innovations. Other viable technologies have been presented to reach the IE5 standard. Synchronous operation motors include permanent magnet synchronous motors (PMSMs) and synchronous reluctance motors (SynRMs). Synchronous motors employ a drive that can also control the speed, and they have introduced a series of improvements in motor drives, such as ease of automation, the possibility of pre-diagnosis, ease of application of intelligent sensors, the possibility of collection and analysis of electrical quantities, etc.

PMSMs, for the same power range (4 kW) and speed (two poles) as those shown in Figure 21 can present a mass/power ratio of approximately 4 kg/kW, with a performance above 93%, even for low power and a power factor above 0.95.

SynRMs, for the same power range (4 kW) and speed (two poles) as those shown in Figure 21 can present a mass/power ratio of approximately 7.5 kg/kW, with a performance above 92.8%, even for low power and a power factor above 0.95.

Synchronous operation electric motors do not have rotor losses, and this is one of the main reasons this equipment can raise the level of performance. Synchronous motors also have a smaller physical volume than traditional SCIMs and are touted as the immediate future of variable-speed motor drives. If the economic factor also becomes an attraction, synchronous motors may also be viable in fixed-speed systems.

For SCIMs to reach IE5, two possibilities are currently considered. One is the use of amorphous materials with high magnetic permeability to reduce core losses. Another is the use of copper to minimize losses in rotors traditionally constructed of aluminium.

The magnetic package of SCIMs can be particularly suited to amorphous laminations, as demonstrated by Hitachi with an 11 kW motor prototype that achieved IE5 efficiency [99]. The Hitachi prototype had a reduced size compared with a traditional SCIM and performance above 93% over a wide load range.

Traditional medium- and low-power SCIMs have a rotor constructed primarily of cast aluminium. However, since 2002, it has been possible to find, for some applications, SCIMs with rotors made with copper [100].

The copper squirrel-cage rotor enables a 15% to 18% reduction in total motor losses (this can represent an efficiency gain of 2 to 4%, depending on the power and number of poles) [101]. A copper rotor is made of electrical steel laminations in which the rotor bars and end rings are made of cast copper instead of cast aluminium. Copper is an excellent material for rotors because it has higher electrical conductivity than aluminium [102].

The use of the copper rotor can also support the resumption of size reduction and overall weight reduction of the motor, since the reduction in losses in the rotor allows the reduction of the total length of the rotor and consequently the stator.

3.4. Research Limitations

In this section, dealing with the limitations of this research, we make suggestions for future research activities on the theme of changes in the performance of electric motors, which will contribute to research in the area:

- Evaluate the changes that have taken place in the forms of SCIM projects, from manual calculations to the use of high-level computer simulation;
- Evaluate the improvements in the copper drawing process and the improvement in the purity of copper (stator) and aluminium (rotor);
- Evaluate the improvements in the design and machining of the ventilation of electric motors;
- Evaluate improvements in the electric drive process and coupling between electric motors and mechanical loads;
- Evaluate advances in metallurgy to produce increasingly accurate cuts, improving the quality of electrical machines.

4. Conclusions

It is common to read in the technical literature that "SCIMs have hardly changed in the last 100 years". However, current SCIMs are significantly different from the SCIM developed by Mikhail Dolivo-Dobrovolsky's team between 1888 and 1890. Therefore, this statement is only valid when referring to the SCIM's working principle. This research showed significant changes in the design, materials, and components that make up the parts of SCIMs.

The present research analysed the performance levels of SCIMs based on the results of tests carried out at the Laboratory of Electrical Machines of IEE/USP in the period between 1945 and 2020. SCIMs with powers of 3.7 kW, 37 kW, and 150 kW were used in a total of 359 electric motors. Regarding the performance levels, the results showed that the SCIMs presented a similar trend, and it was possible to identify three distinct periods in the historical timeline.

Between 1945 and the mid-1960s, SCIMs showed practically constantly increasing gains in performance. This was due to the various technological innovations in the period, mainly the use of oriented grains in the ferromagnetic material, the use of aluminium in the rotor, essential improvements in the projects, and the ventilation of the SCIMs.

Between the 1960s and 1980s, which was a period of cheap energy, manufacturers built cheap and relatively inefficient SCIMs, minimizing the use of materials such as copper, aluminium, and steel. The production of lower-performance, lower-volume SCIMs was made possible by developing insulating materials (particularly varnishes) that could withstand high temperatures. This allowed SCIMs to be designed with higher losses (particularly Joule losses in the stator winding), since the temperature rise due to losses could be transferred to the housing (the varnish is electrically insulating and thermally conductive) without damaging the insulation or reducing the expected motor life (Figure 18). In this period, the reduction in the performance of SCIMs was so high that, in some cases, the performance reached lower levels than for the SCIMs marketed in 1945.

Although these motors had lower start-up costs than previous designs, they used more energy due to their inefficiency.

From the 1980s to 2020, performance improvement dominated the scene again. The central aspect of this performance variation was the technology and materials used to construct the machines. It was possible to observe that the gains were significantly higher for minor power values, due to the large margin for improvements in materials and projects due to the low technical construction rigour.

The reduction of losses in the SCIMs analysed in the period 1945–2020 was in all cases more than 40% for the three analysed output power values (3.7, 37, and 150 kW) and the four possible speeds (two, four, six, and eight poles). In the case of 150 kW SCIMs with a speed corresponding to four poles, the loss reduction in the period reached 72.9%, showing a significant advance.

The 37 kW SCIMs with a speed corresponding to 2 poles had the highest accumulated efficiency gain in the analysed period. They went from 81.5% average yield in 1945 to 94.5% 75 years later (2020), resulting in an absolute 13% performance gain.

The relationship between the mass and power of SCIMs presented two periods in the analysis performed. The first period was the 94% reduction between 1891 and 1984, from 86 kg/kW to 4.8 kg/kW, due to the various technological innovations discussed in this paper. The second period showed a decrease by 112.5% between 2000 and 2020, from 4.8 kg/kW to 10.2 kg/kW on average, due to the need to resume the performance increase.

In conclusion, continuous performance gains occurred during intense technological innovation, showing the importance of performance legislation for SCIMs. In the 1970s and 1980s, the search for lower-cost SCIM manufacturing reduced the equipment's performance. Thus, the self-regulation of the SCIM market, in terms of performance, did not show positive results in periods of low technological innovation. A return of the performance improvement was observed, mainly by the imposition of performance legislation, motivated by a global need to rationalize the final energy use and by sustainable energy considerations.

Author Contributions: Conceptualization, D.F.d.S., F.A.M.S., I.L.S. and A.G.K.; methodology, D.F.d.S., F.A.M.S. and A.G.K.; validation, D.F.d.S., F.A.M.S., H.T. and A.G.K.; investigation, D.F.d.S., F.A.M.S., A.T.d.A. and A.G.K.; resources, F.A.M.S., A.T.d.A., I.L.S. and A.G.K.; data curation, D.F.d.S. and F.A.M.S.; writing—original draft preparation, D.F.d.S., F.A.M.S., I.L.S., H.T., A.T.d.A. and A.G.K.; writing—review and editing, A.T.d.A., F.A.M.S., H.T. and A.G.K.; supervision, H.T. and A.G.K.; funding acquisition, I.L.S. and H.T. All authors have read and agreed to the published version of the manuscript.

Funding: National Electric Energy Agency (ANEEL): Project number 00390-1086/2018 (ENEL)—Integrated assessment of distributed generation, demand management, monitoring, quality, and performance of the network, aiming at optimizing investments and tariff regulation in the underground network, and Project number 00061-0054/2016 (CESP)—Analysis of the efficiency of complementary energy storage with hydroelectric plants, using electrochemical and hydrogen storage technologies; National Council for Scientific and Technological Development (CNPq): Project 870814/1999–0, Process 142323/2020–9.

Institutional Review Board Statement: Not applicable.

Informed Consent Statement: Not applicable.

Data Availability Statement: Not applicable.

Acknowledgments: The authors thank the National Council for Scientific and Technological Development (CNPq) for the scholarship made available to the first author through the project 870814/1999–0, process 142323/2020–9, so that he could dedicate himself to his PhD research in this and related fields. The authors thank the researchers Walter Aguiar Jr. and Kenny Kawaguchi for their support in creating the figures for this paper, and the researcher Jefferson Oliveira for the discussions on electrical machine projects. Furthermore, the workers who worked since the foundation of the Laboratory of Machines of the IEE/USP and throughout their careers generated the database used in this research. The authors thank the researcher Richardson M Abraham-A for improvements in the translation of the manuscript. Moreover, the teacher Lars Eirik Frantzen made the text more friendly. The first author honors Mário Kawaphara and Mateus Rondina from the Federal University of Mato Grosso, for their teaching on SCIMs. The authors would like to thank the researcher José Roberto Cardoso from the Applied Electromagnetism Laboratory (LMAG/USP) for his contributions to the discussion in section (Reduction of Volume and Losses of Ferromagnetic Materials in SCIMs) of this paper. The authors are grateful for the reviewers' criticisms, which contributed to improving this paper.

Conflicts of Interest: The authors declare no conflict of interest.

References

1. Smil, V. *Growth: From Microorganisms to Megacities*; MIT Press: Cambridge, MA, USA, 2020.
2. White, L.A. Energy and the Evolution of Culture. *Am. Anthr.* **1943**, *45*, 335–356. [CrossRef]
3. Zeder, M.A. The Domestication of Animals. *Rev. Anthr.* **1982**, *9*, 321–327. [CrossRef]
4. Mohajan, H. The First Industrial Revolution: Creation of a New Global Human Era. *J. Soc. Sci. Hum.* **2019**, *5*, 377–387.
5. Lamoreaux, N.R.; Levenstein, M.; Sokoloff, K.L. Financing Invention during the Second Industrial Revolution: Cleveland, Ohio, 1870–1920. In *Financing Innovation in the United States, 1870 to the Present*; National Bureau of Economic Research: Cambridge, MA, USA, 2013; Volume 6, pp. 39–84.
6. Thompson, S.P. The Electric Motor and its Applications. *Nature* **1887**, *35*, 410–411. [CrossRef]
7. Baldwin, S.F. The Materials Revolution and Energy-Efficient Electric Motor Drive Systems. *Annu. Rev. Energy* **1988**, *13*, 67–94. [CrossRef]
8. Barkenbus, J. *Roles of Electricity: A Brief History of Electricity and the Geographic Distribution of Manufacturing*; Oak Ridge Associated Universities, Inc.: Oak Ridge, TN, USA, 1987.
9. Devine, W.D. From Shafts to Wires: Historical Perspective on Electrification. *J. Econ. Hist.* **1983**, *43*, 347–372. [CrossRef]
10. Kanefsky, J.; Robey, J. Steam Engines in 18th-Century Britain: A Quantitative Assessment. *Technol. Cult.* **1980**, *21*, 161. [CrossRef]
11. Barnett, R.D. Induction motors: Early development [History]. *IEEE Power Energy Mag.* **2022**, *20*, 90–98. [CrossRef]
12. Heminway, C.M. Automatic safety devices for steam-engines, turbines, and motors. *Proc. Am. Inst. Electr. Eng.* **2014**, *25*, 557–563. [CrossRef]
13. Harford, T. Why Didn't Electricity Immediately Change Manufacturing? *BBC*, 2017. Available online: https://www.bbc.com/news/business-40673694#:~{}:text=Powerwasn\T1\textquoterighttransmittedthrough,thelogicofthedriveshaft(accessed on 15 December 2021).
14. Noe, R. How Did Factories Get Power to Their Machines Before Electricity? *Core77*, 2016. Available online: https://www.core77.com/posts/58982/How-Did-Factories-Get-Power-to-Their-Machines-Before-Electricity#:~{}:text=Factorieshadtheirpowersource,rotatingshaftscalledlineshafts.&text=Beltsranfromthesepulleys,eachmachinethroughanotherpulley(accessed on 13 September 2021).
15. Whitehead, J. 1770's—Water-Powered Textile Mill. Available online: https://www.locallocalhistory.co.uk/brit-land/power/page06.htm (accessed on 16 October 2021).
16. Sens, M.R. *Technological Advances in Electric Motors*; WEG: Jaraguá do Sul, Brazil, 2001; p. 11. (In Portuguese)
17. Neufeld, J.L.; Schurr, S.H.; Burwell, C.C.; Devine, W.D.; Sonenblum, S. *Electricity in the American Economy: Agent of Technological Progress*; Greenwood Press: Westport, CT, USA, 1992; Volume 33, no. 2.
18. Auer, P.L.; Devine, W.D. *An Historical Perspective on Electricity and Energy Use*, 4th ed.; Academic Press: Cambridge, MA, USA, 1983.
19. Edquist, H. Parallel Development? Productivity Growth Following Electrification and the ICT Revolution. *SSRN Electron. J.* **2007**. [CrossRef]
20. Saidur, R. A review on electrical motors energy use and energy savings. *Renew. Sustain. Energy Rev.* **2010**, *14*, 877–898. [CrossRef]
21. De Almeida, J.F.; Fonseca, A.T.; Fernando, P.; Ferreira, J.T.E. EUP Lot 11 Motors—Final. 2008. Available online: https://circabc.europa.eu/sd/d/62415be2-3d5a-4b3f-b29a-d1760f4dc11a/Lot11Motors1-8final28-04-08.pdf (accessed on 7 December 2021).
22. Thangaraj, R.; Pant, M.; Raj, C.T.; Nagar, A.K. In situ efficiency determination of induction motor: A comparative study of evolutionary techniques. *Appl. Artif. Intell.* **2011**, *25*, 116–140. [CrossRef]

23. Ayyappan, G.S.; Nikhil, N.K.; Raja, R.M.; Pandi, R.V.; Angel, T.S.; Babu, R.B. Electrical Motor Maintenance Techniques and Life Cycle Assessment—A Review with Case Studies. In Proceedings of the 2019 2nd International Conference on Power and Embedded Drive Control (ICPEDC), Chennai, India, 21–23 August 2019; Volume 2019, pp. 167–172. [CrossRef]
24. Barnish, T.J.; Muller, M.R.; Kasten, D.J. *Motor Maintenance: A Survey of Techniques And Results*; Rutgers University: New Brunswick, NJ, USA, 1997.
25. Tong, W. *Mechanical Design of Electric Motors*; CRC Press: Boca Raton, FL, USA, 2017.
26. Hughes, A.; Drury, B. *Electric Motors and Drives—Fundamentals, Types and Applications*, 4th ed.; Elsevier Ltd.: Amsterdam, The Netherlands, 2013.
27. Cerra, F. Motor efficiency highly rated. *PACE Process Control Eng.* **2006**, *59*, 21–22.
28. Alger, P.L.; Arnold, R.E. The History of Induction Motors in America. *Proc. IEEE* **1976**, *64*, 1380–1383. [CrossRef]
29. De Almeida, A.T.; Fong, J.; Falkner, H.; Bertoldi, P. Policy options to promote energy efficient electric motors and drives in the EU. *Renew. Sustain. Energy Rev.* **2017**, *74*, 1275–1286. [CrossRef]
30. Aguiar, V.P.B.; Pontes, R.S.T.; Ferreira, F.J.T.E. Technical and economic evaluation of efficiency improvement after rewinding in low-power induction motors: A brazilian case. *Energies* **2018**, *11*, 1701. [CrossRef]
31. Levi, E. Multiphase electric machines for variable-speed applications. *IEEE Trans. Ind. Electron.* **2008**, *55*, 1893–1909. [CrossRef]
32. Reza, C.M.F.S.; Islam, M.D.; Mekhilef, S. A review of reliable and energy efficient direct torque controlled induction motor drives. *Renew. Sustain. Energy Rev.* **2014**, *37*, 919–932. [CrossRef]
33. Kim, Y.; Koo, B.; Nam, K. Induction Motor Design Strategy for Wide Constant Power Speed Range. *IEEE Trans. Ind. Electron.* **2019**, *66*, 8372–8381. [CrossRef]
34. Rautee, J.; Lienesch, F.; Liew, T. Safety improvements of non-sparking and increased safety motors. In Proceedings of the 2008 5th Petroleum and Chemical Industry Conference Europe—Electrical and Instrumentation Applications, Weimar, Germany, 10–12 July 2008. [CrossRef]
35. Clark, P.; Glew, N.; Regan, R. Solutions for MV motors and generators in hazardous locations. In Proceedings of the 1996 IAS Petroleum and Chemical Industry Technical Conference, Philadelphia, PA, USA, 23–25 September 1996; pp. 61–73. [CrossRef]
36. Gunnoe, G.H. Early detection and warning of excessive carbon brush sparking. *IEEE Trans. Power Appar. Syst.* **1972**, *91*, 167–171. [CrossRef]
37. Yang, Z.; Shang, F.; Brown, I.P.; Krishnamurthy, M. Comparative study of interior permanent magnet, induction, and switched reluctance motor drives for EV and HEV applications. *IEEE Trans. Transp. Electrif.* **2015**, *1*, 245. [CrossRef]
38. Le Besnerais, J. Fast prediction of variable-speed acoustic noise due to magnetic forces in electrical machines. In Proceedings of the 2016 XXII International Conference on Electrical Machines (ICEM), Lausanne, Switzerland, 4–7 September 2016; pp. 2259–2265. [CrossRef]
39. Vijayraghavan, P.; Krishnan, R. Noise in electric machines: A review. *IEEE Trans. Ind. Appl.* **1999**, *35*, 1007–1013. [CrossRef]
40. Lopez-Perez, D.; Antonino-Daviu, J. Application of Infrared Thermography to Failure Detection in Industrial Induction Motors: Case Stories. *IEEE Trans. Ind. Appl.* **2017**, *53*, 1901–1908. [CrossRef]
41. Benbouzid, M.E.H. A review of induction motors signature analysis as a medium for faults detection. *IEEE Trans. Ind. Electron.* **2000**, *47*, 984–993. [CrossRef]
42. Bartheld, R.G.; Habetler, T.G.; Kamran, F. Motor Bearing Damage Detection Using Stator Current Monitoring. *IEEE Trans. Ind. Appl.* **1995**, *31*, 1274–1279. [CrossRef]
43. Nandi, S.; Toliyat, H.A. Condition monitoring and fault diagnosis of electrical machines—A review. In Proceedings of the Conference Record of the 1999 IEEE Industry Applications Conference, Thirty-Forth IAS Annual Meeting, Phoenix, AZ, USA, 3–7 October 1999; Volume 1, pp. 197–204. [CrossRef]
44. Wübbeke, J. Rare earth elements in China: Policies and narratives of reinventing an industry. *Resour. Policy* **2013**, *38*, 384–394. [CrossRef]
45. Darcy, J.W.; Bandara, H.M.D.; Mishra, B.; Blanplain, B.; Apelian, D.; Emmert, M.H. Challenges in Recycling End-of-Life Rare Earth Magnets. *JOM* **2013**, *65*, 1381–1382. [CrossRef]
46. Melfi, M.J.; Evon, S.; McElveen, R. Induction versus permanent magnet motors: For power density and energy savings in industrial applications. *IEEE Ind. Appl. Mag.* **2009**, *15*, 28–35. [CrossRef]
47. De Souza, D.F. An Evaluation of the Performance of Three-Phase Induction Electric Motors Sold in Brazil between 1945–2016 and the Impact of Brazilian Legislation. Master's Thesis, University of São Paulo, São Paulo, Brazil, 2018. (In Portuguese)
48. *ISO/IEC 17025*; General Requirements for the Competence of Testing and Calibration Laboratories. International Organization for Standardization (ISO): Geneva, Switzerland, 2017. [CrossRef]
49. De Souza, D.F.; Salotti, F.A.M.; Sauer, I.L.; Tatizawa, H.; Kanashiro, A.G. A Comparison between Reported Values and Measured Values of Power Factor and Efficiency for Electric Induction Motors. *IEEE Lat. Am. Trans.* **2021**, *19*, 173–181. [CrossRef]
50. Boldea, I. Electric generators and motors: An overview. *CES Trans. Electr. Mach. Syst.* **2017**, *1*, 3–14. [CrossRef]
51. Tesla, N. Electro—Magnetic Motor. U.S. Patent 382279, 1 May 1888.
52. Tesla, N. A New System of Alternate Current Motors and Transformers. *Proc. IEEE* **1984**, *72*, 165–173. [CrossRef]
53. Greenwood, P.B. Development of A.C. Motors. In *Electromechanical Prime Movers: Electric Motor Development and Usage*; Palgrave: London, UK, 1972; pp. 9–14.
54. Bowers, B. Scanning Our Past from London: Galileo Ferraris and Alternating Current. *Proc. IEEE* **2001**, *89*, 790–792. [CrossRef]

55. Mitolo, M.; Tartaglia, M. Galileo Farraris—A Life Dedicated to the Electric Sciences [History]. *IEEE Ind. Appl. Mag.* **2016**, *22*, 8–11. [CrossRef]
56. Smil, V. May 1888: Tesla files his patents for the electric motor [Numbers Don't Lie]. *IEEE Spectr.* **2018**, *55*, 24. [CrossRef]
57. Neidhöfer, G. Early three-phase power: Winner in the development of polyphase ac. *IEEE Power Energy Mag.* **2007**, *5*, 88–100. [CrossRef]
58. Guarnieri, M. The development of ac rotary machines [Historical]. *IEEE Ind. Electron. Mag.* **2018**, *12*, 28–32. [CrossRef]
59. Allerhand, A. The Earliest Years of Three-Phase Power-1891–1893. *Proc. IEEE* **2020**, *108*, 215–227. [CrossRef]
60. Allgemeine Elektricitäts-Gesellschaft (AEG). *Little Chronology*; AEG: Berlin, Germany, 2004.
61. Eletrobrás; Procel. *Energy Conservation: Energy Efficiency of Equipment and Facilities*, 3rd ed.; Eletrobrás: Itajubá, Brazil, 2006. (In Portuguese)
62. Smolenskij, I. *Electrical Machines*; Mir Publishers: Moscow, Russia, 1988; Volume 2.
63. WEG. *Module 6—Energy Efficiency in Electric Motor Applications*; WEG Brasil: Jaraguá do Sul, Brazil, 2005. (In Portuguese)
64. Browning, R. Evolution of Induction Motors-Ever-Shrinking Motor. *IEEE Ind. Appl. Mag.* **1997**, *3*, 16. [CrossRef]
65. De Almeida, A.T.; Bertoldi, P.; Leonhard, W. *Energy Efficiency Improvements in Electric Motors and Drives*; Springer: Berlin/Heidelberg, Germany, 1997.
66. Sjöberg, S. Motor Developments and Energy Efficiency. In *Energy Efficiency Improvements in Electric Motors and Drives*; Springer: Berlin/Heidelberg, Germany, 1997; pp. 102–112.
67. Jenkins, J.M.; Parker, J.B.; Clark, R.P. Selection of insulating materials for repair of integral horsepower electrical rotating machinery. In Proceedings of the 1960 EI National Conference on the Application of Electrical Insulation, Chicago, IL, USA, 5–8 December 1960. [CrossRef]
68. Rehder, R.H.; Draper, R.E.; Moore, B.J. How good is your motor insulation system? *IEEE Electr. Insul. Mag.* **1996**, *12*, 8–14. [CrossRef]
69. Mikami, H.; Ide, K.; Shimizu, Y.; Senoo, M.; Seki, H. Historical evolution of motor technology. *Hitachi Rev.* **2011**, *60*, 38–45.
70. Cardoso, J.R.; Salles, M.B.C.; Costa, M.C. *Electromechanical Energy Conversion through Active Learning*; IOP Publishing: Bristol, UK, 2019.
71. Berg, E.J. *Electrical Energy—Its Generation, Transmission, and Utilization*; McGraw Hill: New York, NY, USA, 1908.
72. Steinmetz, C.P.; Berg, E.J. *Theory and Calculation of Alternating Current Phenomena*; Kessinger Publishing, LLC: Whitefish, MT, USA, 1897.
73. Steinmetz, P. On the law of hysteresis. *Trans. Am. Inst. Electr. Eng* **1892**, *9*, 1–64. [CrossRef]
74. Brusso, B.C. One Man's Life Becomes Another's Journey [History]. *IEEE Ind. Appl. Mag.* **2020**, *26*, 8–13. [CrossRef]
75. Steinmetz, C.P. Discussion on 'high-tension testing of insulating materials', 'hysteresis and eddy current exponents for silicon steel', 'commercial problems of transformer design', Schenectady, February 14, 1911. *Trans. Am. Inst. Electr. Eng.* **1911**, *30*, 224–243. [CrossRef]
76. Littmann, M.F. Iron and Silicon-Iron Alloys. *IEEE Trans. Magn.* **1971**, *7*, 48–60. [CrossRef]
77. Krings, A. *Iron Losses in Electrical Machines: Influence of Material Properties, Manufacturing Processes and Inverter Operation*; KTH Royal Institute of Technology: Stockholm, Sweden, 2014.
78. Laithwaite, E.R. The influence of Michael Faraday on power engineering. *Power Eng. J.* **1991**, *5*, 209. [CrossRef]
79. Silveyra, J.M.; Ferrara, E.; Huber, D.L.; Monson, T.C. Soft magnetic materials for a sustainable and electrified world. *Science* **2018**, *362*, 6413. [CrossRef]
80. Cullity, B.D.; Graham, C.D. *Introduction to Magnetic Materials*, 2nd ed.; IEEE Computer Society Press: Washington, DC, USA, 2008.
81. Bozorth, R.M. The physics of magnetic materials. *Electr. Eng.* **2013**, *75*, 134–140. [CrossRef]
82. Mischler, W.R.; Rosenberry, G.M.; Frischmann, P.G.; Tompkins, R.E. Test results on a low loss amorphous iron induction motor. *IEEE Trans. Power Appar. Syst.* **1981**, *100*, 2907–2911. [CrossRef]
83. Duwez, P.; Lin, S.C.H. Amorphous Ferromagnetic Phase in Iron-Carbon-Phosphorus Alloys. *J. Appl. Phys.* **2004**, *38*, 4096. [CrossRef]
84. Yoshizawa, Y.; Oguma, S.; Yamauchi, K. New Fe-based soft magnetic alloys composed of ultrafine grain structure. *J. Appl. Phys.* **1998**, *64*, 6044. [CrossRef]
85. McCoy, G.A.; Douglass, J.G. *Advanced Manufacturing Office Premium Efficiency Motor Selection and Application Guide—A Handbook for Industry Premium Efficiency Motor Selection and Application Guide Disclaimer*; Washington State University Energy Program: Washington, DC, USA, 2014.
86. *ABNT NBR IEC 60085*; Electrical Insulation—Thermal Evaluation and Designation. Brazilian Association of Technical Standards (ABNT): Rio de Janeiro, Brazil, 2017.
87. WEG. Specification Guide—Electric Motors. 2021. Available online: https://static.weg.net/medias/downloadcenter/h32/hc5/WEG-motores-eletricos-guia-de-especificacao-50032749-brochure-portuguese-web.pdf (accessed on 10 January 2022). (In Portugese)
88. *IEC 60034-31*; IEC 60034-31 Ed. 1.0 b:2010—Rotating Electrical Machines—Part 31: Selection of Energy-Efficient Motors Including Variable Speed Applications—Application Guide. International Electrotechnical Commission: Geneva, Switzerland, 2014.
89. *ANSI/NEMA MG 1-2016*; Motors and Generators. The Association of Electrical Equipment and Medical Imaging Manufacturers—NEMA: Arlington, VA, USA, 2016.

90. De Almeida, A.; Fong, J.; Brunner, C.U.; Werle, R.; van Werkhoven, M. New technology trends and policy needs in energy efficient motor systems—A major opportunity for energy and carbon savings. *Renew. Sustain. Energy Rev.* **2019**, *115*, 109384. [CrossRef]
91. Yanti, P.A.A.; Mahlia, T.M.I. Considerations for the selection of an applicable energy efficiency test procedure for electric motors in Malaysia: Lessons for other developing countries. *Energy Policy* **2009**, *37*, 3467–3474. [CrossRef]
92. Waide, P.; Brunner, C.U. *Energy-Efficiency Policy Opportunities for Electric Motor-Driven Systems*; International Energy Agency: Paris, France, 2011; p. 132. [CrossRef]
93. Cepoi, R.D.; Jașcău, F.F.; Szabó, L. Current trends in energy efficient electrical machines. *J. Electr. Electron. Eng.* **2017**, *10*, 13–18. [CrossRef]
94. Lu, S.-M. A review of high-efficiency motors: Specification, policy, and technology. *Renew. Sustain. Energy Rev.* **2016**, *59*, 1–12. [CrossRef]
95. Reine, P. Industrial Motors and Drives: A Global Market Update. In Proceedings of the 9th International Conference on Energy Efficiency in Motor Driven Systems—EEMOD'S15, Helsinki, Finland, 15–17 September 2015; pp. 224–228. [CrossRef]
96. Ferreira, F.J.T.E.; Silva, A.M.; Aguiar, V.P.B.; Pontes, R.S.T.; Quispe, E.C.; de Almeida, A.T. Overview of Retrofitting Options in Induction Motors to Improve Their Efficiency and Reliability. In Proceedings of the 2018 IEEE International Conference on Environment and Electrical Engineering and 2018 IEEE Industrial and Commercial Power Systems Europe (EEEIC/I&CPS Europe), Palermo, Italy, 12–15 June 2018. [CrossRef]
97. De Souza, D.F.; Salotti, F.A.M.; Sauer, I.L.; Tatizawa, H.; de Almeida, A.T.; Kanashiro, A.G. An Assessment of the Impact of Brazilian Energy Efficiency Policies for Electric Motors. *Energy Nexus* **2021**, *5*, 100033. [CrossRef]
98. De Castro, R.A. Feasibility Analysis of Replacement of Oversized Electric Motors and the Influence of Reactive Energy. Master's Thesis, University of Campinas—UNICAMP, Sao Paulo, Brazil, 2008. (In Portuguese)
99. Enomoto, Y.; Tokoi, H.; Imagawa, T.; Suzuki, T.; Obata, T.; Souma, K. Amorphous motor with IE5 efficiency class. *Hitachi Rev.* **2015**, *64*, 480–487.
100. Kotak, V.; Jaiwal, N.K.; Patel, S.N. Improvising strategies for efficiency of IE 4 SCIM 2.2kW through simulation. In Proceedings of the 2016 International Conference on Electrical, Electronics, and Optimization Techniques (ICEEOT), Chennai, India, 3–5 March 2016; pp. 3932–3936. [CrossRef]
101. Lie, S.; di Pietro, C. Copper die-cast rotor efficiency improvement and economic consideration. *IEEE Trans. Energy Convers.* **1995**, *10*, 419–424. [CrossRef]
102. Di Nardo, M.; Marfoli, A.; Degano, M.; Gerada, C.; Chen, W. Rotor Design Optimization of Squirrel Cage Induction Motor-Part II: Results Discussion. *IEEE Trans. Energy Convers.* **2021**, *36*, 1280–1288. [CrossRef]

energies MDPI

Article

Study of the Thermal Conductivity of Soft Magnetic Materials in Electric Traction Machines

Benedikt Groschup *, Alexandru Rosca, Nora Leuning and Kay Hameyer

Institute of Electrical Machines (IEM), RWTH Aachen University, Schinkelstraße 4, 52062 Aachen, Germany; alexandru.rosca@rwth-aachen.de (A.R.); nora.leuning@iem.rwth-aachen.de (N.L.); kay.hameyer@iem.rwth-aachen.de (K.H.)
* Correspondence: benedikt.groschup@iem.rwth-aachen.de

Abstract: The power density of traction drives can be increased with advanced cooling systems or reduced losses. In induction machines with housing and shaft cooling, the produced heat in the stator and rotor winding system needs to be extracted over the rotor and stator lamination. The influence of soft magnetic material parameters, such as texture, thickness or alloy components on the magnetization and loss behavior, are well studied. Studies about influencing factors on the thermal conductivity are hard to find. Within this study, eight different soft magnetic materials are analyzed. An analytical approach is introduced to calculate the thermal conductivity. Temperature-dependent measurements of the electric resistivity are performed to obtain sufficient data for the analytical approach. An experimental approach is performed. The thermal diffusivity, density, and specific heat capacity are determined. An accuracy study of all measurements is performed. The analytical and the experimental approach show good agreement for all materials, except very thin specimens. The estimated measurement error of those specimens has high values. The simplified case study illustrates the significant influence of the different soft magnetic materials on the capability to extract the heat in the given application.

Keywords: induction machines; electrical machines; thermal modeling; soft magnetic material; thermal conductivity

Citation: Groschup, B.; Rosca, A.; Leuning, N.; Hameyer, K. Study of the Thermal Conductivity of Soft Magnetic Materials in Electric Traction Machines. *Energies* **2021**, *14*, 5310. https://doi.org/10.3390/en14175310

Academic Editor: Ryszard Palka

Received: 30 July 2021
Accepted: 20 August 2021
Published: 26 August 2021

Publisher's Note: MDPI stays neutral with regard to jurisdictional claims in published maps and institutional affiliations.

1. Introduction

Increasing the power density of highly utilized traction drives is a frequently discussed research topic. The reduction of losses or the improvement of the heat dissipation capabilities are both potential measures to address this target. A significant influencing factor on the overall efficiency of a traction drive is the selection of the soft magnetic material. The influence of structural material parameters on the efficiency of the electric drive is well studied [1,2]. Eddy losses play a significant role in traction applications, due to their high frequency dependency. In order to reduce this loss share, silicon (Si) and aluminum (Al) can be added as alloy components to the iron matrix. The specific electric resistance ρ_{el} is increased, leading to a reduced loss contribution of the eddy losses [1,2]. A direct dependency between the electron contribution of the thermal conductivity k_e and the specific electric resistance ρ_{el} can be found in the Wiedemann–Franz law:

$$k_e = \frac{L_0 \vartheta}{\rho_{el}}, \tag{1}$$

with the Lorenz number L_0 and the temperature ϑ. As shown within this study, the rule is not fully applicable for alloys, but already indicates a negative impact of increased Si and Al alloy components on the thermal conductivity k. Several influencing thermal parameters, such as the heat transition in the air gap, the interfaces between lamination and housing, the impregnation goodness or the end winding correlation are well studied within the literature [3,4]. A fundamental understanding of the influencing factors of structural soft magnetic parameters on the thermal behavior of electric machines is rare to find. Correlations or validated data for the thermal conductivity of soft magnetic material are not frequently studied. Exact knowledge about the thermal conductivity of soft magnetic materials is crucial for its selection. The selection is especially challenging in the case that the soft magnetic material is placed within the main heat dissipating path. A well suited example for such an application is an Induction Motor (IM) with housing and direct shaft cooling, such as that introduced in [5].

Within this study, an analytical as well as an experimental approach is introduced to obtain data for the thermal conductivity of soft magnetic materials. Eight different soft magnetic materials with different Al and Si content are selected, according to Table 1. The name of the material, an Acronym (Acr.) with the material number from one to eight, the silicon weight content, the aluminum weight content and the nominal thickness are added to the overview. Measurements of the electric resistivity ρ_{el} in dependency of the temperature ϑ are performed to have sufficient data input for the analytical approach. For the experimental approach, the thermal conductivity k_{m} is determined using an indirect measurement technique.

$$k_{\mathrm{m}}(\vartheta) = a(\vartheta) \cdot \rho(\vartheta) \cdot c_{\mathrm{p}}(\vartheta) \tag{2}$$

The thermal diffusivity $a(\vartheta)$ is measured using a Laser Flash Analysis (LFA). The density of the material ρ is measured at room temperature, using the Archimedes principle. A simple model is used to adapt the gained data in dependency of the temperature. A modified model of the Kopp–Neumann law is utilized to determine the specific thermal heat capacity $c_{\mathrm{p}}(\vartheta)$ of the materials. All measurements and models are developed for a temperature range between room temperature and 225 °C. The results of the thermal conductivity in dependency of the temperature $k(\vartheta)$ are compared between the indirect measurement and the analytical approach. The results are used to investigate the influence of the material choice on the thermal heat dissipating capabilities of a traction drive. An IM with direct shaft cooling and housing cooling, as introduced in [5], is selected as a reference for this simplified case study.

Table 1. Alloy weight content and nominal thickness of studied materials.

Name	Acr.	Si in %	Al in %	d in mm
NO1000	M1	0.47	0.03	1
M270-50A	M2	3.38	1.49	0.5
M330-35A	M3	2.6	0.44	0.35
280-30AP	M4	3.64	0.59	0.30
NO30	M5	3	1.067	0.30
NO23	M6	3.64	0.87	0.23
NO20	M7	2.91	1.57	0.20
NO10	M8	6	0	0.10

A study is performed to evaluate the accuracy of the measurement according to [6]. The accuracy is defined as the closeness of agreement between a measured quantity value and a true quantity value of the measurand. The accuracy is not a quantity and cannot be given as a numerical quantity value according to [6]. In order to analyze the accuracy, an estimation of a possible measurement error is performed. This error estimation is aimed to represent the worst possible measurement error. It includes systematic and random measurement errors. The absolute value of the estimated error of a variable x is labeled with Δx. The estimated relative measurement error δx can be expressed with the following:

$$\delta x = \frac{\Delta x}{x}. \quad (3)$$

Please note the difference between accuracy and precision. Precision is the closeness of agreement between indications or measured quantity values obtained by replicate measurements on the same or similar objects under specified conditions [6]. In order to gain good measurement results, both precise and accurate measurement results are necessary. The precision throughout the study shows very good values. The coefficient of variation cv is used for investigating the measurement precision. The cv is calculated by a division of the standard deviation of a measurement sequence and the gained average value.

2. Analytical Formula for the Thermal Conductivity

The two main mechanisms of heat transfer in metallic alloys, such as soft magnetic material, are the phonon k_{p} and the electron thermal conductivity k_{e}. The total value for the analytical approach k_{calc} can be calculated as follows:

$$k_{\mathrm{calc}}(\vartheta) = k_{\mathrm{p}}(\vartheta) + k_{\mathrm{e}}(\vartheta) \quad (4)$$

The mechanism can be subdivided into different scatter processes between phonons, electrons and imperfections. A good overview of the resulting interactions is given in [7,8]. This study mainly uses the correlations as introduced in [8]. The most important assumptions and derivations of the correlations are discussed in the following.

2.1. Phonon Thermal Conductivity

An important quantity for the description of phonon scatter processes is the material-dependent Debye temperature θ_{D}. Several publications address the identification of this variable. Different values are identified in dependency of the used methodology as discussed in [9]. The average value of $\theta_{\mathrm{D}} = 418\,\mathrm{K}$ has developed as the state of the art [8,10,11] and is used in this study. The thermal resistance of the phonon-phonon scatter $W_{\mathrm{p-p}}$ is defined by the following:

$$W_{\mathrm{p-p}}(\vartheta) = \frac{A \cdot \vartheta}{\theta_{\mathrm{D}}}. \quad (5)$$

The constant A is calculated by Julian's modification of the Liebfried–Schlömann equation [8,10,12]. The authors in [10] conclude that the alloy components have a minor influence on the thermal resistance of the phonon–phonon scatter. The values of $A = 0.412\,\mathrm{m\,K/W}$ and $\theta_{\mathrm{D}} = 418\,\mathrm{K}$ for pure α-iron can be used for the calculation. The authors in [11] suggest a correction term for $\vartheta > \theta_{\mathrm{D}}$ to consider the thermal expansion. The temperature range in this study is limited to values of 498 K. The correction term is neglected because the maximum influence is 0.2% in the considered temperature range. The phonon–phonon scatter $W_{\mathrm{p-p}}$ can be simplified using the introduced simplifications as follows:

$$W_{\mathrm{p-p}}(\vartheta) = 9.86 \times 10^{-5}\,\mathrm{m/W} \cdot \vartheta \quad (6)$$

The formulation of the thermal resistance of the phonon–electron scatter $W_{\mathrm{p-e}}$ shows a temperature dependency. A formulation of $W_{\mathrm{p-e}}$ at a temperature that is equal to the Debye temperature θ_{D} is given with the following:

$$W_{\mathrm{p-e}} = 2.69 \times 10^{-2}\,\mathrm{m\,K/W} \quad (7)$$

Additional therms need to be considered for temperatures below the Debye temperature θ_{D} [10,11]. In this study, only small differences from the Debye temperature are considered, and the additional therms are neglected. A similar simplification is used in [8] for alloys. The thermal resistance of the phonon–impurity scatter $W_{\mathrm{p-i}}$ describes the interaction between impurities such as the alloys and the Fe-lattice. The process can be modeled as follows [8]:

$$W_{\mathrm{p-i}} = B \cdot \sum_i c_{\mathrm{a,i}} \cdot \Gamma_{\mathrm{i}}, \tag{8}$$

with an experimental constant B, the impurity parameter Γ_{i} and the atomic content $c_{\mathrm{a,i}}$ of each alloy i. The value of B is given in [8] with $B = 1.36 \times 10^{-2}\,\mathrm{m\,K/W}$. The impurity factor Γ can be calculated based on weighted ratios of the molar masses and the molar volumes between the alloy contents i and the Fe-lattice. The values for nine different alloys are given in [8]. The value for silicon, aluminum, and manganese is equal to 0.59, 0.05, and zero, respectively. This means that the influence of aluminum is almost negligible, and the influence of manganese is not present. The phonon thermal conductivity k_{p} can be calculated as the inverse of a sum of the three influences $W_{\mathrm{p-p}}(\vartheta)$, $W_{\mathrm{p-e}}$, and $W_{\mathrm{p-i}}$:

$$k_{\mathrm{p}}(\vartheta) = \left(9.86 \times 10^{-5}/\mathrm{K} \cdot \vartheta + 2.69 \times 10^{-2} + 1.36 \times 10^{-6} \cdot \sum_i c_{\mathrm{a,i}} \cdot \Gamma_{\mathrm{i}}\right)^{-1} \mathrm{W/(m\,K)}. \tag{9}$$

2.2. Electron Thermal Conductivity

The first process that influences the electron thermal conductivity is the electron-phonon scatter. The process is described by the thermal resistance $W_{\mathrm{e-p}}$ [10]:

$$W_{\mathrm{e-p}}(\vartheta) = \frac{\rho_{\mathrm{el\,Fe}}(\vartheta)}{L(\vartheta) \cdot \vartheta}. \tag{10}$$

The ideal electric resistivity is the electric resistivity of pure iron $\rho_{\mathrm{el\,Fe}}(\vartheta)$ in this application. The value can be calculated utilizing the following [8,13]:

$$\rho_{\mathrm{el\,Fe}}(\vartheta) = \left(-2.4 + 3.65 \times 10^{-2}/\mathrm{K} \cdot \vartheta + 64 \times 10^{-9}/\mathrm{K}^3 \cdot \vartheta^3\right) \mu\Omega\,\mathrm{cm}. \tag{11}$$

Please note that there seems to be a typo in the original source in [13] because the values do not match the measurement data presented in the publication. This typo is corrected in [8]. The values of [8] are used in this study and presented in Equation (11). The Lorenz number $L_0 = 2.443 \times 10^{-8}\,\mathrm{W\,\Omega/K^2}$ is modified by a temperature-dependent therm [10] in Equation (10):

$$L(\vartheta) = L_0 \cdot \left(1 - e^{-\vartheta/159.3\,\mathrm{K}}\right). \tag{12}$$

The second considered electron scattering process is the electron–impurity scatter that is represented by $W_{\mathrm{e-i}}$:

$$W_{\mathrm{e-i}}(\vartheta) = \frac{\rho_{\mathrm{el0}}(\vartheta)}{L_0 \cdot \vartheta}, \tag{13}$$

with the residual electrical resistivity $\rho_{\mathrm{el0}}(\vartheta) = \rho_{\mathrm{el}}(\vartheta) - \rho_{\mathrm{el\,Fe}}(\vartheta)$ as a difference between the electric resistivity of the alloy $\rho_{\mathrm{el}}(\vartheta)$ and the electric resistivity of pure iron $\rho_{\mathrm{el\,Fe}}(\vartheta)$. The electric resistivity of an alloy can be calculated utilizing Matthiessen's rule as follows:

$$\rho_{\mathrm{el}}(\vartheta) = \rho_{\mathrm{el\,Fe}}(\vartheta) + \sum_i \rho_{\mathrm{el_i}} c_{\mathrm{a,i}}, \tag{14}$$

The calculated values utilizing this formula deviate from measurement results as discussed in [10], due to the independency of the electric resistivity $\rho_{\mathrm{el_i}}$ from the temperature. The authors [10] propose an improved formulation:

$$\rho_{\mathrm{el}}(\vartheta) = \rho_{\mathrm{el\,Fe}}(\vartheta) + \sum_i \rho_{\mathrm{el_i}}(\vartheta) c_{\mathrm{a,i}}, \tag{15}$$

The estimation of the necessity of utilizing the improved equation in comparison to the Matthiessen's rule is not possible for this application. The error estimation in [10] is based on values of the electric resistivity of the alloy at 4 K and based on Cr and Ni alloys. While

Equation (14) is preferred, due to the simple availability of the electric resistivity of the alloy components ρ_{el_i}, Equation (15) seems to give more accurate results. Within this study, the electric resistivity of the alloys is measured over the temperature range and compared to the simplified equation. The final value of the electron thermal conductivity $k_e(\vartheta)$ can be calculated with:

$$k_e(\vartheta) = \left(\frac{\rho_{\text{el Fe}}(\vartheta)}{L(\vartheta) \cdot \vartheta} + \frac{\rho_{\text{el0}}(\vartheta)}{L_0 \cdot \vartheta}\right)^{-1}. \tag{16}$$

2.3. *Measurements of the Temperature Dependent Electric Resistivity*

Measurements of the electric resistivity are performed using the measurement setup as depicted in Figure 1. The setup is designed following the recommendations of the standard DIN EN IEC 60404-13 [14] on a smaller scale because the specimens are not available in the recommended size. Probes of 120 mm × 20 mm are used. In order to gain sufficient accuracy, an analysis of the measurement uncertainties is performed. The specimen is inserted into a measurement fixture and placed inside an oven. The electric resistivity $\rho_{\text{el}}(\vartheta)$ is calculated with the following equation:

$$\rho_{\text{el}}(\vartheta) = \frac{U(\vartheta) \cdot d \cdot w}{I \cdot l}. \tag{17}$$

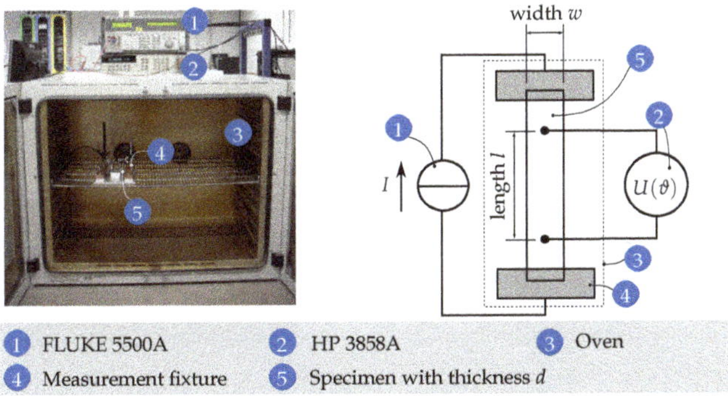

Figure 1. Measurement setup for the electric resistivity.

A DC-current I between 0.2 A and 2 A is introduced into the sample using a FLUKE 5500A Multi-Product Calibrator [15]. The maximum measurement error of this current is given by $\Delta I = 0.64$ mA in the manufacturer data sheet [15]. The voltage is measured with a separate device to improve the accuracy of the measurement. A HP 3458A Multimeter is used for this purpose. The maximum measured voltage during the procedure is around 20 mV. With an maximum measurement error of 16.4 ppm of the reading and 22 ppm of the measurement range, a maximum measurement error for the voltage of $\Delta U = 2.5\,\mu$V is specified in the data sheet of the device [16]. Two measurement tips are placed on the probe for the voltage measurement. The distance between the measurement tips is equal to the measuring length l. The measuring length l as well as the width of the specimen w are measured using a digital caliper from Mitutoyo. The maximum measurement error of the measurement device is given with 0.02 mm. In order to include measurement errors that result from geometrical deviations, a total error estimation of $\Delta l = \Delta w = 0.05$ mm is included in the accuracy calculation. The measurement of the two values is repeated at least three times and the average value is calculated. The thickness of the sheet d is measured after removing the insulation, using an outside micrometer from Mitutoyo. The measurement is repeated six times and the average value is used. The coefficient of variation cv of the iterative geometrical measurement procedure is given in Table 2. cv is well below 1% with one outlier of material M8 for the thickness measurement. The measurements show good precision.

Table 2. Coefficient of variation cv for measurements of geometrical values in %.

Material	$cv(l)$	$cv(w)$	$cv(d)$
M1-NO1000	0.29	0.04	0.08
M2-M270-50A	0.09	0.04	1.00
M3-M330-35A	0.41	0.02	0.91
M4-280-30AP	0.09	0.03	0.63
M5-NO30	0.19	0.03	0.85
M6-NO23	0.06	0.08	0.54
M7-NO20	0.30	0.04	0.72
M8-NO10	0.30	0.06	1.28

The maximum measurement error of the micrometer is given with 0.001 mm. In order to account for geometric errors, a total measurement error estimation of $\Delta d = 0.01$ mm for the thickness d is used for the accuracy evaluation. The measurement error estimation $\Delta\rho_{\mathrm{el}}$ is calculated at room temperature $\vartheta = 293$ K using the following formula:

$$\Delta\rho_{\mathrm{el}}(\vartheta) = \frac{(U(\vartheta) + \Delta U) \cdot (d + \Delta d) \cdot (w + \Delta w)}{(I - \Delta I) \cdot (l - \Delta l)} - \rho_{\mathrm{el}}(\vartheta). \tag{18}$$

The results of the accuracy study ($x = \rho_{\mathrm{el}}$) are depicted in table Table 3. A trend of increasing measurement errors with decreasing specimen thickness can be observed. One exception of this trend is the decreased accuracy of M1 in comparison to M2. This exception is caused by an increased influence of ΔU for this material. Due to the high thickness of this material, the resulting measured voltage U is relatively low at a comparable current I. For decreasing thickness, the error caused by the thickness d dominates the overall influence. For material M8 for instance, the measurement error of the thickness measurement Δd causes about 10.6%, while other influences only contribute by 1.1%. Please note that this influence also could not have been changed utilizing the recommended setup in DIN EN IEC 60404-13 [14], as the thickness of the specimen would have a similar influence. The results of the measurement of the electric resistivity in dependency on the temperature $\rho_{\mathrm{el}}(\vartheta)$ are plotted in Figure 2. The results show a significant difference for the electric resistivity ρ_{el} between the different materials. At first glance, the material d seems to influence the ρ_{el}. This impression is not correct because the alloy components Si and Al are the primary influencing factors. As an example, M1 is the thickest selected material and has a very low silicon and aluminum content. Material M8 is the thinnest material with the highest silicon content. A plausibility check can be performed utilizing the ternary plot of the electric resistivity ρ_{el} as a function of the silicon $c_{\mathrm{w,Si}}$ and aluminum $c_{\mathrm{w,Si}}$ and weight content published in [17]. The eight different materials are added to the plot in Figure 3 based on their silicon and aluminum content. The gained experimental results show very good agreement with this plot gained from the literature at room temperature.

Table 3. Estimated measurement error $\delta\rho_{\mathrm{el}}(\vartheta = 293\,\mathrm{K})$ in %.

Material	$\delta\rho_{\mathrm{el}}$	Material	$\delta\rho_{\mathrm{el}}$	Material	$\delta\rho_{\mathrm{el}}$
M1-NO1000	4.3	M2-M270-50A	2.9	M3-M330-35A	3.9
M4-280-30AP	3.8	M5-NO30	4.3	M6-NO23	4.5
M7-NO20	5.3	M8-NO10	11.7		

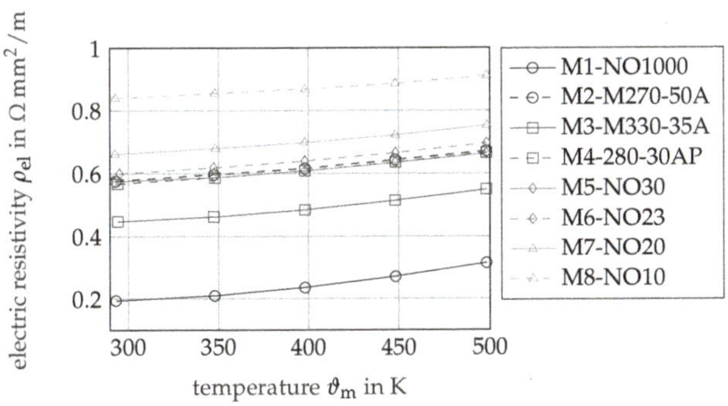

Figure 2. Results of the electric resistivity measurements $\rho_{el}(\vartheta)$.

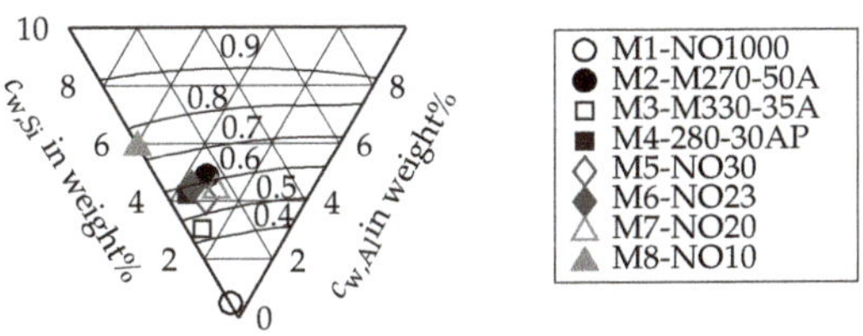

Figure 3. Theoretical values for the electric resistivity $\rho_{el}(\vartheta = 298\,\text{K})$ in $\Omega\,\text{mm}^2/\text{m}$ based on the silicon and aluminum content (Source: Data from [17]).

3. Experimental Evaluation of the Thermal Conductivity

An indirect measurement technique is used for the evaluation of the thermal conductivity $k_m(\vartheta)$, i.e., the thermal diffusivity $a(\vartheta)$ is measured. The thermal conductivity of the measurement k_m is calculated using the following formula:

$$k_m(\vartheta) = a(\vartheta) \cdot \rho(\vartheta) \cdot c_p(\vartheta) \tag{19}$$

The procedure of evaluating the thermal diffusivity $a(\vartheta)$, the mass density $\rho(\vartheta)$, and the specific heat capacity $c_p(\vartheta)$ is introduced in the following.

3.1. Measurements of the Thermal Diffusivity

The measurements of the thermal diffusivity are performed with a Netzsch LFA 427 measurement device. The samples are cut into specimens with a side length of $10\,\text{mm} \pm 0.1\,\text{mm}$. The insulation of the steel sheets is removed with sandpaper with a 500 grit. A thin graphite layer is added on the samples for improved absorption of the laser impulse. A schematic overview as well as some images of the measurement device are depicted in Figure 4. The specimen is inserted into the sample holder. The device is closed, and the sample is purged with argon as a protective gas. A laser pulse is shot at the specimen, and the temperature rise is measured on the back side using an indium antimonide (InSb) infrared detector. An exemplary measurement signal of material M1 at 348 K is depicted in Figure 5. Different models are available for the evaluation of the thermal diffusivity. The first approach is introduced by Parker [18]. The relative maximum signal $s_{max} = 1$ is evaluated. The half time $t_{1/2}$ represents the time, when half $s_{1/2}$ of the maximum signal s_{max} is reached. The half time is used to calculate the thermal diffusivity a according to [18]:

$$a = 0.1388 \frac{d^2}{t_{1/2}}. \tag{20}$$

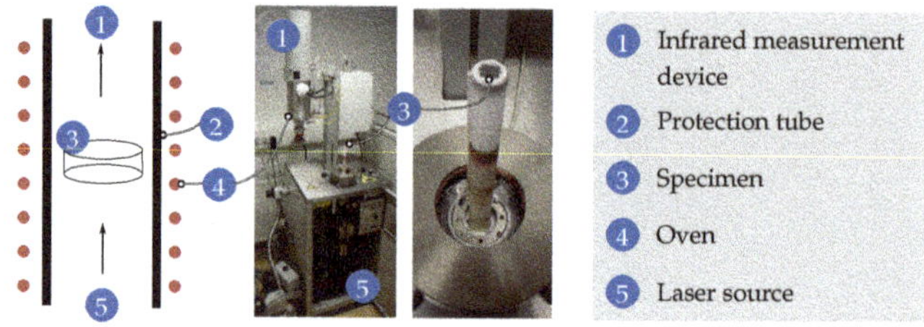

Figure 4. Measurement setup for the thermal diffusivity.

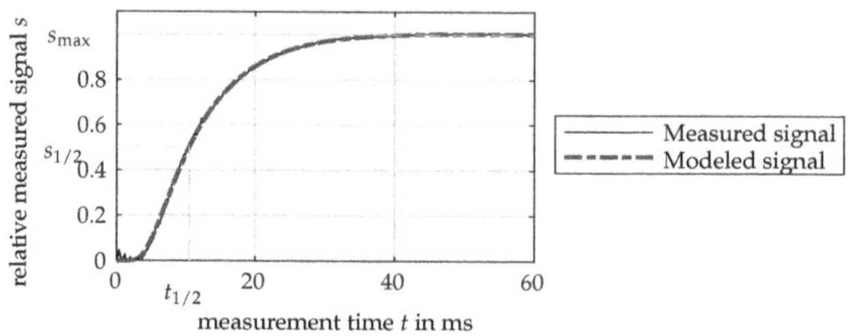

Figure 5. Exemplary measurement signal of material M1 at 348 K.

Parker assumes ideal conditions, such as an instantaneous energy pulse, adiabatic boundary conditions or constant material properties during the temperature rise. Several improvements of this model are performed. The losses on the front and rear end are added by Cowan [19]. Radial losses are added by Cape–Lehman [20]. Within this study, the standard model with a horizontal baseline correction of the NETZSCH Proteus Software Version 7.1.0. is used, i.e., an improved version of the Cape–Lehman formulation. A total of five temperatures are measured for each material, ranging from room temperature up to 498 K. At each temperature, at least six measurements are used for the calculation of the average value and the variation. The coefficient of variation of the thermal diffusivity measurements cv is given in Table 4.

Table 4. Coefficient of variation cv for measurements of thermal diffusivity a in %.

Material	ϑ = 293 K	ϑ = 348 K	ϑ = 398 K	ϑ = 448 K	ϑ = 498 K
M1-NO1000	1.13	0.80	0.62	1.22	0.52
M2-M270-50A	0.90	0.33	0.21	0.19	0.21
M3-M330-35A	0.66	0.45	0.40	0.51	0.40
M4-280-30AP	0.41	0.61	0.44	0.52	0.72
M5-NO30	0.61	0.46	0.60	0.50	0.36
M6-NO23	0.63	0.72	0.59	0.70	0.70
M7-NO20	0.98	0.74	0.56	0.81	0.94
M8-NO10	1.61	1.24	2.65	1.53	0.76

The coefficient of variation is below 1.22% for all materials, except material M8. Material M8 shows a maximum coefficient of variation of 2.7%. These values indicate a good precision of the measurement. Please note that these values only consider the influence of the thermal diffusivity measurement procedure, i.e., the thickness is a constant value and not considered in Table 4. The measurement is repeated five times, and the average value is used for the measurements. For the measurement error estimation, the simplified Parker formula is used [18]. The general accuracy of the NETZSCH LFA is assumed to be ±3% for a 1 mm sample as given in the device data sheet [21]. This value is not included in the accuracy of different thickness measurements, as one constant value is being used during LFA measurements. The thickness is measured with a similar outside micrometer as utilized in the measurement of the electric resistivity with a maximum measurement error of 0.001 mm. A value of $\Delta d = 0.01$ mm is used for the error estimation to include geometrical errors. The estimated error of the resulting thermal diffusivity Δa is calculated as follows:

$$\Delta a = 0.1388 \frac{(d + \Delta d)^2}{t_{1/2}} \cdot x_{\mathrm{m}} \tag{21}$$

The value for x_{m} is selected to be 1%, as for this value, the measurement error of the 1 mm sample of M1 fits to the given data sheet value for the maximum measurement error of 3%. The resulting estimated measurement errors according to Equation (3) for $x = a$ are depicted in Table 5. A strong dependency on the thickness of the specimen is evaluated. The results of the thermal diffusivity measurement a are shown in Figure 6.

The thermal diffusivity a varies in a range from 3.1 $\mathrm{mm^2/s}$ to 13.6 $\mathrm{mm^2/s}$. A significant difference between the values of the eight materials is visible. In particular, material M1 with a very low alloy content (see Table 1) and M8 with a very high alloy content stand out in the comparison. The thermal diffusivity of M1 is about 434% larger than the thermal diffusivity of M8 at 298 K. The thermal diffusivity is expected to play a significant role in the calculation of the thermal conductivity (Equation (19)) and the maximum possible dissipated heat in the application.

Table 5. Estimated measurement error of the thermal diffusivity measurement $x = a$ in %.

Material	$\delta\rho_{\mathrm{el}}$	Material	$\delta\rho_{\mathrm{el}}$	Material	$\delta\rho_{\mathrm{el}}$
M1-NO1000	3.0	M2-M270-50A	5.2	M3-M330-35A	6.8
M4-280-30AP	8.1	M5-NO30	8.0	M6-NO23	9.3
M7-NO20	11.0	M8-NO10	23.4		

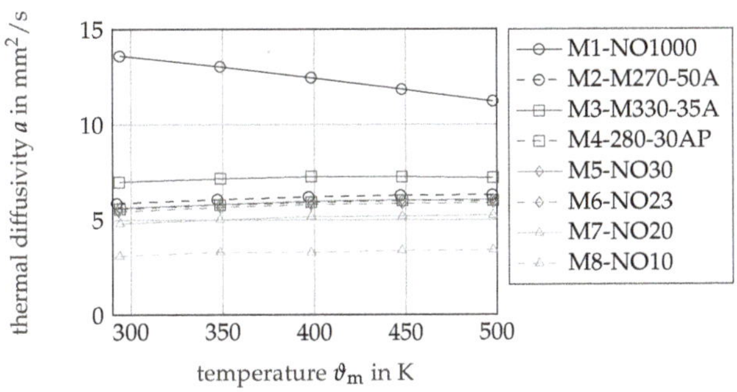

Figure 6. Results of the thermal diffusivity measurements $a(\vartheta)$.

3.2. Measurements of the Density

Two different possibilities to measure the density of the sheets are studied and compared. The first methodology is a geometric-based method, utilizing the dimensions and the mass of the specimens. The LFA specimens are used for this purpose with the length $l \approx 10$ mm, the width $w \approx 10$ mm, and the thickness from $d \approx 0.1$ mm up to $d \approx 1$ mm. The measurement of the three values is repeated five times and an average value is calculated. Similar measurement equipment, as described in the measurements of the electric resistivity, is used for all three quantities. The estimation of the measurement error is $\Delta d = 0.01$ mm and $\Delta l = \Delta w = 0.05$ mm. The weight m of the probe is measured using a Sartorius high precision balance with a maximum error of 0.1 mg. In order to account for possible dust or additional influences, a value of $\Delta m = 1\,\text{mg}$ is used for the accuracy study.

$$\rho = \frac{m}{d \cdot l \cdot w} \tag{22}$$

The error of the procedure can be estimated with the following:

$$\Delta\rho = \frac{m + \Delta m}{(d - \Delta d) \cdot (l - \Delta l) \cdot (w - \Delta w)} \tag{23}$$

The second methodology utilizes the principle of Archimedes. In this measurement technique, no exact cubic probe is necessary. A higher amount of pieces can be utilized to obtain a higher overall measured weight. The insulation of the material is removed by sandblasting. The density measurements are performed with an analytical balance Kern ABT 220-4M. The measurement error of the weight measurement is 0.1 mg. Multiple probes are cut into specimens that fit into the universal immersion basket of the balance. An overview of the measurement equipment is given in Figure 7. The distilled water, used as the reference fluid, is filled into a beaker. The temperature of the reference fluid ϑ_0 is measured, utilizing the thermometer included in the balance equipment. The density of the reference fluid is evaluated from a lookup table $\rho_0 = f(\vartheta_0)$. The error of the mass density of the reference fluid is estimated by a 5 K-deviation in the temperature measurement as follows:

$$\Delta\rho_0(\vartheta) = \rho_0(\vartheta) - \rho_0(\vartheta + 5\,\text{K}) \tag{24}$$

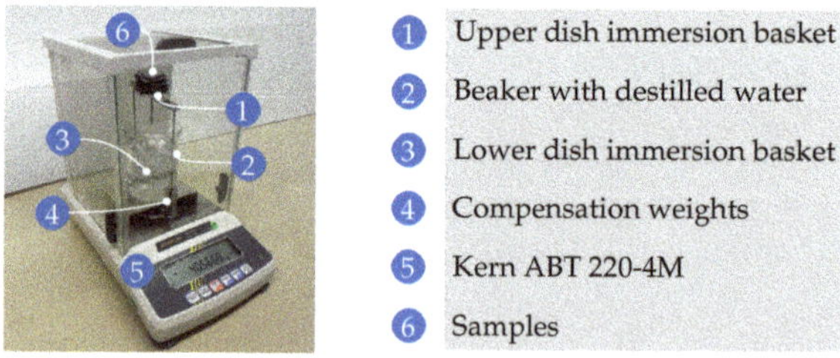

Figure 7. Analytical balance Kern ABT 220-4M with universal immersion basket.

The first measurement is performed with the specimens placed onto an upper sample dish of the immersion basket. The amount of samples is adapted to obtain a total weight of approximately $m_A \approx 40\,\text{g}$. The measured value of m_A is the result of the difference between the buoyancy force of the air and the weight force of the specimen:

$$m_A = (\rho - \rho_{\text{air}}) \cdot V, \tag{25}$$

with the volume of the specimen V, and the density of air ρ_{air}. The influence of the air buoyancy force is neglected in the calculation. An additional factor is considered in the error estimation of Δm_A:

$$\Delta m_A = \left(\frac{\rho_{\text{air}}}{\rho}\right) \cdot m_A + 1\,\text{mg} \tag{26}$$

The measurement is repeated with the samples placed on the lower dish of the immersion basket. The measured weight m_B is equal to the following:

$$m_B = (\rho - \rho_0) \cdot V. \tag{27}$$

The estimated error of the measurement of m_B is assumed to be equal to $\Delta m_B = 1\,\text{mg}$. The combination of Equations (25) and (27) under neglection of the air buoyancy force gives the equation to calculate the mass density ρ of the specimen.

$$\rho = \frac{m_A}{m_A - m_B}\rho_0 \tag{28}$$

A worst-case estimation for the measurement error of the density ρ is performed with the following:

$$\Delta\rho = \frac{m_A + \Delta m_A}{(m_A - \Delta m_A - m_B - \Delta m_B)} \cdot (\rho_0 + \Delta\rho_0) - \rho \tag{29}$$

The entire measurement procedure is repeated three times. After each measurement in the water reference, the samples are dried. A thin rust film develops within seconds and is removed by sandblasting. All values are reevaluated, including the masses m_A and m_B, as the repeated sandblasting also removes some of the material. An average value of the three measurements is calculated.

The results of the measurements as well as the error estimation according to the calculation of Equation (3) for $x = \rho$ are depicted in Table 6. The coefficient of variation cv for the measured values is added. cv of the geometrical values l, d and w shows values below 1.1% for most of the values. Only the thickness measurements show larger values with 2.5% for M8 and 1.86% for M6. The coefficient of variation for the Archimedes principle shows small values below 0.13% for all measured materials. The precision of the Archimedes principle is significantly improved in comparison to the geometrical principle. The accuracy of the geometrical principle is mainly driven by the estimated error of the thickness measurement Δd, which leads to high error estimations for the thin sheets. The error estimation model gives a minimal value of 2.2% for the thickness material M1 and 14.7% for the thinnest material. The estimated error of the Archimedes principle with values around 0.2% is very low and almost equal for all probes. The accuracy is independent from the thickness of the probe. The variation for the three measurement repetitions of the Archimedes principle varies between 0.01% and 0.13%. This value is lower than the predicted values for the measurement error $\delta\rho$ in Table 6. This observation confirms the good precision of the measurement and confirms the estimation of the measurement accuracy being the critical value. The measured density utilizing the geometrical principle is lower than the density evaluated by the Archimedes principle between 3.2% and 4.7%. It is interesting to note that the values are all lower and not spread around the exact values of the Archimedes principle. The difference between the results shows the lowest value for Material M1, which confirms the trend indicated by the accuracy study. Additionally, the value of 3.2% is larger than the predicted error of 2.2% as a sum of the two error estimations. There are obviously some additional systematic errors present. Issues with air bubbles in the Archimedes measurement do not seem to be present, as the variation coefficient of the measurement is low. Air bubbles in the second measurement step would decrease m_B, which leads to an underestimation of the density ρ. This is not the case, because the results of the Archimedes principle are all larger than those of the geometrical probes. A possible explanation for the effect is the cuboid model that is used for the estimation of the volume in the geometric approach. The measured values are the outer dimensions. Irregularities and roughness could lead to a real volume that is lower. This would cause lower values of the density ρ. The temperature dependency is estimated, using a thermal expansion coefficient of $\alpha_{th} = 11.8 \times 10^{-6}/\mathrm{K}$.

$$\rho(\vartheta) = \rho \cdot \frac{1}{1 + 3 \cdot \alpha_{th} \cdot (\vartheta - \vartheta_0)}, \tag{30}$$

with the measurement temperature as the reference temperature ϑ_0.

The density ρ varies in a range from 7479 kg/m to 7834 kg/m. The value for material M1 is only 5% larger than the value of M8 at 298 K. The mass density is expected to play a minor role for the differences in the thermal conductivities of the materials, according to Equation (19) and the maximum possible dissipated heat in the application. In the case of a study with fewer accuracy requirements, an average value of the expected density could be used with a maximum error of the indicated 5%.

Table 6. Measurement results and error estimation of the density measurement at $\vartheta \approx 293\,\mathrm{K}$.

	Geometrical Principle					Archimedes Principle		
Material	$cv(l)$ in %	$cv(w)$ in %	$cv(d)$ in %	ρ in kg/m³	$\delta\rho$ in %	$cv(\rho)$ in %	ρ in kg/m³	$\delta\rho$ in %
M1-NO1000	0.15	0.05	0.25	7581	2.2	0.13	7834	0.21
M2-M270-50A	0.02	0.06	1.05	7288	3.4	0.02	7553	0.21
M3-M330-35A	0.06	0.07	0.57	7398	4.4	0.13	7678	0.20
M4-280-30AP	0.03	1.10	0.38	7222	5.2	0.06	7579	0.21
M5-NO30	0.08	0.07	0.45	7302	5.1	0.04	7565	0.21
M6-NO23	0.16	0.02	1.86	7271	5.9	0.01	7576	0.20
M7-NO20	0.10	0.02	0.56	7162	6.9	0.02	7503	0.24
M8-NO10	0.07	0.20	2.50	6964	14.7	0.12	7479	0.22

3.3. *Evaluation of the Thermal Heat Capacity*

The influence of alloy components on the specific thermal heat capacity c_p can be evaluated utilizing the Kopp–Neumann law. The weight content of each alloy c_{wi} as well as the specific heat capacity of each alloy component $c_{\mathrm{p_i}}$ needs to be known for this law:

$$c_\mathrm{p} = \sum_i c_{\mathrm{wi}} \cdot c_{\mathrm{p_i}}, \tag{31}$$

Both requirements are fulfilled for this study. Temperature-dependent measurement data of the specific heat capacities are available in the literature for silicon [22], aluminum [23], and pure iron [24]. Within this input data, the temperature dependency is considered as used in [25]:

$$c_\mathrm{p}(\vartheta) = \sum_i c_{\mathrm{wi}} \cdot c_{\mathrm{p_i}}(\vartheta) \tag{32}$$

The resulting heat capacities in dependency of the temperature are depicted in Figure 8. The Kopp–Neumann law is, strictly speaking, only valid for composites but the comparison of measured heat capacities of Fe-based alloys with the Kopp–Neumann law can be found in the literature, such as [26]. Further, the difference of the calculated heat capacity of the eight materials shows minor influence on the later thermal conductivity. This can be illustrated with the maximum deviation of the values at room temperature. The c_p of M1 deviates from the M8 value at room temperature by 3.6%. The thermal diffusivity a of M1 is 334% larger than the value of M8 at room temperature. For further validation and estimation of the errors, the calculated values of the heat capacity based on Equation (31) are compared to the results evaluated from the commercial software JMatPro Version 8. The comparison is performed for pure iron and Materials M7 and M8, which are the materials with the highest Al and Si content, respectively. The maximum deviation between the two approaches is around 1%. For the error estimation study, a value of $\delta c_{\mathrm{p_{rel}}} = 2\%$ is used for all materials.

Due to the small differences between c_p of the different materials, only a minor influence on the difference of the thermal conductivities of the materials, according to Equation (19), is expected. A minor influence on the maximum dissipated heat is expected. The value varies between $451\,\mathrm{J/(kg\,K)}$ for M1 and $465\,\mathrm{J/(kg\,K)}$ for M8. In the case of a study with fewer accuracy requirements, an average value of the expected heat capacity could be used with a maximum error of the indicated 3.6%.

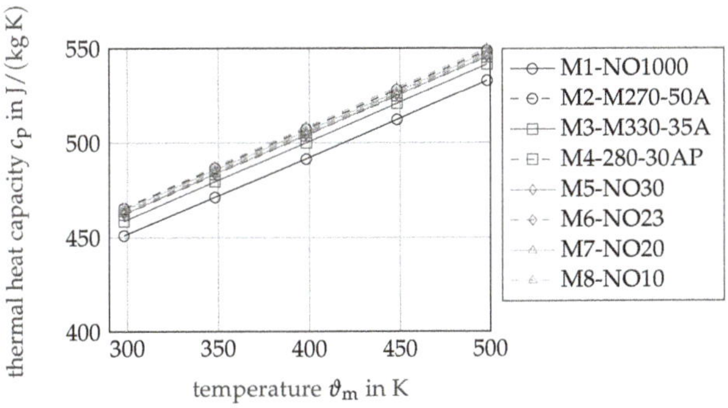

Figure 8. Results of the calculation of the thermal heat capacity $c_p(\vartheta)$.

4. Simplified Case Study

The influence of the thermal conductivity k of the different materials should be studied in a simplified case study. This investigation is performed from a thermal perspective. The used reference design is an induction machine for a commercial truck traction application [5]. The following question is in the focus of the study:

How much loss power can be extracted from the electric machine in dependency of the used soft magnetic material?

Please note that this study does not aim to use a highly accurate thermal model of the machine. The model is kept as simple as possible to understand the fundamental correlations of the material choice. The following assumptions and boundary conditions are made: the rotor and stator of the machine are separated by an ideal thermal insulator, i.e., no heat is transferred through the air gap. The stator is equipped with a housing cooling. The rotor is equipped with a rotor shaft cooling. The stator notch cooling, as found in [5], is not considered. The geometry is simplified by a cylindrical shell model as depicted in Figure 9. The height of the stator yoke and the height of the rotor yoke are kept constant, i.e., the heat transfer path through this part of the lamination is kept constant, and the influence of the teeth is neglected. The model is two-dimensional, i.e., the influence of heat extraction in end windings, bearings and bearing shields is neglected. All stator losses are introduced in the shell of the stator winding, and all rotor losses are introduced in the shell of the rotor bar. The heat conduction of the stator winding shell and the rotor bar shell is infinite. All thermal interface resistances are neglected. The housing and the shaft have the same temperature as the cooling fluid $\vartheta_{\text{fluid}} = 50\,°\text{C}$, i.e., the thermal resistance between the lamination and the housing/shaft, the convectional resistance between wall and fluid, and the heating up of the fluid are neglected. The bending of the shells is neglected, i.e., the shells are modeled as flat plates, utilizing the average diameter of the shell $(d_{\text{out}} + d_{\text{i}})/2$. The thermal resistance of the stator R_1 and rotor R_2 iron are calculated as follows:

$$R_{1/2} = \frac{d_{\text{out},1/2} - d_{\text{i},1/2}}{k_{\text{m}} \cdot \pi \cdot (d_{\text{out},1/2} + d_{\text{i},1/2}) \cdot l_{\text{i}}}, \tag{33}$$

with the outer and inner diameters $d_{\text{out1}} = 282\,\text{mm}$; $d_{\text{out2}} = 168\,\text{mm}$; $d_{\text{i1}} = 214\,\text{mm}$ and $d_{\text{i2}} = 100\,\text{mm}$, the active length of the lamination $l_{\text{i}} = 285\,\text{mm}$, and the evaluated thermal conductivity of the measurements k_{m}. The studied operational point is under steady-state behavior. The two thermal Lumped Parameter Thermal Network (LPTN) circuits are depicted in Figure 10. The maximum allowed temperature of stator winding and rotor bar is $\vartheta_{\text{max}} = 180\,°\text{C}$. The maximum power loss P_{loss} in the steady-state operation that can be extracted from the rotor or the stator is calculated as follows:

$$P_{\text{loss}} = \frac{\vartheta_{\text{max}} - \vartheta_{\text{fluid}}}{R_{1/2}(\vartheta_{\text{avg}})}, \tag{34}$$

with the average temperature of the stator or rotor iron $\vartheta_{avg} = (\vartheta_{max} + \vartheta_{fluid})/2$.

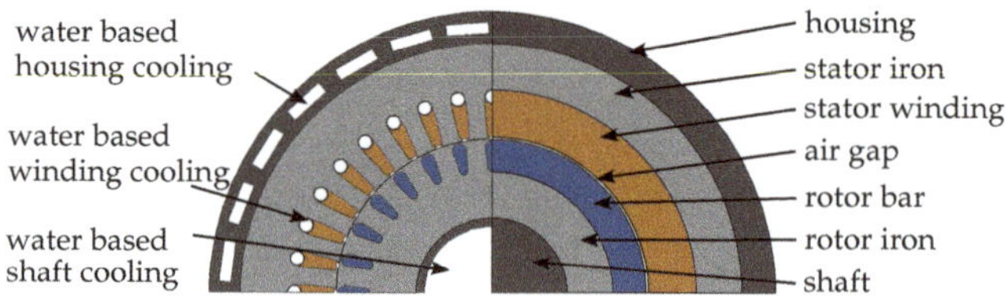

Figure 9. Real geometry (**left**) and simplified thermal shell model of the studied machine (**right**).

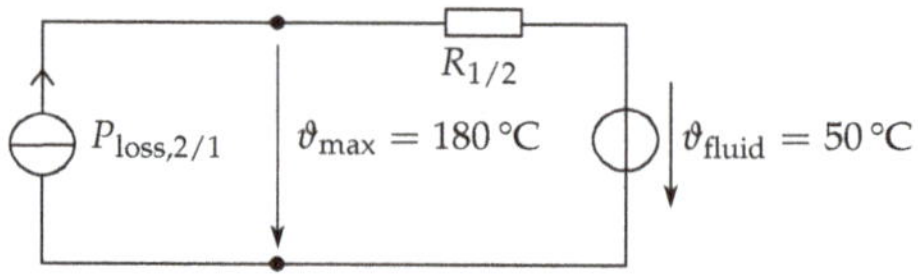

Figure 10. LPTN model of the simplified case study for the stator and the rotor.

5. Results

The thermal conductivity of the measurement approach $k_m(\vartheta)$ is calculated utilizing Equation (19). The results of the more accurate Archimedes principle are used for the density values. The results of the measurement procedure are depicted in Figure 11. The results of the analytical calculation based on Equations (9) and (16) are given in Figure 12. Material M1 has the highest thermal conductivity, while M8 has the lowest thermal conductivity in both approaches. In the measurement approach, the difference in the thermal conductivity k_m mainly results from the different values of the thermal diffusivity measurements a as shown in Figure 6. The differences in the density ρ and the thermal heat capacity c_p seem to have a minor impact on the difference of the thermal conductivity k_m.

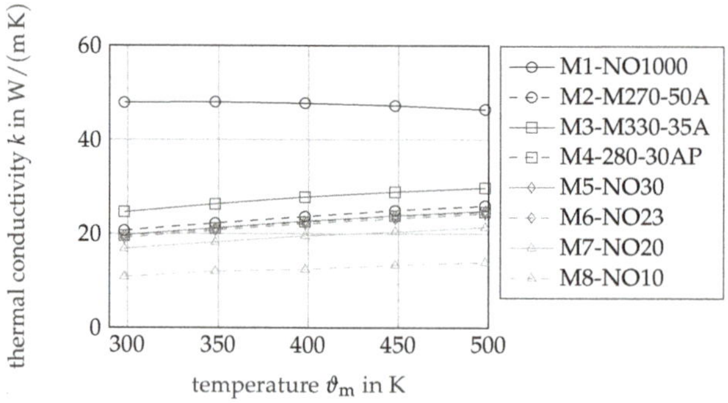

Figure 11. Results of the measurements of the thermal conductivity k_m.

A comparison between the measurement and the analytical approach is performed. The relative difference between $k_{calc}(\vartheta)$ and $k_m(\vartheta)$ is calculated as follows:

$$\delta k_{m-calc}(\vartheta) = \frac{k_{calc}(\vartheta) - k_m(\vartheta)}{k_{calc}(\vartheta)} \tag{35}$$

The value of this difference is plotted in Figure 13. The results of the analytical and the experimental approach show very good agreement for most of the materials. Materials M1 up to M6 show differences smaller than 10% for the entire temperature range. The differences for $\vartheta = \theta_D$ are significantly smaller. Materials M7 and M8 show higher differences, below 18% for material M7 and below 30% for material M8.

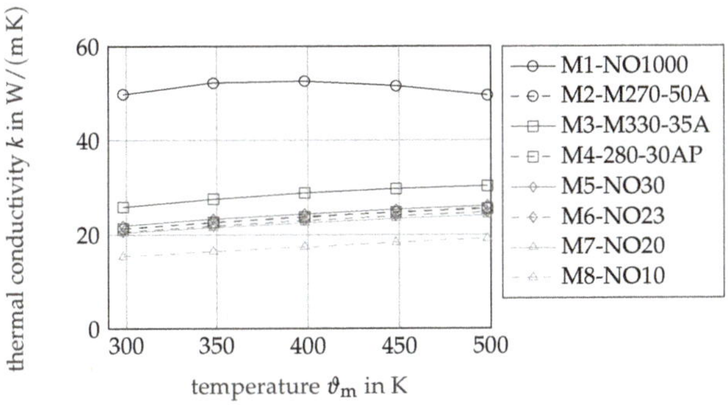

Figure 12. Results of the calculation of the thermal conductivity k_{calc}.

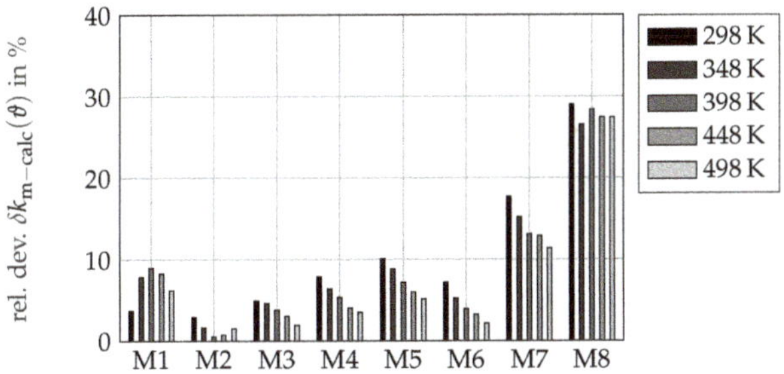

Figure 13. Relative difference between thermal conductivites gained from measurement and analytical formula $\delta k_{m-calc}(\vartheta)$.

A possible reason for the deviation can be evaluated by the analysis of the overall estimated error of the two different procedures. The previously examined errors are multiplied to gain the overall error estimation of the measurement.

$$\delta k_m = \delta \rho \cdot \delta a \cdot \delta c_p \tag{36}$$

For the analytical approach, solely the influence of the measurement of the electric resistivity $\delta \rho_{el}$ is considered.

$$\delta k_{calc}(\vartheta) = \frac{k_{calc}(\vartheta, \rho_{el} \cdot (1 - \delta \rho_{el}))}{k_{calc}(\vartheta, \rho_{el})} - 1 \tag{37}$$

The value is evaluated at $\vartheta = 398$ K. The results of the accuracy studies δk_m and δk_{calc} are depicted in Figure 14. The accuracy study shows a clear dependency of the measurement results on the material thickness d. In particular, the estimated measurement errors of the thermal diffusivity δa has a squared dependency on the material thickness d. The influence of the thickness is also visible for the estimated errors of the electric resistivity measurement $\delta \rho_{el}$, where the estimated thickness error has a linear influence. This linear influence shows some impact on the accuracy of the calculated thermal conductivity value k_{calc}. The deviation between the measurement and calculation results $\delta k_{m-calc}(\vartheta)$ shows a similar trend as the estimated errors δk_m and δk_{calc}. An allocation of the two effects is very likely, but not absolute clearly justifiable in the eyes of the authors. Material M8 is by far the thinnest material $d \approx 0.1$ mm but also has by far the highest silicon content. It is also possible that the used formula has some inaccuracies in predicting such high silicon contents. Due to the estimated measurement errors for M8, a clear separation is not possible. The second material with higher deviations between measurement and calculation is M7. It is the second thinnest material, i.e., the nominal thickness is $d \approx 0.2$ mm. M7 has a significantly lower silicon content than M8, but the highest aluminum content. The alloying contents are close to those of M2 that show the lowest deviation of all materials between the measurement and the analytical approach. This indicates that the formula is accurate for the given alloys, and the differences of M7 occur due to the measurement errors of the material or some other structural influences that are not considered in the given formula.

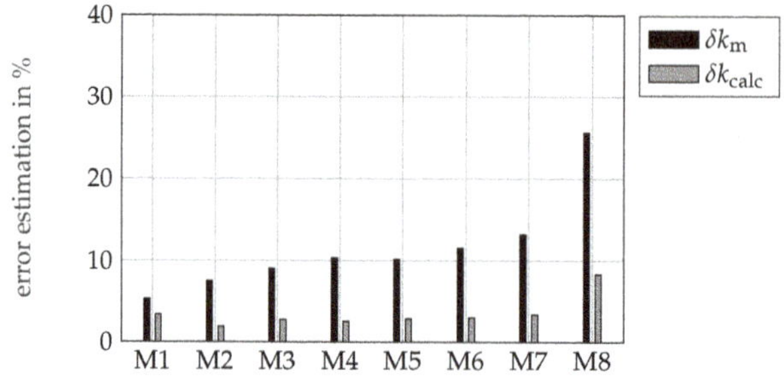

Figure 14. Relative predicted error of the measured values of the thermal conductivity δk.

The influence of the alloys on the phonon and electron thermal conductivity can be analyzed based on the introduced formula. The evaluation is performed at $\vartheta = 398\,\mathrm{K}$. The phonon thermal conductivity k_{p} deviates from $6.8\,\mathrm{W/(m\,K)}$ for M8 up to $15.7\,\mathrm{W/(m\,K)}$ for M1. The electron thermal conductivity k_{e} deviates from $8.5\,\mathrm{W/(m\,K)}$ for M8 up to $34.0\,\mathrm{W/(m\,K)}$ for M1. Both values show significant deviations, whereas the electron thermal conductivity has the higher impact on the overall value. The influence is limited to the electron–impurity scatter $W_{\mathrm{e-i}}$ according to Equation (13) and the phonon–impurity scatter $W_{\mathrm{p-i}}$ according to Equation (8). Other scattering processes are not influenced according to the used formula.

The results of the simplified case study are depicted in Figure 15. Material M1 shows the most preferable thermal properties. With this material, a maximum of 41 kW stator losses and 22 kW rotor losses could be extracted from the motor in the steady-state operation. With material M8, only 10 kW stator losses and 6 kW rotor losses would be allowed to ensure steady-state operation. It is well visible that all materials with high aluminum and silicon content show significant disadvantages regarding their capability for heat extraction. Please note that this estimation is based on some significant simplifications.

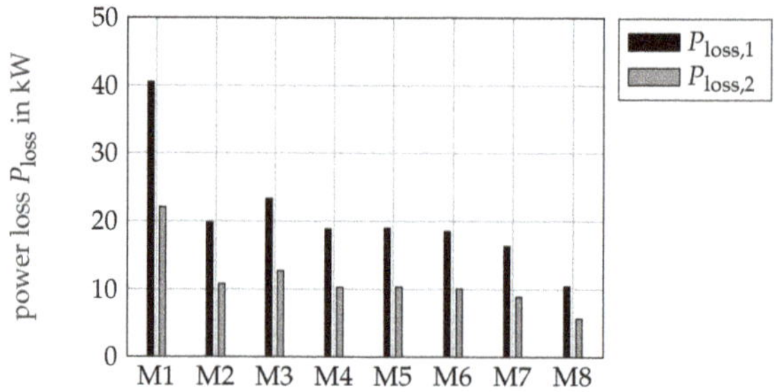

Figure 15. LPTN model of the simplified case study for the stator and the rotor.

6. Discussion

An analytical as well as an experimental approach are performed to study the influence of the silicon and aluminum content of soft magnetic materials on their thermal conductivity. The two approaches show very good agreement for a majority of the studied materials and analyzed temperatures. The thickness of the probes is identified as a crucial factor for the accuracy of the measurements. The validity of the experimental values of the thin samples M8 is restricted. The results indicate a significant influence of the alloying contents on the thermal conductivity of the materials. A simplified case study is performed. The maximum possible heat that can be extracted from the rotor and stator of a reference induction machine varies by a factor of up to four for the different materials. It is obvious that this factor should be considered in the design process of high power density traction drives in the case that a main heat dissipation path is realized over the lamination of the machine.

A measurement of a thicker material with a very high silicon content would be helpful to further identify whether the introduced formula needs to be adjusted for such alloys. In a real application, additional thermal resistances would be present that reduce the calculated impact on the thermal dissipation. The different materials, of course, also influence the magnetization behavior and the produced losses in the material. The selection of the material should consider all these aspects. Detailed measurements of the loss behavior of the studied materials, a thermal model of the studied motor, a test bench evaluation of the entire motor and a simulation study including thermal, loss and magnetization aspects is on the way.

Author Contributions: Conceptualization, B.G, A.R., N.L. and K.H.; methodology, B.G. and A.R.; software, B.G. and A.R.; validation, B.G., A.R. and N.L.; formal analysis, B.G. and A.R.; writing—original draft preparation, B.G.; writing—review and editing, B.G., A.R., N.L. and K.H.; visualization, B.G. and A.R.; supervision, K.H.; project administration, K.H.; funding acquisition, K.H. All authors have read and agreed to the published version of the manuscript.

Funding: This research and development project "Concept-ELV^2(01MY17002B)" is funded by the German Federal Ministry for Economic Affairs and Enegry (BMWi). The authors are grateful to the BMWi for the financing of this collaboration and who are responsible for the contents of this publication.

Institutional Review Board Statement: Not applicable.

Informed Consent Statement: Not applicable.

Acknowledgments: We want to thank the Institute for Materials Applications in Mechanical Engineering (iwm) of RWTH Aachen University and Netzsch Gerätebau GmbH for the excellent consultation and discussion within the project.

Conflicts of Interest: The authors declare no conflict of interest. The funders had no role in the design of the study; in the collection, analyses, or interpretation of data; in the writing of the manuscript, or in the decision to publish the results.

Abbreviations

The following abbreviations are used in this manuscript:

Acr.	Acronym
Al	Aluminum
IM	Induction Motor
InSb	Indium Antimonide
LFA	Laser Flash Analysis
LPTN	Lumped Parameter Thermal Network
SI	Silicon

References

1. Ruf, A.; Steentjes, S.; von Pfingsten, G.; Grosse, T.; Hameyer, K. Requirements on Soft Magnetic Materials for Electric Traction Motors. In Proceedings of the 7th international Conference on Magnetism and Metallurgy (WMM), Rome, Italy, 13–15 June 2016; pp. 111–128.
2. Leuning, N.; Pauli, F.; Hameyer, K. Most appropriate soft magnetic material choice with consideration of structural material parameters for electrical machines. In Proceedings of the 8th International Conference on Magnetism and Metallurgy (WMM), Dresden, Germany, 12–14 June 2018; Universitätsverlag: Freiberg, Germany, 2018.
3. Staton, D.; Boglietti, A.; Cavagnino, A. Solving the More Difficult Aspects of Electric Motor Thermal Analysis in Small and Medium Size Industrial Induction Motors. *IEEE Trans. Energy Convers.* **2005**, *20*, 620–628. [CrossRef]

4. Groschup, B.; Nell, M.; Pauli, F.; Hameyer, K. Characteristic Thermal Parameters in Electric Motors: Comparison between Induction- and Permanent Magnet Excited Machine. *IEEE Trans. Energy Convers.* **2021**, *36*, 2239–2248. [CrossRef]
5. Köller, S.; Uerlich, R.; Westphal, C.; Franck, M. Design of an Electric Drive Axle for a Heavy Truck. *ATZ Heavyduty Worldw.* **2021**, *14*, 20–25.
6. JCGM. *International Vocabulary of Metrology—Basic and General Concepts and Assocciated Terms (VIM)*, 3rd ed.; JCGM: Sevres, France, 2012.
7. Klemens, P.G.; Williams, R.K. Thermal conductivity of metals and alloys. *J. Appl. Phys.* **1986**, *31*, 197–215.
8. Wang, G.; Li, Y. Effects of alloying elements and temperature on thermal conductivity of ferrite. *J. Appl. Phys.* **2019**, *126*, 125118.1–125118.9. [CrossRef]
9. Herbstein, F.H.; Smuts, J. Determination of Debye temperature of alpha-iron by X-ray diffraction. *Philos. Mag.* **1963**, *8*, 367–385. [CrossRef]
10. Williams, R.K.; Yarbrough, D.W.; Masey, J.W.; Holder, T.K.; Graves, R.S. Experimental determination of the phonon and electron components of the thermal conductivity of bcc iron. *J. Phys. Chem. Ref. Data* **1981**, *52*, 5167–5175. [CrossRef]
11. Williams, R.K.; Graves, R.S.; McElroy, D.L. Thermal and electrical conductivities of an improved 9 Cr-1 Mo steel from 360 to 1000 K. *Int. J. Thermophys.* **1984**, *5*, 301–313. [CrossRef]
12. Julian, C.L. Theory of Heat Conduction in Rare-Gas Crystals. *Phys. Rev.* **1965**, *137*, A128–A137. [CrossRef]
13. Bäcklund, N.G. An experimental investigation of the electrical and thermal conductivity of iron and some dilute iron alloys at temperatures above 100 °K. *J. Phys. Chem. Solids* **1961**, *20*, 1–16. [CrossRef]
14. DIN German Institute for Standardization. *Magnetic Materials—Part 3: Methods of Measurement of the Magnetic Properties of Electrical Steel Strip and Sheet by Means of a Single Sheet Tester (IEC 60404-3:1992 + A1:2002 + A2:2009)*; Beuth Verlag GmbH: Berlin, Germany, 2010.
15. Corporation Fluke. *Datasheet: Service Manual 5500A Multi-Product Calibrator*; Fluke Corporation: Everett, WA, USA, 1995.
16. Hewlett-Packard Company (HP). *Datasheet: 3458A Multimeter Operating, Programming and Configuration*; Manual, Hewlett-Packard Company (HP): Palo Alto, CA, USA, 1988.
17. Wijn, H.P.J. *Magnetic Alloys for Technical Applications. Soft Magnetic Alloys, Invar and Elinvar Alloys*; Springer: Berlin/Heidelberg, Germany, 1994; Volume 19i1. [CrossRef]
18. Parker, W.J.; Jenkins, R.J.; Butler, C.P.; Abbott, G.L. Flash Method of Determining Thermal Diffusivity, Heat Capacity, and Thermal Conductivity. *J. Appl. Phys.* **1961**, *32*, 1679–1684. [CrossRef]
19. Cowan, R.D. Pulse Method of Measuring Thermal Diffusivity at High Temperatures. *J. Appl. Phys.* **1963**, *34*, 926–927. [CrossRef]
20. Cape, J.A.; Lehman, G.W. Temperature and Finite Pulse—Time Effects in the Flash Method for Measuring Thermal Diffusivity. *J. Appl. Phys.* **1963**, *34*, 1909–1913. [CrossRef]
21. NETZSCH. *Laser Flash Apparatus LFA 427: Brochure*; Netzsch-Gerätebau GmbH: Seib, Germany, 2021.
22. Desai, P.D. Thermodynamic Properties of Iron and Silicon. *J. Phys. Chem. Ref. Data* **1986**, *15*, 967–983. [CrossRef]
23. Chao, J.; Phillips, E.W.; Sinke, G.C.; Hu, A.T.; Prophet, H.; Syverud, A.N.; Karris, G.C.; Wollert, S.K. *JANAF Thermochemical Tables Addendum*; U.S. Department of Commerce: Washington, DC, USA, 1966.
24. Chase, M.W.; Curnutt, J.L.; Downey, J.R.; McDonald, R.A.; Syverud, A.N.; Valenzuela, E.A. JANAF Thermochemical Tables, 1982 Supplement. *J. Phys. Chem. Ref. Data* **1982**, *11*, 695–940. [CrossRef]
25. Novikov, V.V. Debye–Einstein model and anomalies of heat capacity temperature dependences of solid solutions at low temperatures. *J. Therm. Anal. Calorim.* **2019**, *138*, 265–272. [CrossRef]
26. Hafenstein, S.; Werner, E.; Wilzer, J.; Theisen, W.; Weber, S.; Sunderkötter, C.; Bachmann, M. Influence of Temperature and Tempering Conditions on Thermal Conductivity of Hot Work Tool Steels for Hot Stamping Applications. *Steel Res. Int.* **2015**, *86*, 1628–1635. [CrossRef]

energies
MDPI

Article

Procedure for the Accurate Modelling of Ring Induction Motors

Krzysztof Tomczyk [1,*], Tomasz Makowski [1], Małgorzata Kowalczyk [2], Ksenia Ostrowska [2] and Piotr Beńko [3]

[1] Faculty of Electrical and Computer Engineering, Cracow University of Technology, Warszawska 24, 31-155 Krakow, Poland; tomasz.makowski@pk.edu.pl
[2] Mechanical Faculty, Cracow University of Technology, Jana Pawła II 37 Avenue, 31-864 Krakow, Poland; malgorzata.kowalczyk@pk.edu.pl (M.K.); ksenia.ostrowska@pk.edu.pl (K.O.)
[3] Faculty of Environmental Engineering and Energy, Cracow University of Technology, 31-155 Krakow, Poland; piotr.benko@pk.edu.pl
* Correspondence: krzysztof.tomczyk@pk.edu.pl; Tel.: +48-12-628-2543

Abstract: This paper proposes a procedure for the accurate modelling of the ring induction motors (RIMs), based on the Monte Carlo (MC) method and the relations presented in the relevant metrology guidelines. Modelling was carried out based on the measured data for the torque-slip characteristic (TSC) and using the equivalent circuit for the RIM. The parameters included an extended Kloss equation (EKE) and the associated uncertainties were determined using the MC method. The polynomial procedure was applied as a numerical tool to complement the MC method to determine the power losses in the stator iron and the relevant uncertainty. This is in line with international standards for the theory of uncertainty application in the field of engineering. The novelty of this paper refers to the accurate modelling of the RIMs obtained by determining the corresponding uncertainties. The procedure presented in this paper was developed based on the assumption that the parameters of the equivalent circuit are independent of the temperature, influence of core saturation, and the phenomenon of current displacement. Our procedure can be successfully used for both the theoretical calculations related to the modelling of the RIMs, and in practical applications involving detailed measurements and the corresponding uncertainties. The use of the MC method allowed for significant improvement in the modelling results, in terms of both the TSC and EKE.

Keywords: modelling of ring induction motors; Monte Carlo method; accurate modelling

Citation: Tomczyk, K.; Makowski, T.; Kowalczyk, M.; Ostrowska, K.; Beńko, P. Procedure for the Accurate Modelling of Ring Induction Motors. *Energies* **2021**, *14*, 5469. https://doi.org/10.3390/en14175469

Academic Editor: Armando Pires

Received: 5 August 2021
Accepted: 31 August 2021
Published: 2 September 2021

1. Introduction

The ring induction motors are a group of electrical machines mainly used in electric drives with heavy starting. The electrical power of these machines varies over a wide range, from a fraction of a kilowatt to several megawatts, which allows for an extensive range of applications [1]. They are characterised by a simple structure, which implies low construction costs, easy operation and maintenance, and high levels of operational reliability [2].

One of the most important curves used to describe the RIMs is the torque-slip characteristic, which represents the relationship between the electromagnetic moment and the slip [3,4]. This is determined by the constant value of the stator supply voltage and the variable rotational speed of the motor, which is related to the slip by a simple and well-known formula. It is most often described using a simplified Kloss equation [3], which approximates the corresponding measured points with a high uncertainty value, however, most often in the case of very low-power machines [5]. An extended version of this equation [6–8], denoted here as EKE, was therefore developed based on the theory and practice of electric machines, for which much lower values of the approximation uncertainty of the mechanical characteristic of the motor can be obtained. This equation is necessary for the analytical determination of machine acceleration times [9], an analysis of the transient regimes [10], frequency control of inductive electric drives under conditions of overload [11], and for the testing of electric motors, for example in Tesla vehicles [12].

Using a simplified and extended version of the Kloss equation, the value of maximum (critical) motor torque and the corresponding value of maximum (critical) slip can be determined [13]. For when the slip is equal to one, the value of the starting torque of the motor can be determined. Knowledge of the critical and starting torques is extremely important from the point of view of evaluating the mechanical properties of the RIM [14]. The first type of torque allows us to determine the possibility of short-term motor overload [15], while the second one represents the possibility of starting a drive system including both a motor and a generator [16].

All of the procedures for determining the TSC involve the accurate measurements of the corresponding points [17], and then their approximation with a minimum value of uncertainty [18–20]. The accurate determination of the maximum torque and slip is only possible by applying this approach to modelling. When the mathematical formula representing the measured points of the TSC is known, the satisfactory approximation results and the associated uncertainties can be obtained by applying the MC method [21–23]. This method should involve a pseudorandom number generator with a uniform distribution [24,25]. However, it is only possible to determine the maximum values of the torque and slip by applying the MC method. A third parameter related to the EKE also needs to be determined. This can be obtained by performing indirect calculations based on the equivalent circuit of the RIM [26–28] and by using the additional numerical method to obtain values of the power losses in the stator iron. Additionally, it should be emphasized that the MC method has so far been applied to the modelling of the RIMs, by analysis of their thermal behaviour and the detection of corresponding faults, based on the stator current measurements [29,30].

In Section 2 of this paper, we present a detailed discussion of the issues related to the determination of the equivalent circuit parameters corresponding to the RIM, based on measurements of the motor idle and short-circuit states. The use of polynomial approximation [31–33] to determine the power losses are also proposed as the additional numerical method. Section 3 describes the use of the MC method to model the RIM by using the TSC, while Section 4 presents an example of the use of MC-based modelling and the verification of the corresponding results.

The solution presented in this paper represents a new approach to modelling the RIM based on the measurement points of the TSC and the EKE. This is obtained by an application of the MC method and the polynomial approximation which allows us to determine the values of the maximum moment and slip with the associated uncertainties. Therefore, it is an example of accurate modelling of the RIM which is worked out according to the guidelines in [18,21] and can be applied in the field of the precise elaboration of measurement results. The lack of uncertainty analysis in the modelling of the RIMs can be considered a weakness of the research so far in the field of electrical motors.

The proposed method can be used for accurate modelling and appropriate control over the mechanical properties of the RIMs in order to ensure the correct operation of both single motors and complex electrical drive systems.

2. Theoretical Basis for the Accurate Modelling of the RIM

The EKE is defined by the following formula:

$$T = \frac{T_{\max}(2 + \beta \cdot s_{\max})}{\frac{s}{s_{\max}} + \frac{s_{\max}}{s} + \beta \cdot s_{\max}}, \tag{1}$$

where T, $T_{\max}$, s, and $s_{\max}$ are the motor torque, maximum torque, slip, and maximum slip, respectively. The coefficient β is expressed as follows:

$$\beta = 2\frac{R_s C_s}{R_r K_V{}^2}, \tag{2}$$

where R_s, R_r , and K_V denote the stator resistance, rotor resistance, and voltage ratio, respectively [6–8]. The voltage ratio is calculated by the formula:

$$K_V = \frac{1}{2}\left(\frac{V_{s1}}{V_{rm}} + \frac{V_{sm}}{V_{rm}}\right), \tag{3}$$

where V_{s1} is the stator supply volge, which is lower than the rated voltage V_{sn}, while V_{rm} is the maximum voltage produced between any two rotor phases, and V_{sm} is the maximum voltage produced between any two stator phases when the rotor is supplied by the voltage V_{rm} [27]. The coefficient C_s is defined by:

$$C_s = \frac{X_{mg}}{X_{\sigma s} + X_{mg}}, \tag{4}$$

where X_{mg} and $X_{\sigma s}$ denote the magnetisation reactance and the stator phase leakage reactance, respectively, and are calculated using the following formulae:

$$X_{mg} = \frac{V_{Sn}}{I_{mg}} \tag{5}$$

and

$$X_{\sigma s} = (V_{Sn} - K_V{\cdot}V_{rm})/I_{s0}, \tag{6}$$

where I_{mg} is the magnetisation current, I_{s0} is the stator current under idle conditions, and V_{rm} is the voltage induced in the rotor [6–8]. Figure 1 shows a circuit model of the RIM, which is valid when the measurements are made in the idle state.

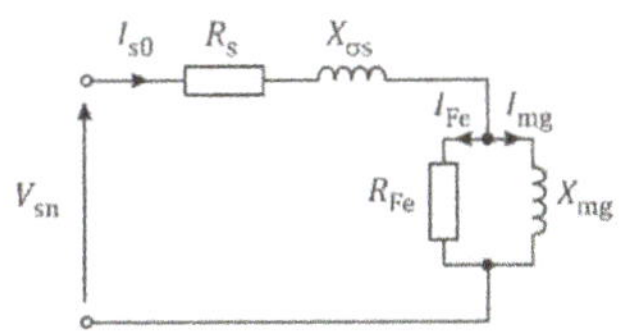

Figure 1. Circuit model of the RIM in the idle state.

The magnetisation current is

$$I_{mg} = \sqrt{{I_{s0}}^2 - {I_{Fe}}^2}, \tag{7}$$

where I_{Fe} denotes the current corresponding to the power losses in the stator iron, and is calculated using the simple formula:

$$I_{Fe} = \frac{V_{sn}}{R_{Fe}}. \tag{8}$$

Here, R_{Fe} denotes the resistance of the iron, and is defined by the following equation:

$$R_{Fe} = \frac{3{V_{sn}}^2}{\Delta P_{Fe}}, \tag{9}$$

while ΔP_{Fe} denotes the power losses in the stator iron.

The complete equivalent circuit of the RIM is shown in Figure 2, where R'_r and $X'_{\sigma r}$ denote the rotor phase resistance and the reactance transformed to the stator side.

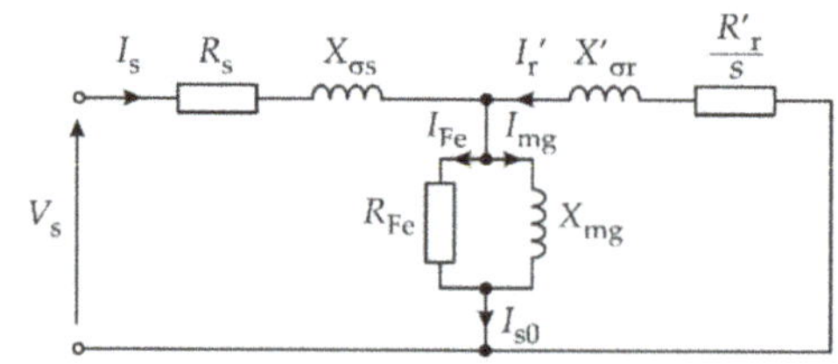

Figure 2. Complete equivalent circuit of the RIM.

The quantities R_{Fe}, I_{Fe}, X_{mg}, and I_{mg}, given in Equations (5), (7)–(9) are calculated based on the complete phase equivalent diagram (the left side of Figure 2), omitting the voltages related to the stator resistance and the stator winding leakage reactance.

The values of the parameters R'_r and $X'_{\sigma r}$ can be determined based on the circuit model of the RIM in the short-circuit state, as shown in Figure 3, where V_{sk} denotes the current in this state.

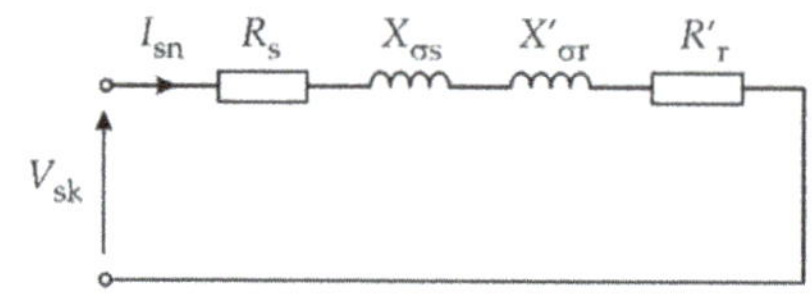

Figure 3. Circuit model of the RIM in the short-circuit state.

The transverse branch of the circuit model shown in Figure 2 is omitted in Figure 3, due to the significant value of its impedance compared to the impedance of the longitudinal branch. The short-circuit current I_{sk} is also assumed to be equal to the rated stator current I_{sn} [6–8].

Based on the circuit model shown in Figure 3, we have:

$$R_k = R_s + R'_r = R_s + R_r K_V{}^2 \tag{10}$$

and

$$X_k = X'_{\sigma r} + C_s X_{\sigma s} = X'_{\sigma r} + \frac{X_{mg} X_{\sigma s}}{X_{mg} + X_{\sigma s}} \tag{11}$$

while

$$X'_{\sigma r} = X_{\sigma s} K_V{}^2. \tag{12}$$

Based on Ohm's law, we obtain:

$$Z_k = \frac{V_{sk}}{I_{sn}}, \tag{13}$$

and applying the impedance triangle gives

$$R_k = \sqrt{Z_k{}^2 - X_k{}^2}. \tag{14}$$

By transforming Equation (10), we obtain the rotor resistance

$$R_r = (R_k - R_s)/K_V{}^2 \tag{15}$$

We can express the coefficient β in terms of the measured values of the parameters of the RIM. We can obtain this relation by substituting Equations (3)–(13) into Equation (2) to give [1,6–8]:

$$\beta = \frac{2R_s}{\left(\frac{\gamma_2\gamma_3}{\gamma_4}+1\right)\left[\sqrt{\left(\frac{V_{sk}}{I_{sn}}\right)^2-\left(\frac{\gamma_1{}^2\gamma_3}{8\cdot I_{s0}}+\frac{V_{sn}\cdot\gamma_3}{\gamma_4+\gamma_2\gamma_3}\right)^2}-R_s\right]} \quad (16)$$

where the auxiliary parameters are

$$\gamma_1 = \frac{V_{S1}}{V_{rm}} + \frac{V_{sm}}{V_{rm}},\ \gamma_2 = \sqrt{I_{s0}{}^2 - \left(\frac{\Delta P_{Fe}}{3V_{sn}}\right)^2}, \quad (17)$$
$$\gamma_3 = 2V_{sn} - \gamma_1 V_{rm},\ \gamma_4 = 2V_{sn} I_{s0}.$$

The uncertainties associated with the quantities given in Equations (3)–(16) can be calculated using the formula:

$$u(x) = \sqrt{\sum_{j=1}^{J}\left[\frac{\partial x}{x_j}u(x_j)\right]^2}, \quad (18)$$

where x denotes the quantity under consideration, and J is the number of indirect quantities necessary to determine the value of x. The relative uncertainty associated with the quantity x is defined by the equation:

$$\delta(x) = u(x)/x. \quad (19)$$

For both analogue and digital measuring instruments, the uncertainty $u(x)$ is determined by the formula:

$$u(x) = \Delta(x)/\sqrt{3}, \quad (20)$$

where $\Delta(x)$ is the absolute error, while the value of the denominator results from the probability density function of a uniform distribution, which is valid for both analogue and digital instruments.

For analogue instruments, the error is determined on the basis of the static accuracy class κ, according to the equation:

$$\Delta_a(x) = \frac{\kappa Y_m}{100\%} \quad (21)$$

where Y_m denotes the measurement range for the quantity to be measured. For digital instruments, the following formula is usually applied:

$$\Delta_d(x) = aY + cY_m, \quad (22)$$

where Y denotes the value of the quantity to be measured, and a and c are constant parameters that are typical for the particular instrument and are included in the corresponding datasheet.

The power losses in the stator iron ΔP_{Fe} are determined using a graphical method for the rated stator voltage V_{sn}, as shown in Figure 4.

The quantity P_0 is the active power consumed by the motor during idling, and is equal to the sum of the losses in the stator iron ΔP_{Fe}, and the mechanical power losses ΔP_m.

Based on the measured points for the active power P_0, it is easy to determine the linear characteristic $P_0 = f(V_s{}^2)$ and the associated uncertainty. This can be done by applying the polynomial method, using the formula:

$$P_0\left(V_s{}^2\right) = a_0 + a_1\cdot V_s^2 + \varepsilon, \quad (23)$$

where a_0 and a_1 are the polynomial coefficients, and ε denotes the error of approximation.

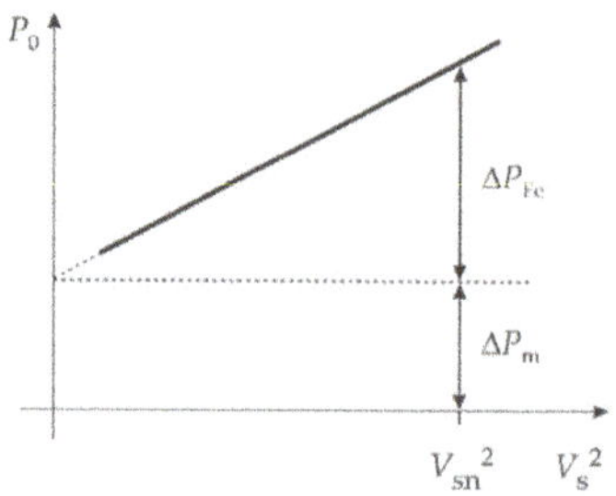

Figure 4. Graphical method for determining the power losses ΔP_{Fe}.

Estimates $\tilde{a}$ of the polynomial coefficients are obtained using the following matrix equation:

$$\tilde{\mathbf{A}} = \left(\mathbf{\Phi}^T\mathbf{\Phi}\right)^{-1}\mathbf{\Phi}^T\mathbf{\Lambda}, \tag{24}$$

where

$$\mathbf{\Phi} = \begin{bmatrix} 1 & \left(V_{\mathrm{s}}^2\right)_0 \\ \vdots & \vdots \\ 1 & \left(V_{\mathrm{s}}^2\right)_N \end{bmatrix}, \quad \mathbf{\Lambda} = \begin{bmatrix} \varepsilon_0 & \varepsilon_1 & \ldots & \varepsilon_N \end{bmatrix}^T, \tag{25}$$

and N denotes the number of measured points for the characteristic $P_0 = f\left(V_{\mathrm{s}}^2\right)$.

The uncertainty of approximation is denoted as the error ε, and is given by the following formula:

$$u\left(P_0\left(V_{\mathrm{s}}^2\right)\right) = \sqrt{\frac{\left(\mathbf{\Phi}\tilde{\mathbf{A}} - \mathbf{\Lambda}\right)^T\left(\mathbf{\Phi}\tilde{\mathbf{A}} - \mathbf{\Lambda}\right)}{N-3}} \tag{26}$$

The standard uncertainty associated with the coefficients a_0 and a_1 is

$$u(a_i) = u\left(P_0\left(V_{\mathrm{s}}^2\right)\right)\sqrt{\Theta_{i,i}}, \tag{27}$$

where

$$\mathbf{\Theta} = \left(\mathbf{\Phi}^T\mathbf{\Phi}\right)^{-1}, \tag{28}$$

and $i = 0,\ 1,\ 2$ [29–31].

The relative uncertainties associated with the coefficients a_0 and a_1 are calculated as follows:

$$\delta(a_i) = \frac{u(a_i)}{a_i}100\%. \tag{29}$$

The values of the power losses in the stator iron ΔP_{Fe} and the associated uncertainty $u(\Delta P_{\mathrm{Fe}})$ are calculated using the expressions:

$$\Delta P_{\mathrm{Fe}} = P_0\left(V_{\mathrm{sn}}^2\right) - P_0(0) \tag{30}$$

and

$$u(\Delta P_{\mathrm{Fe}}) = u(a_0) + u(a_1)P_0\left(V_{\mathrm{sn}}^2\right). \tag{31}$$

The corresponding relative uncertainty is:

$$\delta(P_{\mathrm{Fe}}) = \frac{u(\Delta P_{\mathrm{Fe}})}{\Delta P_{\mathrm{Fe}}}100\%. \tag{32}$$

The procedure for determining the values of the parameters included in Equation (1) and the corresponding uncertainties is discussed in detail in the section below.

3. Monte-Carlo-Based Modelling of the RIM

We now present the example of the application of the MC method in the accurate modelling of the RIM, which involves determining the corresponding parameters of the EKE and the associated uncertainties. This procedure is based on an intuitive method of determining the parameters T_{maxi} and s_{maxi} for the possible ranges of variability of the parameters T_{max} and s_{max}, which are included in the EKE [21–24]. Figure 5 shows the typical TSC which describes the RIM for its motor work and covers the stable range of this characteristic. Figure 5 also shows examples of the variability ranges of the parameters T_{max} and s_{max}.

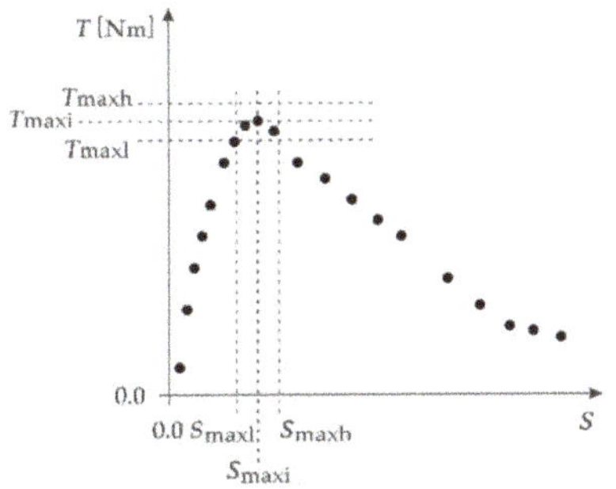

Figure 5. The typical TSC for the motor work of the RIM.

The index 'i' denotes the intuitive values of both parameters, while the indexes 'h' and 'l' represent the high and low assumed values of these parameters. The high and low values are assumed in advance to ensure that the estimated values of the parameters T_{max} and s_{max} are within these selected ranges. A suitable selection of these ranges constitutes the first step in this method.

The parameter β and the associated uncertainty $u(\beta)$ are determined based on the procedure discussed in Section 2. The low and high values, β_{l} and β_{h}, are determined as follows:

$$\beta_{\text{l}} = \beta - u(\beta),\ \beta_{\text{h}} = \beta + u(\beta). \tag{33}$$

The second step in our MC-based procedure involves the choice of the type of pseudorandom number generator. Taking into account the analogous probability of the occurrence of the optimal value of the estimates $\widetilde{T}_{\text{max}}$, $\widetilde{s}_{\text{max}}$, and $\widetilde{\beta}$ for any value from the above intervals, we are justified in choosing the pseudorandom number generator with a uniform distribution. The above estimates should accurately map the parameters T_{max}, s_{max}, and β, which requires an approximation of the TSC with minimal uncertainty.

In the third step, we determine the number of MC trials. According to the recommendations given in the guide [20], the optimal number of trials should be greater than $10^4/(1-v)$, where v denotes the coverage probability.

In the fourth step, the following matrix is determined

$$\mathbf{\Psi} = \begin{bmatrix} T(s_0)_0 & \cdots & T(s_0)_{M-1} \\ \vdots & \ddots & \vdots \\ T(s_{N-1})_0 & \cdots & T(s_{N-1})_{M-1} \end{bmatrix}, \tag{34}$$

based on Equation (1), where N and M denote the number of measured points for the TSC and the number of MC trials, respectively [22–25]. The matrix $\mathbf{\Psi}$ is determined by substituting the values of the parameters ${T_{\text{max}}}^m$, ${s_{\text{max}}}^m$, and β^m into Equation (1), as obtained for a sequence of MC trials $m = 0, 1, \ldots, M-1$. The value of each slip s_n is substituted into Equation (1) for each MC trial m, where $n = 0, 1, \ldots, N-1$. In the fifth step, the matrix of approximation errors for the measured points of the TSC is determined as follows:

$$\mathbf{\Delta}_{\mathrm{e}} = \begin{bmatrix} \widetilde{T}(s_0)_0 & \cdots & \widetilde{T}(s_0)_{M-1} \\ \vdots & \ddots & \vdots \\ \widetilde{T}(s_{N-1})_0 & \cdots & \widetilde{T}(s_{N-1})_{M-1} \end{bmatrix}, \tag{35}$$

where $\widetilde{T}(s_n)_m = T(s_n)_m - T(s_n)$. The next rows of the matrix $\mathbf{\Delta}_{\mathrm{e}}$ correspond to the approximation uncertainties obtained for each value of the slip s_n.

The sixth step in our MC-based modelling process involves the determination of the vector

$$\mathbf{\Delta}_{\mathrm{c}} = \sum\nolimits_m (\mathbf{\Delta}_{\mathrm{e}})^2, \tag{36}$$

in which each element is the sum of the squared errors calculated for each column of the matrix $\mathbf{\Delta}_{\mathrm{e}}$ [22–25].

In the next step, the minimum value $\left(\mathbf{\Delta}_{\mathrm{c}}{}^{\min}\right)$ of the vector $\mathbf{\Delta}_{\mathrm{c}}$ and the corresponding number of trials $\left(m^{\min}\right)$ are determined. The parameters $T_{\max}{}^{\mathrm{opt}}$, $s_{\max}{}^{\mathrm{opt}}$, and β^{opt} corresponding to the value $\mathbf{\Delta}_{\mathrm{c}}{}^{\min}$ are assumed to represent the optimal solution to the MC-based model. These parameters correspond to the estimates $\widetilde{T}_{\max}$, $\widetilde{s}_{\max}$, and $\widetilde{\beta}$ as defined above.

We then determined the uncertainty associated with the MC method using the following formula:

$$u(\mathrm{MC}) = \sqrt{\frac{1}{M(M-1)} \sum\nolimits_{m=0}^{M-1} \left[\mathbf{\Delta}_{\mathrm{c}m} - \overline{\mathbf{\Delta}}_{\mathrm{c}}\right]^2}, \tag{37}$$

where

$$\overline{\mathbf{\Delta}}_{\mathrm{c}} = \frac{1}{M} \left(\sum\nolimits_{m=0}^{M-1} \mathbf{\Delta}_{\mathrm{c}m}\right). \tag{38}$$

The last step in our MC-based procedure involves the determination of the uncertainties associated with the parameters $T_{\max}$, $s_{\max}$, and β using the formula:

$$u(x) = \sqrt{\frac{1}{M(M-1)} \sum\nolimits_{m=0}^{M-1} [x_m - \overline{x}]^2}, \tag{39}$$

where

$$\overline{x} = \frac{1}{M} \sum\nolimits_{m=0}^{M-1} x_m. \tag{40}$$

The last two formulae are valid for all the parameters in the MC model.

4. Example Application and Verification of Results

Let us present below the experimental results referring to the modelling of the real RIM with the following rated data: $P_{\mathrm{n}} = 3.3$ kW (rated power), $V_{\mathrm{sn}} = 400$ V (rated stator voltage), $I_{\mathrm{sn}} = 9.5$ A (rated stator current), $f_{\mathrm{n}} = 50$ Hz (rated frequency), $n_{\mathrm{n}} = 940$ rpm/min (rated rotational speed), Δ/y (winging connections), $\cos\varphi_{\mathrm{n}} = 0.89$ (rated power factor), and $\eta_{\mathrm{n}} = 0.87$ (rated efficiency). Single-phase resistances for the stator and rotor are 2.9 Ω and 0.1 Ω, respectively.

Table 1 shows the measured results for the magnetisation characteristic obtained in the idle state. The value of the current $I_{\mathrm{s}0}$, obtained for the rated voltage V_{sn}, was measured with a digital instrument, and is equal to 5.60 A. The uncertainty associated with this current was calculated using Equations (20) and (22), and is equal to 0.22 A.

Figure 6 shows the results obtained for the power loss in the stator iron ΔP_{Fe} using the method shown in Figure 4 and Equations (23)–(30). The value of this power loss is 155.8 W, and the linear equation that approximates the measurement points is

$$P_0\left(V_{\mathrm{s}}{}^2\right) = a_0 + a_1 \cdot V_{\mathrm{s}}{}^2 = 41.5 + 9.74 \cdot 10^{-4} \cdot V_{\mathrm{s}}{}^2. \tag{41}$$

The uncertainty associated with this approximation is $u\left[P_0\left(V_s^2\right)\right] = 3.52$ W.

Table 1. Measured results for the magnetisation characteristic in the idle state.

No.	1	2	3	4	5	6	7
V_s [V]	160	180	200	220	240	260	280
P_0 [W]	70	70	81	85	95	110	120
No.	8	9	10	11	12	13	
V_s [V]	300	320	340	360	380	400	
P_0 [W]	130	147	150	165	180	200	

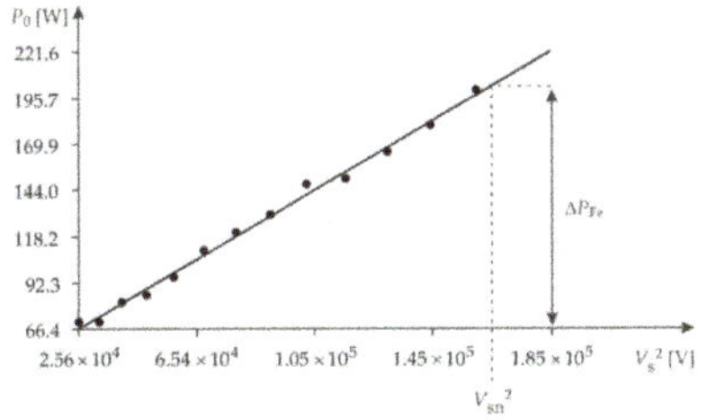

Figure 6. Results for the power loss in the stator iron ΔP_{Fe}.

The uncertainties associated with the parameters in the linear equation are $u(a_0) = 2.17$ and $u(a_1) = 2.31 \cdot 10^{-5}$, while the corresponding relative uncertainties are $\delta(a_0) = 5.24\%$ and $\delta(a_1) = 2.38\%$.

The uncertainty and relative uncertainty associated with the power losses, obtained using Equations (31) and (32), respectively, are $u(\Delta P_{Fe})$ =2.18 W and $\delta(P_{Fe}) = 1.40\%$. The measured results for the short-circuit state of the RIM are shown in Table 2. These results enabled us to determine the parameters included in the circuit model shown in Figure 3, using Equations (10)–(14).

Table 2. Measurement results for the short-circuit state.

V_{sk} [V]	I_{sn} [A]
168	9.5

The values of the voltages V_{sn}, V_{s1}, and V_{sm} were determined using analogue voltmeters with an accuracy and measurement range of 0.5% and 400 V, respectively. The values of the quantities K_V, R_{Fe}, I_{Fe}, I_{mg}, X_{mg}, $X_{\sigma s}$, and C_s, calculated using Equations (3)–(9), are 4.167 V/V, 3.08 kΩ, 0.13 A, 5.59 A, 71.45 Ω, 2.23 Ω and 0.971, respectively. The corresponding uncertainties $u(K_V)$, $u(R_{Fe})$, $u(I_{Fe})$, $u(I_{mg})$, $u(X_{mg})$, $u(X_{\sigma s})$ and $u(C_s)$ are 0.001 V/V, 55 Ω, 0.020 A, 0.22 A, 14.23 Ω, 0.22 Ω and 0.0070, respectively.

Table 2 shows the measured results obtained for the short-circuit state. Based on these measurements, we can calculate the values of the parameters R_k, X_k, and Z_k using Equations (11), (13) and (14), respectively; we can then determine the corresponding uncertainties using Equations (18)–(22). These measurements were made using a voltmeter and ammeter with accuracy and measurement ranges of 0.50%, 200 V and 0.50%, 10 A, respectively.

The values of the parameters R_k, X_k, and Z_k and the associated uncertainties $u(R_k)$, $u(X_k)$ and $u(Z_k)$ are 36.90 Ω, 40.92 Ω, 17.68 Ω, 1.62 Ω, 2.36 Ω and 1.08 Ω, respectively.

Based on the above parameters, the coefficient β and the associated uncertainty $u(\beta)$ were calculated using Equations (16) and (18) as 0.152 and 0.141, respectively. The value of the relative uncertainty $\delta(\beta)$ is 92.8%. The high value of this uncertainty was due to the significant complexity of Equation (16), which depends on eight indirectly measured quantities.

Table 3 shows the measured results for the torque-slip characteristic of the RIM.

Table 3. Measured points for the torque-slip characteristic of the RIM.

No.	1	2	3	4	5	6	7	8	9	10	11
s	0.004	0.018	0.036	0.054	0.074	0.095	0.128	0.154	0.181	0.213	0.253
T [Nm]	0.00	2.55	3.63	4.41	5.00	5.59	6.18	6.47	6.75	6.83	6.67
No.	12	13	14	15	16	17	18	19	20	21	
s	0.312	0.379	0.445	0.510	0.567	0.681	0.760	0.833	0.893	0.961	
T [Nm]	6.38	6.08	5.69	5.30	5.00	4.41	4.02	3.92	3.83	3.42	

The values of the parameters T_{maxi} and s_{maxi} were determined intuitively, as shown in Figure 5, as 7.00 Nm and 0.200, respectively. The values of the parameters T_{maxl}, T_{maxh}, s_{maxl}, and s_{maxh} were assumed in advance around the above parameters. These values define the draw ranges for the parameters T_{max} and s_{max}. The draw range for the coefficient β is determined based on the associated uncertainty $u(\beta)$ by Equation (33), as follows:

$$\beta_{\mathrm{l}} = 0.152 - 0.141 = 0.011 \text{ and } \beta_{\mathrm{h}} = 0.152 + 0.141 = 0.293.$$

A total of 2×10^5 MC trials were carried out using the pseudo-random number generator with a uniform distribution. Equations (34)–(40) were applied to the execution of the relevant numerical calculations using the MathCad 15 program, and the total computation time was 3 h and 24 min. The calculations were performed on a PC with the following parameters: Inter® Core™, Duo CPU E8400, processor ×64, 3.00 GHz, 4.00 GB RAM.

The minimum value $\Delta_{\mathrm{c}}{}^{\mathrm{min}}$ for the vector Δ_{c} and the corresponding number of trials m^{min} were 3.581 and 95360, respectively. The values of the parameters $T_{\mathrm{max}}{}^{\mathrm{opt}}$, $s_{\mathrm{max}}{}^{\mathrm{opt}}$, and β^{opt} corresponding to quantity $\Delta_{\mathrm{c}}{}^{\mathrm{min}}$ are 7.3861 Nm, 0.19721, and 0.28927, respectively. The uncertainty $u(\mathrm{MC})$ associated with the MC method is 0.032. The uncertainties $u(T_{\mathrm{max}})$, $u(s_{\mathrm{max}})$, and $u(\beta)$, associated with the parameters T_{max}, s_{max}, and β are $7{\cdot}10^{-4}$ Nm, $1.3{\cdot}10^{-4}$ and $1.82{\cdot}10^{-4}$, respectively.

The relative uncertainty $\delta(\beta)$ associated with the coefficient β is $u(\beta)/\beta$ =0.07%. Based on these results, it should be noted that the value of the uncertainty $\delta(\beta)$ was reduced from 92.8% (obtained from analytical calculations) to 0.07% (obtained using our MC-based procedure).

Figure 7 shows the results from our MC-based model of the RIM for the example of the TSC characteristic and using the EKE.

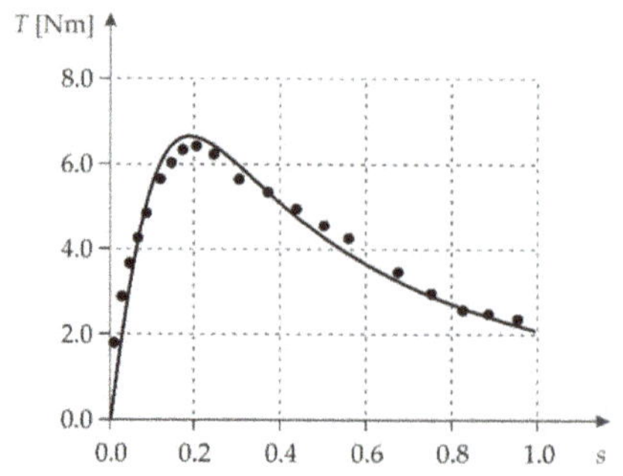

Figure 7. Results from our MC-based model of the RIM.

Figure 8 shows the distribution of the approximation uncertainty $u(T)$ for the particular values of the slip s.

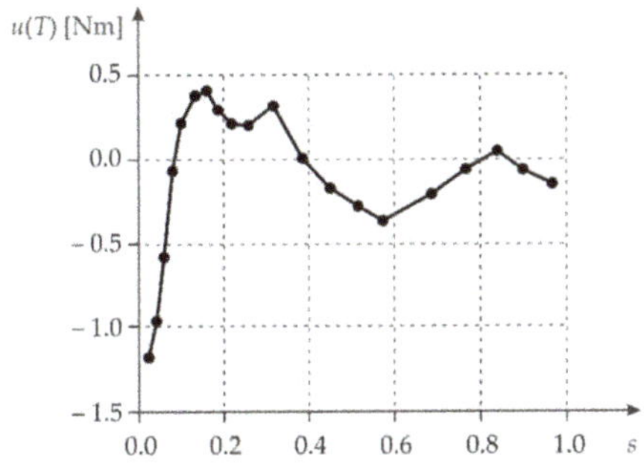

Figure 8. Distribution of the approximation uncertainty for the TSC characteristic.

The highest value of the approximation uncertainty was obtained for the slip s within the maximum slip value s_{max}, as well as for the slip s with a value of about 0.85.

We now verify the implementation of our MC-based procedure by examining the influence of the number of MC trials on the value of the uncertainty u(MC). The results for the values of the parameters included in the EKE are given in Table 4.

Table 4. Results from our Monte Carlo procedure.

No. MC Trials	10^3	2×10^3	10^4	2×10^4	10^5	2×10^5
u(MC)	0.4291	0.3214	0.1463	0.1038	0.0464	0.0327
T [Nm]	7.3313	7.3479	7.4250	7.3624	7.3800	7.3861
s	0.20024	0.19865	0.19741	0.19827	0.19748	0.19721
β	0.25231	0.27863	0.29152	0.28259	0.28993	0.28927

It can be seen from Table 4 that the values for the uncertainty u(MC) decrease as the number of MC trials increases. The values of the EKE parameters obtained for 2×10^5 MC trials (the lowest number of MC trials recommended by the corresponding guide), were assumed to represent the optimal solution to the modelling task for this example.

The results obtained in the section above show that the application of the MC method and the polynomial procedure in particular allows for a significant increase in the accuracy of the RIM modelling compared to other methods, which do not include analysis of the modelling uncertainty.

5. Conclusions

This paper presents a procedure that allows us to assess the accuracy of modelling of the RIMs on the example of the RSC and EKE, by applying the MC method. Based on the numerical simulations and calculations performed for an example of the RIM, it has shown that the effect of the uncertainty on the results of measurements is significant. The proposed method is based on the corresponding guidelines for the implementation of accurate measurements, and can significantly reduce the values of the uncertainties associated with the parameters in the ECE. For example, for the coefficient β, the application of our MC-based numerical modelling procedure reduced the value of corresponding uncertainty from 92.8% (analytical calculations) to 0.07% (MC-based procedure). It should also be emphasised that a further increase in the modelling uncertainty of both the TSC and the other parameters in the equivalent circuit of the RIM can be obtained by using more accurate measuring instruments, and by applying modern measurement techniques based on specialised computer software such as LabVIEW.

The solutions presented in this paper can be used in other applications in the field of electric machines and electric drives, for example in the precise determination of the characteristics of all types of motors, or for the development of accurate measurement reports.

Author Contributions: Conceptualisation, K.T. and T.M.; data curation, T.M., M.K., K.O. and P.B.; writing—original draft, K.T.; formal analysis, T.M., M.K., K.O. and P.B.; methodology, K.T. and T.M.; writing—review and editing, K.T., T.M., M.K., K.O. and P.B.; software, K.T. and T.M. All authors have read and agreed to the published version of the manuscript.

Funding: This research was conducted at the Faculty of Electrical and Computer Engineering, Krakow University of Technology, and was financially supported by the Ministry of Science and Higher Education, Republic of Poland (grant No. E-3/2021).

Institutional Review Board Statement: Not applicable.

Informed Consent Statement: Not applicable.

Data Availability Statement: MDPI Research Data Policies.

Conflicts of Interest: The authors declare no conflict of interest.

References

1. Ćalasan, M.; Alqarni, M.; Rosić, M.; Koljčević, N.; Alamri, B.; Abdel Aleem, S. A novel exact analytical solution based on Kloss equation towards accurate speed-time characteristics modeling of induction machines during no-load direct startups. *Appl. Sci.* **2021**, *11*, 5102. [CrossRef]
2. Ejiofor, O.; Nwosu, C.; Nnadi, D.; Okoro, O.I. Performance study of three-phase induction motor driving a load. *Discovery* **2019**, *55*, 279–290.
3. Aree, P. Analytical determination of speed-torque and speed-current curves of single-cage induction motor under supply voltage and frequency variations. *COMPEL* **2018**, *37*, 2279–2298. [CrossRef]
4. Malagoli, J.A.; Camacho, J.R.; Ferreira da Luz, M.V. Optimal electromagnetic torque of the induction motorgenerated automatically with Gmsh/GetDP software. *Int. Trans. Electr. Energy Syst.* **2020**, *3*, 1–20.
5. Gärtner, J.; Halámka, Z.; Pavelka, J. The speed control of the induction motors by the change of supply voltage. In Proceedings of the Silnoproudá Elektrotechnika a Elektroenergetika, Brno, Czech Republic, 26 April 2001; pp. 41–44.
6. Cociu, V.R.; Cociu, L. Consideration about current-speed expression of the induction machine. *Bul. Inst. Politeh. Iaşi* **2020**, *66*, 39–50.
7. Gerling, D. *Electrical Machines*; Springer: Berlin, Germany, 2015; pp. 477–480, ISBN 978-3-642-17584-8.
8. Rieciciarova, E.; Nanasi, T. Static and dynamic characteristics of asynchronous motor. In Proceedings of the Annals of DAAAM and Proceedings of the International DAAAM Symposium, Vienna, Austria, 23–26 November 2011; Volume 22, pp. 747–748.
9. Bitar, Z.; Al Jabl, S. Studying the performances of induction motor used in electric car. *Energy Procedia* **2014**, *50*, 342–351. [CrossRef]
10. Abdelati, R.; Mimouni, M.F. Loss minimization of induction machines during torque transients. *Proc. Inst. Mech. Eng. Part I J. Syst. Control. Eng.* **2019**, *234*, 314–329. [CrossRef]
11. Kodkin, V.L.; Anikin, A.S.; Shmarin, Y.A. Effective frequency control for induction electric drives under overloading. *Russ. Electr. Engin.* **2014**, *85*, 641–644. [CrossRef]
12. Un-Noor, F.; Padmanaban, S.; Mihet-Popa, L.; Mollah, M.N.; Hossain, E. A comprehensive study of key electric vehicle (EV) components, technologies, challenges, impacts, and future direction of development. *Energies* **2017**, *10*, 1217. [CrossRef]
13. Abdel-Wahab, R.R.; Abdo, T.M.; Hanafy, H.H. Dynamic investigation of Georges phenomenon related to three-phase wound-rotor induction motor. *JESIT* **2020**, *7*, 1–21.
14. Guo, Z.; Zhang, J.; Zheng, C.; Sun, Z. Dynamic Performance analysis of the induction motor drive fed by current-source based on Ansoft. *AMSE J. Ser. Model. A* **2016**, *89*, 118–129.
15. Elavenil, P.E.; Kalaivani, R. Overload protection and speed monitoring of induction motor using ZigBee wireless sensor networks and GSM technology. *Springer Lect. Notes Electr. Eng.* **2014**, *326*, 1007–1016.
16. Cowder, R. *Electric Drives and Electromechanical Systems*, 2nd ed.; Elsevier: London, UK, 2019; pp. 1–322, ISBN 9780081028841.
17. Bodolai, T. Computer-aided measurement method for determination of torque-speed characteristics of electrical motors used in practical education. *J. Phys. Conf. Ser.* **2010**, *238*, 1–8. [CrossRef]
18. Evaluation of measurement data—Supplement 2 to the Guide to the Expression of Uncertainty in Measurement—Extension to any Number of Output Quantities. *JCGM* **2011**, *102*, 1–80.
19. Wu, S.; Hannig, J.; Lee, T.C.M. Uncertainty quantification for principal component regression. *Electron. J. Statist.* **2021**, *15*, 2157–2178. [CrossRef]
20. Beck, K. Drivers of structural convergence: Accounting for model uncertainty and reverse causality. *Entrep. Bus. Econ. Rev.* **2021**, *9*, 189–208.
21. Evaluation of measurement data—Supplement 1 to the Guide to the Expression of Uncertainty in Measurement—Propagation of Distributions Using a Monte Carlo Method. *JCGM* **2008**, *104*, 1–90.
22. Kubisa, S.; Warsza, L.Z. Identification of parameters of the capacitor equivalent scheme Using Monte Carlo methods. *Springer Adv. Intell. Syst. Comput.* **2017**, *550*, 165–177.

23. Harris, P.M.; Cox, M.G. On a Monte Carlo Method for measurement uncertainty evaluation and its implementation. *Metrologia* **2014**, *51*, 176–182. [CrossRef]
24. Tomczyk, K. Influence of Monte Carlo generations applied for modelling of measuring instruments on maximum distance error. *IEEE Trans. Instrum. Meas. Control.* **2019**, *41*, 74–84. [CrossRef]
25. Kowalczyk, M.; Tomczyk, K. Procedure for determining the uncertainties in the modeling of surface roughness in the turning of NiTi alloys using the Monte Carlo method. *Materials* **2020**, *13*, 4338. [CrossRef] [PubMed]
26. Rolek, J.; Utrata, G. An identification procedure of electromagnetic parameters for an induction motor equivalent circuit including rotor deep bar effect. *Arch. Electr. Eng.* **2018**, *67*, 279–291.
27. Weinreb, K. Diagnostics of an induction-motor rotor by the spectral analysis of stator currents. *Therm. Eng.* **2013**, *60*, 1006–1023. [CrossRef]
28. Sulowicz, M.; Ludwinek, K.; Tulicki, J.; Depczynski, W.; Nowakowski, L. Practical adaptation of a low-cost voltage transducer with an open feedback loop for precise measurement of distorted voltages. *Sensors* **2019**, *19*, 1071. [CrossRef] [PubMed]
29. Bouheraoua, M.; Atig, M.; Bousbaine, A.; Benamrouche, N. Electro-Thermal Coupled Modeling of Induction Motor Using 2D Finite Element Method. *Adv. Electr. Comput. Eng.* **2021**, *21*, 33–40. [CrossRef]
30. Trachi, Y.; Elbouchikhi, E.; Choqueuse, V.; Benbouzid, M. Induction Machines Fault Detection Based on Subspace Spectral Estimation. *IEEE Trans. Ind. Electron.* **2016**, *63*, 1–11. [CrossRef]
31. Tomczyk, K.; Piekarczyk, M.; Sokal, G. Radial basis functions intended to determine the upper bound of absolute dynamic error at the output of voltage-mode accelerometers. *Sensors* **2019**, *19*, 4154. [CrossRef]
32. Ostertagova, E. Modelling using polynomial regression. *Procedia Eng.* **2021**, *48*, 500–506. [CrossRef]
33. Sun, B.; Liu, H.; Zhou, S.; Li, W. Evaluating the performance of polynomial regression method with different parameters during color characterization. *Math. Probl. Eng.* **2014**, *2014*, 1–7. [CrossRef]

energies

Article

Multi-Stage Optimization of Induction Machines Using Methods for Model and Parameter Selection

Martin Nell *, Alexander Kubin and Kay Hameyer

Institute of Electrical Machines—IEM, RWTH Aachen University, 52062 Aachen, Germany; alexander.kubin@rwth-aachen.de (A.K.); kay.hameyer@iem.rwth-aachen.de (K.H.)
* Correspondence: martin.nell@iem.rwth-aachen.de

Abstract: Optimization methods are increasingly used for the design process of electrical machines. The quality of the optimization result and the necessary simulation effort depend on the optimization methods, machine models and optimization parameters used. This paper presents a multi-stage optimization environment for the design optimization of induction machines. It uses the strategies of simulated annealing, evolution strategy and pattern search. Artificial neural networks are used to reduce the solution effort of the optimization. The selection of the electromagnetic machine model is made in each optimization stage using a methodical model selection approach. The selection of the optimization parameters is realized by a methodical parameter selection approach. The optimization environment is applied on the basis of an optimization for the design of an electric traction machine using the example of an induction machine and its suitability for the design of a machine is verified by a comparison with a reference machine.

Keywords: induction machine; electromagnetic models; model selection; optimization; artificial neural networks; pattern search; evolutionary strategy; simulated annealing

Citation: Nell, M.; Kubin, A.; Hameyer, K. Multi-Stage Optimization of Induction Machines Using Methods for Model and Parameter Selection. *Energies* **2021**, *14*, 5537. https://doi.org/10.3390/en14175537

Academic Editor: Ryszard Palka

Received: 30 July 2021
Accepted: 1 September 2021
Published: 4 September 2021

1. Introduction

Optimization methods are increasingly used in the electromagnetic design and re-design process of electrical machines. The selection of the electromagnetic machine model and the optimization methods has a great influence on the computational effort and the convergence of the optimization.

Optimization methods can be divided into deterministic and stochastic methods. A well-known deterministic method is Pattern Search (PS) [1,2]. Among the most commonly used stochastic methods are evolutionary algorithms such as Genetic Algorithms (GA) [3,4] or Evolution Strategy (ES) [5], Particle Swarm Optimizations [6], and Simulated Annealing (SA) [4]. Both deterministic and stochastic optimization methods are applied in the field of electric machine optimization. In addition, couplings of deterministic and stochastic methods, such as GA coupled with PS [7], are applied.

In the field of electrical machines, multi-objective optimization is often considered, which allows the machine to be optimized with respect to several objective functions. In these optimizations, stochastic methods such as GA or ES and the Design of Experience (DoE) are used. Other methods used are the sequential optimization method [8] and the multi-level or multi-stage optimization [3,9–11]. In some of these referred multi-stage optimizations, a successive two stage optimization using one optimization method is conducted. The optimization parameters are divided into significant and less significant parameters and the parameter groups thus defined are varied or kept constant depending on the current stage. Examples of such methods can be found in [3,9] using a GA and [10,11] using DoE as the optimization method. An overview and further literature on optimization methods in the field of electrical machines can be found in [12,13].

The machine models can be divided into direct and indirect machine models. Direct models include numerical ones such as the Finite Element Method (FEM) and analytical ones such as the Equivalent Circuit Diagram (ECD) model. They differ in their range of values of the modeled effects, their level of detail and their computational effort. In the field of optimizing synchronous machines [14,15] and synchronous reluctance machines [16], machine modeling is often performed using the FEM. This is possible due to the negligible transient effects and the resulting lower computational effort compared to the consideration of the FEM in the optimization of an Induction Machine (IM) [17,18]. For the IM, transient effects can no longer be neglected without accepting a significant reduction of the level of detail, resulting in a high computational effort for the FEM. Therefore, lower order models like analytical ECD models are applied [19–22].

In machine optimization environments, a very high number of machine simulations can be required. In this case, the FEM and other analytical methods can lead to very high computational effort. To reduce this high computational effort indirect machine models like the Response Surface Model (RSM) [8], Kriging Model [8,10] or Artificial Neural Network (ANN) are used. These surrogate models replace the machine model and estimate the output parameters of the machine based on the input parameters.

This paper presents a multi-stage optimization environment for IM design optimization that combines the advantages of several of the described optimization methods. The methods used are SA by [23], ES by [24], and PS by [1]. While the stochastic SA method has good global convergence with low local convergence speed, the ES method is known for stable convergence in the local group [25]. PS, as a deterministic method, provides a tool for fast local convergence [26]. Both the application of the successive ES-PS optimization and the previously executed stage of SA improve the convergence behavior in this case. In all these stages, direct machine models are used for electromagnetic modeling. The successive ES-PS method also reduces the computational effort compared to the single-stage ES-PS method. The increased computational effort due to the use of SA is compensated for by the application of an indirect machine model in the form of an ANN. This leads to a multi-stage optimization environment that combines the advantages of deterministic and stochastic optimization methods and those of direct and indirect model building. The selection of electromagnetic machine models and optimization parameters in each stage is methodically performed using the model selection and parameter selection procedure approach presented in [27]. Thus, in each stage, the model can be adjusted according to the desired range of values and level of detail.

The presented optimization environment is exemplary used to design an IM as a traction drive for a small vehicle. The aim of the optimization is to minimize the losses occurring over a given driving cycle while at the same time minimizing the required installation space. Starting with a rough design of the machine, it is optimized using the optimization methods presented, considering geometric and thermal constraints. The resulting machine is compared with a reference machine, which is used as a benchmark. The simulation results of this example show a good robustness of the multi-stage optimization environment including SA, a successive ES-PS method and the use of an ANN. Compared to classical multi-stage optimizations using one optimization method and one type of machine model, as in [3,9–11], it shows an improved convergence behavior. It can therefore be used for the design process and the design optimization of IM.

2. Optimization Environment and Optimization Methods

The classical optimization methods can be categorized into stochastic and deterministic methods. While deterministic methods realize a fast convergence to a local optimum, stochastic methods enable the search for a global optimum. Furthermore, stochastic methods offer other advantages, including easy consideration of constraints and numerical stability avoiding the use of derivatives [25]. A detailed review of optimization methods in the context of electrical machines is presented in [12,28].

In this paper, with the SA, the ES and the PS, three different optimization methods are combined in one optimization environment, thus exploiting synergies. While the stochastic SA method exhibits good global convergence with low local convergence speed, the ES method is known for stable convergence in the local group [25]. PS, as a deterministic method, provides a tool for fast local convergence [26]. In the following, the multi-stage optimization environment used in this work is presented. First, the overall structure of the optimization environment is explained and then the individual parts of it are described in more detail.

2.1. Structure of the Optimization Environment

The optimization environment used in this work is shown in Figure 1 and consists of several optimization steps and further functionalities to reduce the simulation effort.

The input of the optimization environment consists of the problem description, the electromagnetic IM models used in the respective optimization stages and the optimization parameters that are varied during the optimization. The problem description includes the physical problem including its constraints and decision parameters used to define the objective function and the fitness function respectively. The electromagnetic machine models used in the respective optimization stages are defined using a methodical model selection approach presented in [27]. Based on the problem description, this determines the most suitable problem-specific IM model for the individual stage of the optimization process. This results from the fact that, in a global search, if necessary, a lower level of detail of the model is sufficient for a rough estimate of the fitness value, whereas in a local search a high precision is required to converge to the actual minimum. The optimization parameters are also defined methodically. Building on the model selection methodology, the methodological parameter selection approach presented in [27] is used to determine the variable optimization parameters based on their influence on the optimization problem. For these parameters, possible lower and upper bounds are described.

The optimization itself consists of five optimization stages. The result of each stage represents the initial solution of the further stages. The first stage is performed using the SA method. This features good global convergence, but has a low local convergence speed. The SA optimization is used to identify a local group of solutions. The following stages involve a successive hybrid optimization method with a faster local convergence. A hybrid optimization method is a combination of a stochastic and a deterministic search method. In this paper, the combination of the ES method with the PS method forms such a hybrid method. In the ES method, both the required population size and the required number of generations increase for stable convergence of the method with an increasing number of optimization parameters and thus the dimensions of the solution space. This relationship can be explained as a function of the number of optimization variables n by $O(n^2)$ [24]. To reduce the solution effort, the hybrid optimization is performed successively in two consecutive steps. In the first step, significant optimization parameters are varied and less significant optimization parameters are set to constant values. In the second step, the parameters optimized in the first step are assumed to be constant and the less significant parameters are varied. The classification of the optimization parameters into significant and less significant parameters is done by the user.

The hybrid successive optimization approach results in a reduction of the solution space and thus also of the computational effort within the individual stages. Since there is no holistic global search for the optimum, but the stages act independently of each other, this is a heuristic approach to solve the optimization problem.

To further reduce the simulation effort, an ANN is introduced. This is used to determine the objective function without the need for electromagnetic simulation. The database for the training, selection and testing instances of the ANN is set up during the first stage of the optimization environment. Once the minimum size of the database has been reached, the ANN is constructed and applied in the further course of the optimization.

In the following, the individual parts of this multi-stage optimization environment are explained in more detail. First, the model and parameter selection approach are introduced. Second, the used optimization methods and the fitness function used by all optimization methods are explained. Finally, the ANN used is described.

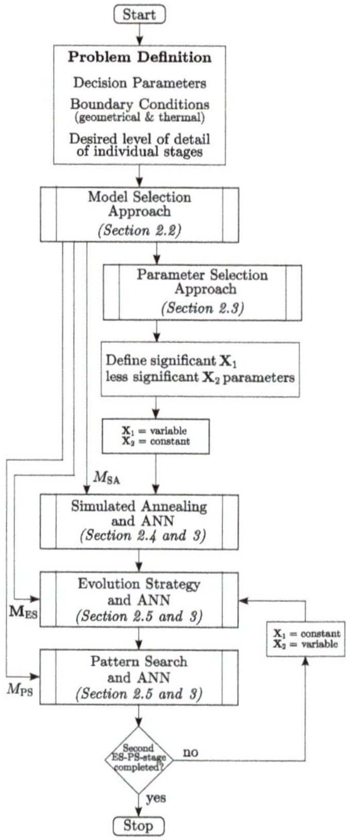

Figure 1. Flow chart of the multi-stage optimization environment.

2.2. Model Selection Approach in the Optimization

Whereas in the global search a lower level of detail of the model may be sufficient for a rough estimation of the fitness value, in the local search, a high precision is required to converge to the true minimum. For this reason, the model selection approach presented in [27] is used to methodically assign different machine models of the IM to each step of the optimization environment. In the context of the optimization environment, the output variables and effects to be considered describe those variables, which influence the decision parameters of the optimization problem. The use of the model selection approach will allow different levels of detail to be considered and defined in the individual optimization steps, allowing both precision and solution effort to be flexibly adjusted.

The model selection approach assigns a constant model to the SA optimization. Since SA implements a global search to identify a suitable local group, a lower level of detail may be sufficient in this step, which in turn reduces the solution cost of the resulting model.

For the ES method, a model vector of arbitrary length and increasing level of detail is used. The desired level of detail is specified and the most suitable model which fulfils this is determined by the model selection approach and added to the model vector. In general, the model selection does not distinguish between the first and second stage in the case of the successive optimization. The ES optimization starts in both stages with the first model in the model vector. If the minimum fitness value $f(\vec{x}_{\mathrm{min,k}})$ of the generation k has not changed more than a tolerance ε over a given number of generations n, thus

$$|f(\vec{x}_{\mathrm{min,k-n}}) - f(\vec{x}_{\mathrm{min,k}})| < \varepsilon \quad (1)$$

is satisfied, the next most accurate machine model in the model vector is selected. This detects convergence to a local minimum and supports it with a more precise model. Provided that the criterion from (1) is met again, the next model in the model vector is selected, and so on.

Since the PS method is a deterministic and thus local optimization method, a high level of detail of the machine model of the IM is required. Otherwise, the geometry changes required for convergence to the actual minimum may not be adequately represented. This may even lead to a degradation of the solution. The model selection approach assigns a constant model to the PS optimization, which is also independent of the stage of successive optimization. This model is already used in the last iteration of the ES method in order to support the local convergence there by a more precise model and thus to prepare the use of the PS optimization.

2.3. *Parameter Selection Approach in the Optimization*

The optimization parameters of the machine geometry optimization represent those geometry parameters which are varied during the optimization process in order to find a better solution. These parameters are intended to cover a high degree of possible degrees of freedom of the geometry. With a higher number of optimization parameters, the number of dimensions of the search space increases and so does the required solution effort. It is therefore desirable to choose those geometry parameters that have the greatest influence on the searched output variables and the lowest degree of redundancy. Therefore, the choice of optimization parameters in this work is done by using the parameter selection method presented in [27]. The approach for the parameter selection is problem-specific and is also influenced by the selected system model. This is a result of the approach for the model selection, which must be applied accordingly in advance.

In the approach, the sensitivities and the elasticities with respect to the output variables relevant for the optimization are examined for each geometry parameter. The optimization parameters are sorted according to their elasticity and the parameters with the greatest influence on the optimization problem are identified. The sensitivity and elasticity analysis used is briefly described in the following.

2.3.1. Sensitivity Analysis

The sensitivity analysis is based on the approach of the local sensitivity analysis in [9,10] and is carried out on the basis of the given initial design of the IM. For this purpose, the individual geometric parameters of the machine are varied by user defined factor r_{sens}. A change that is too small leads to numerical instabilities and changes that are too large possibly lead to inconsistent machine geometries. If the sensitivity to a geometry parameter is to be examined, the influence of this parameter on other geometry variables must also be taken into account. These influences are described using a correlation matrix, which describes for each possible variable geometry parameter which other geometry parameter still has to be changed. Both machines, the initial design and the slightly changed machine geometry, are simulated and the influence on the objective function is assessed by comparing the output variables.

2.3.2. Elasticity Analysis

The elasticity ϵ describes the ratio of the change of an output quantity, related to the change of the input quantity, and therefore results in the case of the sensitivity analysis for a physical quantity to

$$\epsilon_{\text{phys}} = \frac{r_{\text{phys}}}{r_{\text{sens}}}, \tag{2}$$

where r_{phys} describes the relative change of the physical quantity between the two simulated machine models. This allows for capturing for each model how much a certain quantity changes when the parameter under investigation is modified.

2.4. Simulated Annealing

The basic principle of the methaheuristic optimization method of SA according to [23] describes the quality of a possible solution or an individual x by means of a fitness function f. The goal of the optimization is to minimize the fitness $f(x)$. Starting from an initial solution x_0, a random selection of new individuals x_{k+1} within the solution space is performed. A new individuum is accepted and not discarded if one of the constraints

$$f(x_{k+1}) \leq f(x_k) \tag{3}$$

$$\text{or} \quad P(f(x_{k+1}) - f(x_k)) = e^{-\frac{f(x_{k+1}) - f(x_k)}{T_k}} \tag{4}$$

is satisfied. Thereby (4) describes, based on Boltzmann statistics, the probability P with which an intermediate result of the optimization may deteriorate. This realizes the possibility to leave a local optimum in the search for the global optimum. For this purpose, a temperature T_k is introduced, which is successively reduced over the iterations. The lower the temperature, the lower the probability of accepting a worse solution. Accordingly, the algorithm converges with the "cooling" of the solution space.

Application of SA for the IM Optimization

The initial solution x_0 of the SA process in the presented IM optimization process is a roughly designed machine geometry. The rough design is based on problem-specific design parameters, such as the rated power, and is computed according to [29]. An individual of the SA optimization is defined by a chromosome. This contains all geometry parameters described in the optimization variables. The optimization variables are determined by the parameter selection approach in [27].

In the SA, as well as in the ES and PS method, the parameters of the chromosome are changed, resulting in new designs of the IM. The new machine designs are calculated, like the initial solution according to the chromosome and [29]. The evaluation of an individual based on its chromosome is done by a fitness function, which will be discussed in Section 2.6.

In this work, new individuals are determined using a normally distributed random vector $\vec{X}$, whose normalization results in the random vector $\vec{y}$

$$\vec{X} \sim \mathcal{N}(0,1) \in \mathbb{R}^n \tag{5a}$$

$$\vec{y} = \frac{\vec{X}}{||\vec{X}||}, \tag{5b}$$

with an expected value of zero and a standard deviation of one. The number of dimensions is equal to the number of optimization parameters n. From the chromosome $\vec{x}_k$ of the current individual and the updated temperature T_k, the new individual is determined by

$$\vec{x}_{k+1} = \vec{x}_k + \sqrt{T_k}\frac{\vec{x}_k}{||\vec{x}_k||} * \vec{y}, \tag{6}$$

where operator $(*)$ describes an element-wise vector multiplication. The temperature adaptation is performed according to the Boltzmann annealing by

$$T_k = \frac{T_0}{\ln k}, \tag{7}$$

since this method guarantees global convergence for sufficiently large starting temperatures T_0 [30]. However, since only very high starting temperatures ensure global convergence, but these result in low convergence rates, the technique of Very Fast Re-Annealing (VFR) is used. This increases the temperature after a given number of iterations to avoid the method converging to local minima [31].

If one or more dimensions of the resulting chromosome lie outside the solution domain, they are set to the upper or lower boundary conditions, depending on which boundary was violated. This results in the chromosome $\vec{x}^*_{k+1}$ which, in a convex combination with the current chromosome, realizes a valid solution

$$\vec{x}_{k+1} = \alpha\vec{x}^*_{k+1} + (1-\alpha)\vec{x}_k \tag{8}$$

via a random variable α evenly distributed between zero and one. The from (6) or (8) resulting chromosome can subsequently be accepted or rejected analogous to the criteria described in Section 2.4.

By considering the step-wise cooling temperature in the description of the new chromosome, local convergence is realized. The termination criterion of the SA is a given number of iterations at which the change in fitness of the best individual is less than a defined tolerance.

2.5. *Hybrid Optimization: Evolution Strategy and Pattern Search*

The stochastic method used in the hybrid optimization is the ES according to [24]. The method is based on populations of individuals in the solution space, in which a generation with μ parents creates λ children via crossover and mutation. A distinction is made between the plus strategies (μ + λ), in which a new generation can be composed of parents and children, and the comma strategies (μ, λ), which only consider the best descendants for the next generation of parents [24]. The method used in this work is based on the comma strategy, as this introduces a maximum "lifetime" of the individuals and thus counteracts premature local convergence. The method is structured in an initialization, a selection, a mutation, a crossover and an inheritance process. The method, especially in the case of mutation and crossover, is problem-specific. Gaussian distributions can be used as a statistical basis. Their variance is flexibly adapted as a function of various parameters in order to achieve good local convergence while at the same time enabling a global search. This achieves the already mentioned stable convergence in the local group. Due to the generation principle, there is the inherent possibility of parallelization, which results in a reduced computing time compared to other stochastic optimization methods [32].

The deterministic method of the hybrid optimization in this work is the PS method. In contrast to other deterministic optimization methods, PS as a direct optimization method does not require a gradient of the fitness function. In addition to the associated numerical stability avoiding the use of derivatives, a local search for problems that are neither continuous nor differentiable is enabled [1]. In this way, a fast local convergence can be realized especially for complex optimization tasks with problem-specific boundary conditions. In this work, the method according to [1] is used.

2.5.1. Application of the Hybrid Optimization Method for the IM Optimization

The implementation of the two optimization methods is as previously described. In the following, the implemented crossover and mutation of individuals based on random distributions in the ES optimization will be discussed. For this purpose, normal distributions are used in the presented optimization environment, whose standard deviations and expected values are adjusted via various parameters depending on the situation.

Crossover

Crossover of individuals is performed using the (2,1)-strategy. From two parents separated by their chromosomes $\vec{x}_{p1}$ and $\vec{x}_{p2}$, by means of the convex function

$$\vec{x}_c = \vec{x}_{p1} + \left(\vec{x}_{p2} - \vec{x}_{p1}\right) * \vec{X} \tag{9}$$

with the normally distributed random vector

$$\vec{X} \sim \mathcal{N}(\mu, \sigma) \in \mathbb{R}^n, \tag{10}$$

a descendant with the chromosome $\vec{x}_c$ is generated. The expected value μ and the standard deviation σ are thereby influenced by different, parent- and population-specific factors. The aim of the crossover is to project the properties of the parent with the lower fitness value $f(\vec{x}_p)$, more onto the descendant. For this purpose, a fitness factor:

$$FF = \begin{cases} \frac{f(\vec{x}_{p1})}{f(\vec{x}_{p2})}, & f(\vec{x}_{p1}) < f(\vec{x}_{p2}) \\ \frac{f(\vec{x}_{p2})}{f(\vec{x}_{p1})}, & f(\vec{x}_{p2}) < f(\vec{x}_{p1}) \end{cases} \quad (11)$$

is introduced, which is derived from the fitness values of the parents and shifts the expected value to the parent with the lower fitness value. Since a larger distance between the two parents increases the uncertainty of how the fitness behaves in the solution space between the parent chromosomes, a distance factor

$$DF = 1 - \frac{d_{p12}}{d_{max}} \quad (12)$$

is introduced. This describes the normalized relative deviation d_{p12} of the two parents relative to the maximum occurring value d_{max} of all parent pairs in the current generation. This factor shifts the expected value toward the parent with the lower fitness when the parents are far apart and reduces the standard deviation, counteracting the uncertainty in the space between. The closer an individual's fitness is to the minimum fitness $f(\vec{x}_{min})$ of a parent in the current generation, the lower the variation of the descendant should be to allow local convergence. This is achieved by an overall fitness factor:

$$OFF = \begin{cases} 1 - \frac{1}{2}\frac{f(\vec{x}_{min})}{f(\vec{x}_{p1})}, & f(\vec{x}_{p1}) < f(\vec{x}_{p2}) \\ 1 - \frac{1}{2}\frac{f(\vec{x}_{min})}{f(\vec{x}_{p2})}, & f(\vec{x}_{p2}) < f(\vec{x}_{p1}) \end{cases} \quad (13)$$

which affects both the expected value and the standard deviation. The expected value of the random vector $\vec{X}$ is given by

$$\mu = \begin{cases} \frac{1}{2} \cdot FF \cdot DF \cdot OFF, & f(\vec{x}_{p1}) < f(\vec{x}_{p2}) \\ 1 - \frac{1}{2} \cdot FF \cdot DF \cdot OFF, & f(\vec{x}_{p2}) < f(\vec{x}_{p1}) \end{cases}, \quad (14)$$

which shifts the expected value of the descendant's chromosome toward the parent with the lower fitness value as a function of the factors introduced. The resulting standard deviation

$$\sigma = \sigma_{crossover} \cdot DF \cdot OFF \quad (15)$$

considers an adjustable maximum standard deviation $\sigma_{crossover}$ in addition to the factors described. This should be set as a function of the elasticities of the optimization parameters in order to respond to large elasticities with a lower variance, thereby counteracting larger jumps through the solution space and thus improving the convergence behavior in the local group.

Mutation

Mutation of a selected parent with the chromosome $\vec{x}_p$ is performed using a normally distributed random vector analogous to (10). The descendant's chromosome $\vec{x}_c$ is calculated by

$$\vec{x}_c = \vec{x}_p * \left(1 + \vec{X}\right), \quad (16)$$

where operator $(*)$ describes an element-wise vector multiplication. Here, the random vector has an expected value of $\mu = 0$, whereas the standard deviation, analogous to the crossover of individuals, is adapted depending on various parameters. Through these factors, local convergence is to be realized in particular. For this purpose, the overall fitness factor

$$OFF = 1 - \frac{1}{2}\frac{f(\vec{x}_{\mathrm{min}})}{f(\vec{x}_{\mathrm{p}})} \tag{17}$$

analogous to (13) is introduced. This reduces the variance as a function of the fitness value. In addition to this, with

$$GF = 1 - \frac{g_{\mathrm{k}}}{g_{\mathrm{max}}} \tag{18}$$

a generation factor is defined for the current generation g_{k}, relative to the maximum number of generations g_{max}. This factor reduces the standard deviation across generations, which also supports local convergence of mutant descendants.

The resulting standard deviation of the random vector describing the mutation follows accordingly with

$$\sigma = \sigma_{\mathrm{Mutation}} \cdot GF \cdot OFF. \tag{19}$$

The adjustable maximum standard deviation of the mutation $\sigma_{\mathrm{mutation}}$ should be adapted as in the case of the crossover depending on the elasticity of the optimization parameters.

2.6. Fitness Function

The fitness function is identical for all stages of the optimization environment. It assigns an individual x a fitness value $f(x)$ based on its chromosome, considering geometry and thermal constraints as well as a given driving cycle. This fitness value thereby also categorizes unacceptable machine geometries, depending on the number of fulfilled boundary conditions and the achievable points of the driving cycle. By this categorization, inadmissible solutions can be of different fitness, which improves the convergence behavior of the optimization methods and, in particular, realizes a faster search for admissible solutions. The maximum fitness of an individual is given by

$$f_{\mathrm{max}} = 2 \cdot n_{\mathrm{BC}} \cdot n_{\mathrm{DC}}, \tag{20}$$

where n_{BC} represents the number of constraints to be considered and n_{DC} represents the number of points in the driving cycle. This allows a clear differentiation and thus categorization of the fitness values. The procedure to evaluate the fitness of an individual is based on a sequential process:

1. Creation of the machine geometry based on the chromosome and the necessary design parameters
2. Verification of geometric boundary conditions. In case of violations of the boundary conditions, the resulting fitness value of the individual as a function of the number of satisfied geometry boundary conditions $n_{\mathrm{BC,Geo}}(x)$ is given by

$$f(x) = \frac{f_{\mathrm{max}}}{n_{\mathrm{BC,Geo}}(x)}. \tag{21}$$

3. Operating map calculation of the geometrically permissible individual using the selected machine model.
4. Verification of the thermal boundary conditions using a thermal simulation. If the given maximum temperatures are exceeded, the fitness results in

$$f(x) = \frac{f_{\mathrm{max}}}{n_{\mathrm{BC,Geo}}}, \tag{22}$$

with the number of all geometry boundary conditions $n_{\mathrm{BC,Geo}}$.

5. Determination of the individual fitness of an admissible individual by means of

$$f(x) = \frac{1}{\sum_i w_\mathrm{i}} \sum_j w_\mathrm{j} \frac{p_\mathrm{j,indiv}}{p_\mathrm{j,ref}}. \tag{23}$$

where $p_\mathrm{j,indiv}$ are weighted problem-specific decision parameters, $p_\mathrm{j,ref}$ are the decision parameters of a reference machine, and w_i is the sum of the weighting factors. In case of an invalid solution due to the non-achievement of several operating points in the driving cycle, an additional penalty term

$$f_\mathrm{penalty}(x) = \frac{f_\mathrm{max}}{2n_\mathrm{BC}} v_\mathrm{DC}(x) + \Delta_\mathrm{offset} \tag{24}$$

is considered. Here, $v_\mathrm{DC}(x)$ stands for the share of the not reachable operating points in the total number of operating points of the driving cycle and Δ_offset for an offset to separate the invalid solutions from the allowed ones.

3. Reduction of the Solution Effort

In order to realize a stable convergence of the optimization environment with a solution space with sufficiently many degrees of freedom for the variation of the machine geometry, many iterations and large populations are required due to the used stochastic optimization methods. This yields a long computation time, particularly for machine models with a high number of degrees of freedom. Depending on the required level of detail of the optimization problem, which influences the model resulting from the model selection methodology, the problem may thus not be solvable within a few days. A reduction of the solution effort is desirable. The solution effort can be reduced by a preselection of the individuals to be simulated. In this work, the preselection is realized by estimating the fitness values of the individuals using an ANN . This estimates for each individual those decision parameters of the optimization environment that require a time-intensive operating map simulation, such as the mean losses over a given drive cycle. Geometry-dependent decision parameters, such as the volume or mass of the machine, on the other hand, are computed in a regular manner, so that a fast estimate of an individual's fitness can be obtained from the estimate of the ANN as well as from the individual geometry parameters. Based on these estimates, solutions are discarded and only those individuals promising better fitness than the current minimum are simulated. In the following, the ANN and its application will be discussed.

Artificial Neural Networks

ANN consist of several artificial McCulloch–Pitts neurons. The structure of such a neuron is shown in Figure 2b. The input values x_i of the neuron are thereby multiplied by weighting factors w_i. The result of the summation of the weighted input values is adjusted with an offset b (*also: bias*) and projected to the output y via a transfer function f [33]. This results according to

$$y = f(\vec{w}^\mathsf{T}\vec{x} + b). \tag{25}$$

Various transfer functions exist whose suitability depends on the problem. An overview of numerous transfer functions and their respective characteristics is given in [34].

Based on the representation of a single artificial neuron, an ANN can be realized, and various constructions are possible. In this work, the feed-forward architecture, which is shown in Figure 2a is considered. This is described by several layers, each of which is fully connected to the following layer. These are the input and output layer, which describe the input and output variables of the network, and one or more hidden layers. The latter are composed of an arbitrary number of artificial neurons [35].

The weighting factors w_i and offsets b of the individual neurons are determined when training the network using a training method. An overview of various training methods is given in [36].

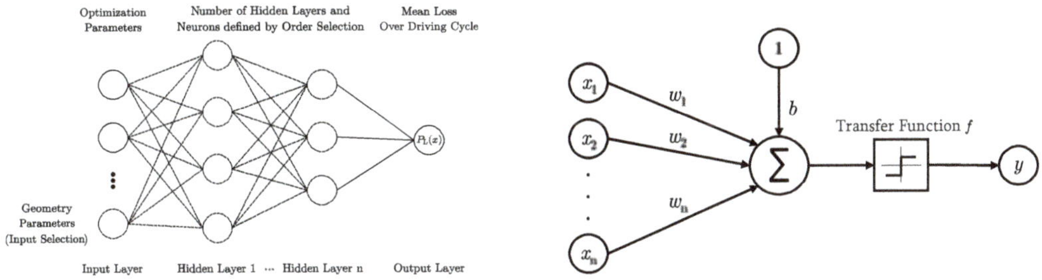

(**a**) Representation of the used feed-forward ANN structure.

(**b**) Illustration of an artificial neuron.

Figure 2. Representation of the used feed-forward ANN and artificial neuron structure.

Application of the ANN for the IM Optimization

The ANN implemented in this work is based on a fully connected feed-forward architecture of McCulloch–Pitts neurons and shown in Figure 2a. As a transfer function, a sigmoidal function

$$f(x) = \frac{1}{1 + e^{-x}} \tag{26}$$

is chosen, which is characterized by a simple and fast calculation of the derivative [34]. Moreover, its range of values includes only positive values, which meaningfully describes the estimation of physical quantities, such as the electromagnetic losses. The inputs of the ANN are the optimization parameters that are the geometry parameters of the actual IM design. Within the input selection, which will be described in the following, these will be supplemented by further geometry parameters. Two common problems in the construction of ANN are under-fitting and over-fitting, which, due to insufficiently complex or over-complex architectures, can cause imprecise estimates of the network. This can be counteracted by a multi-stage method in which different network architectures are analyzed by varying the number of neurons and layers. From these possible architectures, the most appropriate one is selected. For this purpose, the following sequential, five-step flow is introduced.

1. Database: For the training and the use of the ANN, a database of individuals is created, which covers the solution space representative. The database is built in parallel to the optimization from the already simulated individuals, which result during the SA and ES method. Here, the database is composed exclusively of geometrically and thermally admissible solutions that reach all points of the driving cycle and stores the chromosomes and the fitness values of the individuals.
2. Construction: As soon as the database reaches an adjustable minimum number of individuals, the construction of the ANN starts. The database is first divided into training, selection and testing instances. Based on the operating map-dependent decision parameters of the training instances, which are derived from the respective fitness values, increasingly complex neural networks are generated. Starting with a hidden layer and an adjustable number of neurons, this number is successively increased up to a maximum value. A predefined set of further layers with an increasing number of neurons is added step by step. Here, the input variables of the networks are composed of the optimization parameters. The choice of the input variables of

the ANN is essential, since they determine the precision significantly. Therefore, two phenomena must be considered in their selection:

- Input Correlation: The input variables should be minimally correlated with each other, and accordingly have as little redundancy as possible.
- Input Target Correlation: The correlation between output and input variables should be maximized, thereby having the greatest possible influence on the output.

Both aspects can be ensured by the introduced sensitivity analyses. As a training procedure the Levenberg–Marquart method is used in this paper.

3. Order Selection: In the context of the order selection algorithm, the architecture with the lowest selection error is selected from all created network constructions on the basis of the selection instances. For this purpose, the constructed neural networks are used to estimate the operating map-dependent decision parameters of the individual selection instances, and the estimates of the individual fitness values are thus derived from the respective geometry parameters, which are used to determine the selection error.
4. Input Selection: Based on the optimization parameters, successively additional geometry parameters, starting with the highest elasticity to realize the highest possible input-target correlation, are added as input variables to the selected network architecture. For each new geometry parameter to be added, the correlation with the already existing input variables must be checked, since in terms of input correlation the input variables should have as few redundancies as possible. Therefore, if a correlation exists, the geometry parameter to be added is discarded in this work. To realize the growing inputs algorithm, new geometry parameters are added as additional input variables until the selection error starts to increase.
5. Characterization: The problem-specific constructed ANN follows from the order selection and the subsequent input selection. Analogous to the selection instances, this neural network is used to estimate the individual fitness values of the testing instances. Based on this estimation, the error distribution of the testing error can be determined, which is essential for the further application of the ANNin the optimization environment.

The steps for network construction, order and input selection, and characterization are repeated at predetermined intervals once the required number of individuals in the database has been reached. Thus, new individuals added to the database are considered in the ANN, further increasing the estimation accuracy.

The output or solution of the network are the mean loss P_L over the driving cycle of the considered individual. To ensure that the preselection of individuals by the ANN does not discard solutions that would otherwise represent a new optimum, a threshold f_{th} is introduced. Here, individuals whose estimated fitness is above the threshold are discarded and otherwise simulated. To determine the threshold, a delta $\Delta \geq 0$ is introduced. This results from the given estimation reliability γ of the ANN as well as the error distribution of the testing error to

$$\int_{e=-\infty}^{\Delta} p(e)\,\mathrm{d}e = \gamma. \tag{27}$$

The relationship is visualized in Figure 3a. The threshold, as shown in Figure 3b, is obtained as a function of the current optimum $f(\vec{x}_{\min})$ to

$$f_{th} = f(\vec{x}_{\min}) + \Delta. \tag{28}$$

Estimates above the resulting threshold consequently lie within the γ confidence interval with respect to the null hypothesis that these individuals do not represent a new optimum. In addition to the threshold, the relative deviation or distance of the chromosome of the solution to be estimated from the individuals in the database is also considered. If the distance to the nearest individual exceeds a predefined value, the solution is simulated and not discarded because in that case the estimation of the ANN is based on an extrapolation whose validity is not guaranteed.

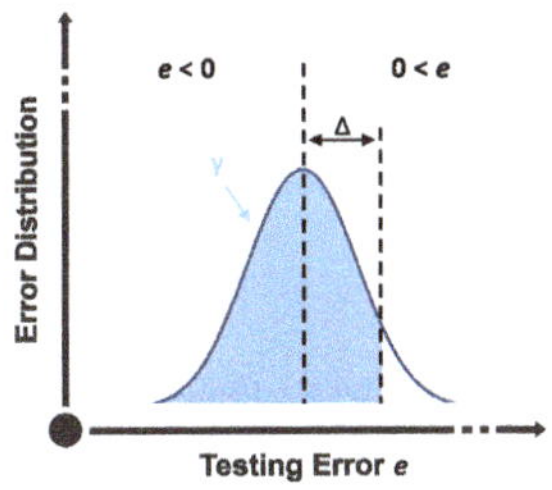

(**a**) Delta Δ based on the distribution of the testing error.

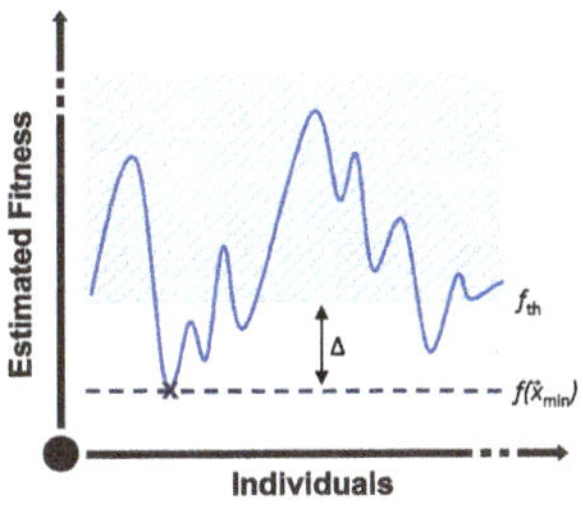

(**b**) Threshold f_{th} for discarding solutions.

Figure 3. Representation of the delta based on the distribution of the testing error (**a**) as well as the resulting threshold (**b**) for the preselection of solutions by means of a ANN.

4. Exemplary Design Optimization of an IM

The presented optimization environment is used to design an IM as a traction drive for a small vehicle. The aim of the optimization is to minimize the losses occurring over a driving cycle while at the same time minimizing the required installation space. Starting with a rough design of the machine, it is optimized using the presented multi-stage optimization environment, considering geometric and thermal constraints. The resulting machine is compared with a reference machine, which is used as a benchmark. To model the IM, different electromagnetic machine models with different value ranges and level of detail are used. The thermal behavior of the machine necessary for the evaluation of the thermal constraints is modeled with an equivalent thermal network with four nodes, as presented in [37].

4.1. Description of the Multiphysics Problem

Boundary conditions and design parameters of the IM are defined on the basis of a reference geometry, whose cross sectional area is shown in Figure 4a. This is a four-pole IM with a rated power of 41.5 kW. An unchorded three-phase copper winding is inserted into the stator, which is connected in delta. The rotor of the machine is a squirrel cage rotor with bars and rings made of a die-cast aluminum with an electrical conductivity of $28 \cdot 10^6$ S/m. The stator and rotor laminations are made of electrical steel sheets of type M400-50A, the designation of which is given by DIN EN 10027-1. The main rated and geometrical parameters are summarized in Table 1. The geometry of the reference machine was designed electromagnetically in an experience-based, iterative process. By solving the problem using the presented optimization environment, a machine geometry with identical boundary conditions and design parameters is to be realized starting from the roughly designed IM in an automated process. The geometry properties and operating behavior of the resulting optimized machine should be similar to those of the reference machine. The cross sectional area of the roughly designed machine is shown in Figure 4b.

To evaluate the machine behavior, two differently weighted decision parameters $p_{j,\text{indiv}}$ are introduced, which are used to determine the fitness value of a geometry relative to the reference machine according to (23).

The first decision parameter is the average losses over a given driving cycle. For this purpose, the Worldwide Harmonized Light Vehicles Test Cycle (WLTC) class 3 driving cycle is considered, which is a part of the Worldwide Harmonized Light Vehicles Test Procedure (WLTP). Based on this driving cycle, the associated speed and torque combinations of the WLTC 3 can be derived for an example small car defined by the vehicle parameters from

Table 2 using the vehicle model according to [38]. To ensure that the driving cycle is fully represented by the reference machine, the resulting speeds and torques are additionally scaled by the factor 0.7. This is due to the fact that the vehicle data of the reference machine are not known. However, this has no influence on the methodical optimization of the machine. Using these speed and torque combinations, the average losses over the driving cycle can be determined via the operating map of the machine.

The second decision parameter is the installation space of the machine. The installation space is not only used as a decision parameter but is further restricted by boundary conditions. The installation space is weighted by a factor of three less than the losses over the driving cycle.

The reference machine is also be used to define the other design-relevant parameters of the multiphysics problem, on the basis of which the rough design of the IM is carried out.

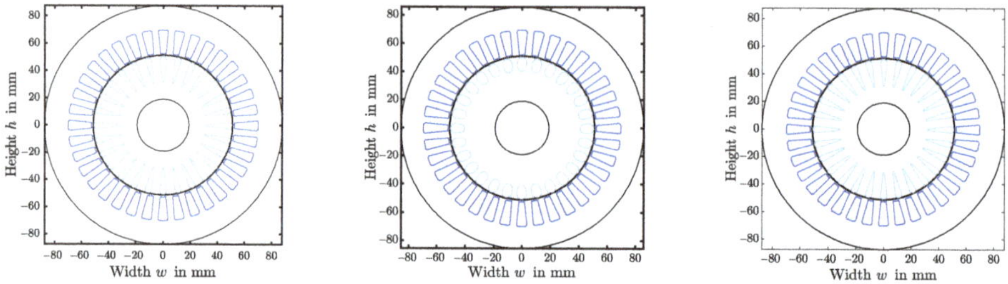

(**a**) Reference. (**b**) Roughly designed. (**c**) Optimized.

Figure 4. Representation of the cross sectional area of the reference (**a**), roughly designed (**b**) and optimized machine (**c**).

Table 1. Rated and geometrical parameters of the reference machine.

Parameter	Value	Unit	Parameter	Value	Unit
Rated power	41.5	kW	Stator outer diameter	175	mm
Rated current	400	A	Active iron length	0.2	m
Rated voltage	90	V	Air gap width	0.5	mm
Number of phases	3	–	Volume (incl. housing)	0.0112	m^3
Number of pole pairs	2	–	Mass (excl. housing)	38.15	kg
Winding connection	Delta	–			

Table 2. Parameters of the example vehicle.

Parameter	Value	Unit
Vehicle weight	1550	kg
Rolling resistance coefficient	0.01	–
Cross span area	2.2	m^2
Drag coefficient	0.29	–
Mass moment of inertia wheel	1.2	$kg\,m^2$
Gearbox efficiency	0.95	–
Wheel radius	0.3065	m

4.2. Methodological Optimization

Based on the problem description, the most suitable machine model is derived for the individual optimization stages using the model selection methodology. The optimization parameters are then derived in dependence of this model using the parameter selection methodology. The used electromagnetic IM models, the procedure and the results of the model and parameter selection for the given problem are briefly explained. Furthermore, the adjustable convergence parameters of the optimization environment are discussed.

4.2.1. Electromagnetic Machine Models

For the electromagnetic simulation of the IM, the machine models described in [27] are used. These are the Fundamental Wave Model (FWM), the Harmonic Wave Model (HWM) and the Extended Harmonic Wave Model (E-HWM), three analytical models, and the Time Harmonic Finite Element Model (TH-FEM) and the Transient Finite Element Model (T-FEM), two numerical models. The models differ in their range of values and level of detail. The following is a brief description of the models.

The fundamental wave model is based on a single-phase ECD, also presented in [39]. For the calculation of the elements of the ECD, which can be derived exclusively from the machine geometry and constant parameters, reference is made to the literature [40].

The HWM and the E-HWM are based on the harmonic wave theory of the IM presented in [41–43]. The HWM is the implementation under the assumption of an infinite permeability of the stator and rotor iron. The E-HWM is an extension of the HWM, where the influence of saturation is modeled by multiplying the flux densities by an air gap conductance function. The circumferential location dependent air gap conductance function [40] is a description of the effective air gap of the machine. The air gap is is increased on average by a saturation factor $k_h > 1$ as a result of the main field saturation. In the region of large iron saturation, i.e., in the maximum of the air gap flux density B, the air gap is further enlarged by an additional saturation factor k_{h1}, and reduced at zero crossings. The air gap conductance function defined in this way moves synchronously with the fundamental wave field. The air gap flux density flattened by the saturation follows from multiplying the air gap flux density calculated with the HWM by the air gap conductance function.

The TH-FEM is based on the state of the art FEM for time-harmonic simulation of electromagnetic components including a slip transformation [44–46]. To consider for the nonlinear material behavior of the stator and rotor laminations, an iterative procedure is used in the field solution. For this purpose, the successive substitution approach or the Newton method can be used.

For the computation of the T-FEM, an in-house state-of-the-art Finite Element (FE) solver called *i*MOOSE/*py*MOOSE [47] is used. To reduce the computational effort of the T-FEM, a hybrid simulation approach presented in [48] is applied.

The FWM, HWM, E-HWM, and TH-FEM are implemented in Matlab®, whereas the T-FEM is implemented in python™ and `C++`. An operating map simulation needs 15 min using the TH-FEM on an PC with an i7 3.6 GHz Processor and 16 GB RAM and 12 h using the T-FEM on the compute cluster of the RWTH Aachen University.

4.2.2. Model Selection Approach

Based on the decision parameters presented, the output variables or effects to be investigated are defined on the basis of those variables of the machine model which influence the decision parameters. This is the case for the diverse loss types of the IM, since they determine the average losses over the driving cycle. For this reason, all loss types are included as output variables to be considered. However, other output variables and effects are not defined in this use case, since none of the other variables has an influence on either of the decision parameters.

In Table 3, the output variables to be investigated are shown with the required levels of detail. Here, identical precision is required for all stages of the optimization environment, since the resulting higher computation times are acceptable in the context of this work. The illustrated levels of detail must be achieved for all operating points of the operating point matrix, so that the precision of the map calculation is ensured. The required level of detail of the ohmic losses is obtained assuming a measurement deviation of 5%, a deviation of the electromagnetic reference model of 5%, and a scatter of the optimum identified by the optimization environment of 5%. A deviation of 25% is allowed for the iron losses since they are not dominant compared to the ohmic losses.

From the model selection approach, the TH-FEM follows as the model that can represent the required levels of detail at all operating points while minimizing the computational effort. Consequently, in the context of the optimization problem considered in this work, the TH-FEM simulation is used in all stages of the optimization environment to describe the behavior of the IM.

Table 3. Output variables of the machine model to be investigated.

Output Variables and Effects		Level of Detail
Output variable	Ohmic losses	15%
	Stator iron losses	25%
	Rotor iron losses	25%
Effects	–	–

4.2.3. Parameter Selection Approach

Based on the defined problem-specific output variables and the machine model of the IM resulting from the model selection methodology, the parameter selection approach is carried out. With this methodology, seven optimization parameters, which have no geometric correlation among each other, are determined. Possible optimization parameters include all geometry parameters of the IM.

Using the approach described in [27], the optimization parameters presented in Table 4 with descending elasticity are derived. Consideration of the specified lower and upper bounds on the variables reduces the size of the solution space and thus the computational effort required. A large part of the bounds results from experience. However, the upper bounds on the rotor diameter and active length can be estimated by considering the installation space limitations in combination with the diameter or length increase of the housing and, in the case of the active length, the winding head length. The lower limit of the outer diameter of the rotor follows from the shaft diameter. In the context of successive optimization, the top four optimization parameters shown are chosen as significant parameters due to their higher elasticity, and the remaining three variables are chosen as less significant parameters.

Table 4. Optimization parameters with associated lower and upper bounds.

Optimization Parameter	Lower Bounds	Upper Bounds
Number of rotor slots	6	50
Number of pole pairs	1	6
Number of slots per pole and phase	1	10
Outer rotor diameter	38 mm	175 mm
Active iron length	1 mm	202 mm
Rotor tooth width	1 mm	20 mm
Stator yoke height	1 mm	40 mm

4.2.4. Convergence Parameters of the Optimization Environment

For the different stages of the optimization environment, numerous adjustable convergence parameters result, such as variances, tolerances, population sizes, or the maximum number of iterations. These parameters can be used to adapt the convergence behavior of the individual stages to the problem-specific specifications of precision and solution effort. In addition, the behavior of the ANN can be influenced by settings for the network construction and the database. The definition of these convergence parameters is thereby experience-based.

4.3. Simulation Results

The simulation results of the presented exemplary IM design optimization are divided in optimization methods results, describing the convergence, robustness, and stability of the presented multi-stage optimization environment, and the IM optimization results, analyzing the optimized machine design.

4.3.1. Optimization Methods Results

Verification of the Multi-Stage Optimization

The analysis of the convergence behavior of the multi-stage optimization environment is performed in two steps. In the first step, the simulation results of the successively applied hybrid ES-PS method are compared with the standard single-stage ES-PS method. In the second step, the use of SA as a preliminary stage to the ES-PS method is compared with the standard single-stage ES-PS method. This approach allows for evaluating separately the advantages of the second hybrid optimization stage and those of the pre-connected SA stage. For this comparison, the described exemplary design optimization is simulated 60 times with each of the different methods and the medians, means, dispersions, and variances of the resulting fitness functions are analyzed.

In Figure 5a, the distribution of the fitness values of the single-stage ES-PS method are plotted in comparison with the two-stage ES-PS method. In the single-stage method, all optimization parameters from Table 4 are varied simultaneously, and the number of generations is $g = 50$ and the population size is $p = 200$. In the two-stage method, the number of optimization parameters in each stage is less than in the one-stage method. Due to this resulting reduction of the solution space in the both stages, the number of generations and populations in the ES method, and thus the computation effort, can be reduced compared to the single-stage method. For the two-stage method, it is $g = 20$ and $p = 50$ in both stages. The reduction of the solution space also causes a reduction of the dispersion and variance, as can be seen in Figure 5a, with a smaller mean value and median. This shows that using the two-stage ES-PS method results in more stable convergence behavior while increasing the speed of convergence.

In Figure 5b, the results of a second single-stage ES-PS method are plotted in comparison to the combination of SA and the single-stage ES-PS method. The single-stage ES-PS method is performed with the same settings as in the previous comparison. This allows an indication of the robustness of the convergence behavior of this method. As shown in Figure 5a,b, the median, the mean value, the dispersion, and the variance are in a very similar range in both cases, suggesting a robust convergence behavior of the method. Combining this method with the SA results in a significant reduction of the median and mean value as shown in Figure 5b. The dispersion and variance also decrease when SA is used. This shows that the previously conducted global search using SA also improves the convergence behavior of the optimization compared with the standard single-stage ES-PS method. The computational effort increases with the use of SA. However, this can be counteracted by ANN as explained in the following section. Since the multi-stage optimization environment runs sequentially, an improvement in the convergence behavior by the SA and the two-stage ES-PS method compared to the single-stage one also means that the convergence behavior of the entire optimization environment is improved.

Verification of the ANN

The verification of the ANN is done by comparing the calculated and the estimated fitness of individuals that arise in the context of the SA optimization following the construction of the network and are to be discarded due to a estimated value above the threshold. An example comparison is shown in Figure 6a. The mean deviation of the estimated from the calculated fitness of the discarded individuals in this case is 2.7%, but much of the estimation is more precise. However, some estimated fitness values have a relatively high deviation from the calculated fitness, which is a consequence of a large distance to the closest individual in the database. Accordingly, the average precision can be increased if the allowed maximum distance is further reduced. It can also be seen that no individual was discarded whose fitness is better than the current optimum $f(\vec{x}_{\min})$.

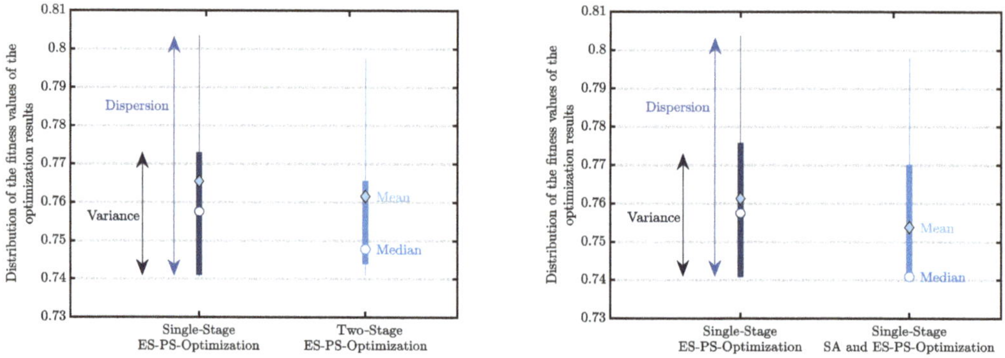

(**a**) Single-stage vs. two-stage ES-PS method.

(**b**) Single-stage ES-PS vs. single stage SA and ES-PS method

Figure 5. Comparison of the distribution of results for different optimization methods.

Reduction of the Solution Effort

By using the ANN, the types of fitness determination can be divided into three different variants:

- Invalid: Neither estimation nor simulation is required. Fitness can be described directly using (21).
- Estimation: Individuals for which the ANN estimates a fitness value above the threshold and are discarded accordingly.
- Simulation: Individuals that are neither invalid nor discarded by the ANN. These require a simulation of the operating map.

In Figure 6b, the percentage breakdown of fitness determination of individuals in these three types is shown for the SA method for the optimization of the example machine. During the optimization, 150 simulations are performed using the TH-FEM to build the database of the ANN. In the SA stage, additional 1130 evaluations of individuals are performed. In each of the hybrid stages, 35 generations were evaluated using the ES method and 20 evaluations were calculated using the PS method. While a large proportion of individuals in the SA stage have invalid geometries, approximately 20.4% of the solutions can be discarded by neural networks and only 1.4% require time-consuming simulation in this use case. This results in a total of 166 simulations performed for the SA stage. Without the use of the ANN, all solutions that would otherwise be discarded due to the estimated fitness also require a simulation. The required simulations would thus increase to 397. In contrast to the SA optimization, the ANN only has a small impact on the solution effort of the ES-PS method. This is a consequence of the local convergence of this hybrid method, since it searches in a vicinity of the optimum where the estimated fitness values often do not exceed the required threshold. The simulation of each population in the 35 generations of the ES method is fully parallelized for all individuals. With the assumption that all valid individuals are electromagnetically simulated in the hybrid stage, the number of simulations performed by the ES method thus increases by $2 \times 35 = 70$ and by the PS method by $2 \times 20 = 40$. The performed total electromagnetic simulations using the ANN are 276 and without the ANN 507. This means a reduction of the simulation effort by 45.5%. This reduction is achieved assuming the same models in each stage. With the use of different models in the individual stages, this reduction factor will vary.

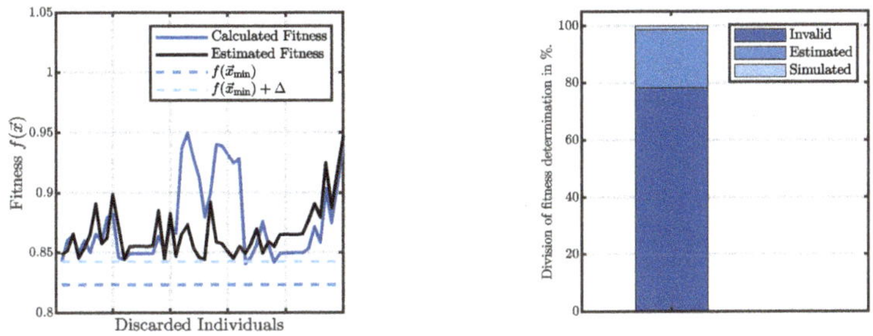

(a) Exemplary comparison of calculated and estimated fitness.

(b) Division of types of fitness determination for SA optimization.

Figure 6. Exemplary comparison of calculated and estimated fitness (**a**) and the breakdown of fitness determination (**b**).

4.3.2. IM Optimization Results

Starting from the cross-section of the rough designed machine shown in Figure 4b, the geometry optimization of the IM is performed. The cross section of the resulting machine geometry of the optimization is shown in Figure 4c. The basic design is similar to that of the reference machine. The optimized geometry differs primarily by two additional rotor bars, a rotor diameter larger by approximately 10% and a shortened active length of $l_{Fe} = 154\,\text{mm}$.

The evaluation of the quality of a solution is done with the fitness described in Section 2.6. It is calculated using the decision parameters defined problem-specifically in Section 4.1 by means of the TH-FEM simulation and related to the reference machine. The resulting fitness values of the roughly designed IM, the optimized machine, and the reference machine are presented in Table 5 and divided into the fractions of the volume as well as the mean losses over the WLTC 3. It can be seen that the optimization environment improves the fitness of the roughly designed IM by approximately 20%. Both the mean losses over the test cycle and the volume are lower in the optimized machine. Compared to the reference machine, the optimized geometry has a lower volume due to the shortened active length, but higher mean losses over the drive cycle. This results in a by 2.6% worse fitness. This is a consequence of the insufficient coverage of all possible degrees of freedom of the machine geometry by the seven optimization parameters, which leads to the fact that the reference machine cannot be completely reproduced. In addition, it is possible that the optimization method has not converged to the global minimum. As shown in Figure 5a, the optimum identified by the optimization environment has a dispersion of approximate 6%.

Table 5. Fitness values and decision parameters resulting from the TH-FEM.

	Roughly Designed IM	Optimized IM	Reference IM
Fitness	1.221	1.026	1
Mean losses	361 W	299 W	276 W
Volume	$0.0107\,\text{m}^3$	$0.0095\,\text{m}^3$	$0.0112\,\text{m}^3$

For further verification of the optimized machine geometry, it is modeled by means of T-FEM simulations and compared with the reference machine. The operating maps of the total losses of the reference machine and the optimized machine resulting from the T-FEM are shown in Figure 7a,b. Here, the optimized geometry exhibits higher total losses, especially at the borders of the operating map, but the losses in the WLTC 3 driving cycle are of a similar order of magnitude to those of the reference machine.

The resulting fitness values related to the transiently simulated reference machine as well as the resulting proportions of the decision parameters are shown in Table 6. The increase in fitness of the optimized geometry from 1.026 for the TH-FEM to 1.059 in the case of a a T-FEM is thereby within the level of detail required by the model selection methodology. Since the deviation of the fitness values between the optimized geometry and the reference machine with approximately 6% is near the range of the assumed accuracy of the transient FEM of approximately 5%, the optimized geometry derived automatically from the rough design and the reference machine can thus be assumed to be similarly suitable solutions of the multiphysics problem.

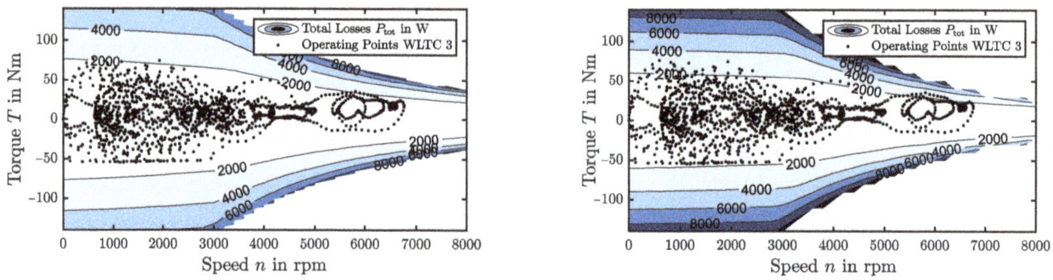

(**a**) Loss operating map of the reference IM.

(**b**) Loss operating map of the optimized IM.

Figure 7. Loss operating maps of the reference (**a**) and optimized machine (**b**).

Table 6. Fitness values and decision parameters resulting from the T-FEM.

	Optimized IM	**Reference IM**
Fitness	1.059	1
Mean losses	541 W	479 W
Volume	0.0095 m^3	0.0112 m^3

5. Discussion and Conclusions

The focus of this work is the development of a multi-stage optimization environment for the design of a IM. In the individual stages, the advantages of two stochastic and one deterministic optimization method are combined by successively applying SA, ES and PS. The search for the optimum starts in the SA stage in a global solution space and continues locally in the successive hybrid use of the ES and PS methods. In the first successive stage, significant optimization parameters are varied and less significant parameters are kept constant. In the second stage, significant parameters are then assumed to be constant and less significant parameters are varied. This successive implementation of the hybrid ES and PS method improves the convergence behavior in terms of a lower mean value, dispersion and variance. In addition, the reduction of the optimization parameters in the individual stages compared to a single-stage hybrid ES-PS method results in a reduction of the computational effort. Using the SA method as a global search performed before the successive hybrid optimization method also improves the convergence behavior in terms of lower mean and median, and lower dispersion and variance of the optimization results. A disadvantage, however, is an increased computational effort due to the additional introduction of another optimization stage. This disadvantage is compensated by the application of an indirect machine model in the form of an ANN. By the ANN individual parts of the objective function, which otherwise require a computationally expensive simulation, are estimated. As a result, the computational cost of the multi-stage optimization environment in the presented application can be reduced by 45% by using the ANN. This value is achieved when the same electromagnetic machine model is used in each stage. The use of different models leads to smaller reductions.

If a precise estimate via the ANN is not possible, direct machine models are used for electromagnetic computation of the IM in the optimization stages. Using a model selection approach in each stage different levels of detail can be considered and defined in each optimization step. Thus, a model of lower level of detail can be used in the global search and models of increasing level of detail can be used in the local search. This procedure results in a high degree of flexibility with respect to the accuracy and the solution effort of the optimization environment.

A methodical approach to parameter selection is used to determine the optimization parameters. For each geometry parameter, the sensitivities and elasticities are studied with respect to the output variables relevant to the optimization. The optimization parameters are sorted by their elasticities and the parameters with the greatest impact on the optimization problem are identified. Sorting by elasticities also allows a systematic division of the parameters into variable and constant values depending on the optimization level.

The optimization environment using the model and parameter selection procedure is applied to the design of a traction machine. The objective of the optimization is to minimize the design space while minimizing the mean electromagnetic losses over the WLTC. The optimization is performed using the TH-FEM of the IM determined by the model selection approach. The quality of the result is determined based on the fitness of the optimized machine and a reference machine. For comparison, both machines are then simulated again using the model with the highest level of detail, the T-FEM. The fitness value of the optimized machine is about 6% higher than that of the reference machine. Since this deviation is within the range of the assumed accuracy of the T-FEM of about 5%, it can be assumed that the optimized geometry automatically derived from the rough design and the reference machine are similarly suitable solutions for the multiphysics problem. Thus, the use of the presented optimization environment as a tool for the design of the machine is verified.

A further verification of the optimization environment with further machine designs for different applications still has to be performed. In addition, the individual machine models can be further improved with respect to their level of detail. Not integrated in the optimization is the simulation of external components, such as inverter or battery. In addition, active cooling of the machine is not considered in the current status. For further improvement, a more detailed thermal model of the machine can be considered.

Regarding the uncertainty of the parameters, the proposed method may be extended using neutrosophic statistics as future research, but this issue is not the content of the study presented here.

Author Contributions: Conceptualization, M.N. and A.K.; methodology, M.N. and A.K.; software, M.N. and A.K.; validation, M.N.; formal analysis, M.N.; investigation, M.N. and A.K.; writing—original draft preparation, M.N.; writing—review and editing, M.N., A.K. and K.H.; visualization, M.N. and A.K.; supervision, K.H. All authors have read and agreed to the published version of the manuscript.

Funding: This research received no external funding.

Institutional Review Board Statement: Not applicable.

Informed Consent Statement: Not applicable.

Conflicts of Interest: The authors declare no conflict of interest.

Abbreviations

The following abbreviations are used in this manuscript:

ANN	Artificial Neural Network
BC	Boundary Condition
DC	Drive Cycle
DoE	Design of Experience
E-HWM	Extended Harmonic Wave Model
ES	Evolution Strategy
FE	Finite Element
FEM	Finite Element Method
FWM	Fundamental Wave Model
GA	Genetic Algorithm
HWM	Harmonic Wave Model
IM	Induction Machine
MDPI	Multidisciplinary Digital Publishing Institute
PS	Pattern Search
SA	Simulated Annealing
TFEM	Transient Finite Element Model
THFEM	Time Harmonic Finite Element Model
VFR	Very Fast Re-Annealing
WLTC	Worldwide harmonized Light vehicles Test Cycle
WLTP	Worldwide harmonized Light vehicles Test Procedure

References

1. Hooke, R.; Jeeves, T.A. "Direct Search"-Solution of Numerical and Statistical Problems. *J. ACM (JACM)* **1961**, *8*, 212–229. [CrossRef]
2. Gradinaru, V.; Tutelea, L.; Boldea, I. 25 kW, 15 krpm, 6/4 PMSM: Optimal Design and Torque Pulsation Reduction via FEM. In Proceedings of the 2008 11th International Conference on Optimization of Electrical and Electronic Equipment, Brasov, Romania, 22–24 May 2008; pp. 249–256. [CrossRef]
3. Wang, S.; Meng, X.; Guo, N.; Li, H.; Qiu, J.; Zhu, J.G.; Guo, Y.; Liu, D.; Wang, Y.; Xu, W. Multilevel Optimization for Surface Mounted PM Machine Incorporating With FEM. *IEEE Trans. Magn.* **2009**, *45*, 4700–4703. [CrossRef]
4. Han, S.I.; Muta, I.; Hoshino, T.; Nakamura, T.; Maki, N. Optimal Design of Superconducting Generator Using Genetic Algorithm and Simulated Annealing. *IEE Proc.-Electr. Power Appl.* **2004**, *151*, 543–554. [CrossRef]
5. Kauder, T.; Hameyer, K.; Belgrand, T. Design Strategy and Simulation of Medium-Frequency Transformers for a Three-Phase Dual Active Bridge. In Proceedings of the IECON 2018—44th Annual Conference of the IEEE Industrial Electronics Society, Washington, DC, USA, 21–23 October 2018; pp. 1158–1163. [CrossRef]
6. Hasanien, H.M. Particle Swarm Design Optimization of Transverse Flux Linear Motor for Weight Reduction and Improvement of Thrust Force. *IEEE Trans. Ind. Electron.* **2011**, *58*, 4048–4056. [CrossRef]
7. Lee, D.; Lee, S.; Kim, J.W.; Lee, C.G.; Jung, S.Y. Intelligent Memetic Algorithm Using GA and Guided MADS for the Optimal Design of Interior PM Synchronous Machine. In Proceedings of the Digests of the 2010 14th Biennial IEEE Conference on Electromagnetic Field Computation, Chicago, IL, USA, 9–12 May 2010; p. 1. [CrossRef]
8. Lei, G.; Shao, K.R.; Guo, Y.; Zhu, J.; Lavers, J.D. Sequential Optimization Method for the Design of Electromagnetic Device. *IEEE Trans. Magn.* **2008**, *44*, 3217–3220. [CrossRef]
9. Xiang, Z.; Zhu, X.; Quan, L.; Du, Y.; Zhang, C.; Fan, D. Multilevel Design Optimization and Operation of a Brushless Double Mechanical Port Flux-Switching Permanent-Magnet Motor. *IEEE Trans. Ind. Electron.* **2016**, *63*, 6042–6054. [CrossRef]
10. Lei, G.; Liu, C.; Zhu, J.; Guo, Y. Techniques for Multilevel Design Optimization of Permanent Magnet Motors. *IEEE Trans. Energy Convers.* **2015**, *30*, 1574–1584. [CrossRef]
11. Lei, G.; Xu, W.; Hu, J.; Zhu, J.; Guo, Y.; Shao, K. Multilevel Design Optimization of a FSPMM Drive System by Using Sequential Subspace Optimization Method. *IEEE Trans. Magn.* **2014**, *50*, 685–688. [CrossRef]
12. Lei, G.; Zhu, J.; Guo, Y.; Liu, C.; Ma, B. A Review of Design Optimization Methods for Electrical Machines. *Energies* **2017**, *10*, 1962. [CrossRef]
13. Duan, Y.; Ionel, D.M. A Review of Recent Developments in Electrical Machine Design Optimization Methods With a Permanent-Magnet Synchronous Motor Benchmark Study. *IEEE Trans. Ind. Appl.* **2013**, *49*, 1268–1275. [CrossRef]
14. Fatemi, A.; Demerdash, N.A.O.; Nehl, T.W.; Ionel, D.M. Large-Scale Design Optimization of PM Machines Over a Target Operating Cycle. *IEEE Trans. Ind. Appl.* **2016**, *52*, 3772–3782. [CrossRef]
15. Zarko, D.; Stipetic, S.; Martinovic, M.; Kovacic, M.; Jercic, T.; Hanic, Z. Reduction of Computational Efforts in Finite Element-Based Permanent Magnet Traction Motor Optimization. *IEEE Trans. Ind. Electron.* **2018**, *65*, 1799–1807. [CrossRef]

16. Pellegrino, G.; Cupertino, F.; Gerada, C. Automatic Design of Synchronous Reluctance Motors Focusing on Barrier Shape Optimization. *IEEE Trans. Ind. Appl.* **2015**, *51*, 1465–1474. [CrossRef]
17. Buschbeck, J.; Vogelsberger, M.A.; Orellano, A.; Schmidt, E.; Bazant, M. Multi-Physics Optimization of High Power Density Induction Machines for Railway Traction Drives. In Proceedings of the 2015 IEEE International Conference on Industrial Technology (ICIT), Seville, Spain, 17–19 March 2015; pp. 2656–2661. [CrossRef]
18. Buschbeck, J.; Vogelsberger, M.; Orellano, A.; Schmidt, E. Pareto Optimization in Terms of Electromagnetic and Thermal Characteristics of Air-Cooled Asynchronous Induction Machines Applied in Railway Traction Drives. *IEEE Trans. Magn.* **2016**, *52*, 1–4. [CrossRef]
19. György, T.; Biró, K.A. Genetic Algorithm Based Design Optimization of a Three-Phase Induction Machine with External Rotor. In Proceedings of the 2015 Intl Aegean Conference on Electrical Machines & Power Electronics (ACEMP), 2015 Intl Conference on Optimization of Electrical & Electronic Equipment (OPTIM) & 2015 Intl Symposium on Advanced Electromechanical Motion Systems (ELECTROMOTION), Side, Turkey, 2–4 September 2015; pp. 462–467. [CrossRef]
20. Arko, D. Design of Premium Efficiency (IE3) Induction Motors Using Evolutionary Optimization and Scaling Laws. *PrzeglĄd Elektrotechniczny* **2016**, *1*, 183–186. [CrossRef]
21. Çunkaş, M.; Akkaya, R. Design Optimization of Induction Motor by Genetic Algorithm and Comparison with Existing Motor. *Math. Comput. Appl.* **2006**, *11*, 193–203. [CrossRef]
22. Duan, Y.; Harley, R.G. A Novel Method for Multiobjective Design and Optimization of Three Phase Induction Machines. *IEEE Trans. Ind. Appl.* **2011**, *47*, 1707–1715. [CrossRef]
23. Kirkpatrick, S.; Gelatt, C.D.; Vecchi, M.P. Optimization by Simulated Annealing. *Science* **1983**, *220*, 671–680. [CrossRef]
24. Schwefel, H.P. *Evolution and Optimum Seeking Programs*; Wiley: New York, NY, USA, 1995.
25. Kasper, M.; Hameyer, K.; Kost, A. Automated Optimal Design of a Permanent Magnet DC Motor Using Global Evolution Strategy and FEM. *Int. J. Appl. Electromagn. Mech.* **1995**, *1995*, 367–376.
26. Torczon, V. On the Convergence of Pattern Search Algorithms. *SIAM J. Optim.* **1997**, *7*, 1–25. [CrossRef]
27. Nell, M.; Kubin, A.; Hameyer, K. Approach for the Model and Parameter Selection for the Calculation of Induction Machines. *Energies* **2021**, under review.
28. Stipetic, S.; Miebach, W.; Zarko, D. Optimization in Design of Electric Machines: Methodology and Workflow. In Proceedings of the 2015 Intl Aegean Conference on Electrical Machines & Power Electronics (ACEMP), 2015 Intl Conference on Optimization of Electrical & Electronic Equipment (OPTIM) & 2015 Intl Symposium on Advanced Electromechanical Motion Systems (ELECTROMOTION), Side, Turkey, 2–4 September 2015; pp. 441–448. [CrossRef]
29. Müller, G.; Vogt, K.; Ponick, B. *Berechnung Elektrischer Maschinen*; 6., Völlig neu Bearbeitete Auflage; Wiley-VCH-Verl.: Weinheim, Germany, 2008; Volume Bd. 2.
30. Geman, S.; Geman, D. Stochastic Relaxation, Gibbs Distributions, and the Bayesian Restoration of Images. *IEEE Trans. Pattern Anal. Mach. Intell.* **1984**, *PAMI-6*, 721–741. [CrossRef] [PubMed]
31. Ingber, L. Very Fast Simulated Re-Annealing. *Math. Comput. Model.* **1989**, *12*, 967–973. [CrossRef]
32. Bäck, T.; Hoffmeister, F. Basic Aspects of Evolution Strategies. *Stat. Comput.* **1994**, *4*. [CrossRef]
33. Rosenblatt, F. The Perceptron: A Probabilistic Model for Information Storage and Organization in the Brain. *Psychol. Rev.* **1958**, *65*, 386–408. [CrossRef] [PubMed]
34. Ding, B.; Qian, H.; Zhou, J. Activation Functions and Their Characteristics in Deep Neural Networks. In Proceedings of the 2018 Chinese Control In Addition, Decision Conference (CCDC), Shenyang, China, 9–11 June 2018; pp. 1836–1841. [CrossRef]
35. Lippmann, R. An Introduction to Computing With Neural Nets. *IEEE ASSP Mag.* **1987**, *4*, 4–22. [CrossRef]
36. Cömert, Z.; Kocamaz, A. A Study of Artificial Neural Network Training Algorithms for Classification of Cardiotocography Signals. *Bitlis Eren Univ. J. Sci. Technol.* **2017**, *7*, 93–103. [CrossRef]
37. Nell, M.; Groschup, B.; Hameyer, K. Using scaled FE solutions for an efficient coupled electromagnetic–thermal induction machine model. *COMPEL-Int. J. Comput. Math. Electr. Electron. Eng.* **2021**, *40*, 267–279. [CrossRef]
38. Mitschke, M.; Wallentowitz, H. *Dynamik der Kraftfahrzeuge*; Springer: Wiesbaden, Germany, 2014; [CrossRef]
39. Nell, M.; Lenz, J.; Hameyer, K. Scaling laws for the FE solutions of induction machines. *Arch. Electr. Eng.* **2019**, *68*, 677–695. [CrossRef]
40. Binder, A. *Elektrische Maschinen und Antriebe: Grundlagen, Betriebsverhalten*; Springer: Berlin/Heidelberg, Germany, 2012; [CrossRef]
41. Oberretl, K. Allgemeine Oberfeldtheorie für ein- und dreiphasige Asynchron-und Linearmotoren mit Käfig unter Berücksichtigung der mehrfachen Ankerrückwirkung und der Nutöffnungen. Teil I: Theorie und Berechnungsverfahren. *Arch. Für Elektrotechnik* **1993**, *76*, 111–120. [CrossRef]
42. Oberretl, K. Allgemeine Oberfeldtheorie für ein- und dreiphasige Asynchron-und Linearmotoren mit Käfig unter Berücksichtigung der mehrfachen Ankerrückwirkung und der Nutöffnungen. Teil II: Resultate, Vergleich mit Messungen. *Arch. für Elektrotechnik* **1993**, *76*, 201–212. [CrossRef]
43. Oberretl, K. Losses, Torques and Magnetic Noise in Induction Motors With Static Converter Supply, Taking Multiple Armature Reaction and Slot Openings into Account. *IET Electr. Power Appl.* **2007**, *1*, 517–531. [CrossRef]
44. Hameyer, K.; Belmans, R. *Numerical Modelling and Design of Electrical Machines and Devices*; WIT Press: Southampton, UK, 1999; Volume 1.

45. De Geserm, H.; Mertens, R.; Kay, H. Comparison of Time-Harmonic and Transient Finite Element Models for Asynchronous Machines. In Proceedings of the Internationel Conference on Electrical Machines (ICEM00), Helsinki, Finland, 28–30 August 2000; Volume 1, pp. 66–70.
46. Schuhmann, T.; Cebulski, B.; Paul, S. Comparison of Time-Harmonic and Transient Finite Element Calculation of a Squirrel Cage Induction Machine for Electric Vehicles. In Proceedings of the 2014 International Conference on Electrical Machines (ICEM), Berlin, Germany, 2–5 September 2014; pp. 1037–1043. [CrossRef]
47. van Riesen, D.; Monzel, C.; Kaehler, C.; Schlensok, C.; Henneberger, G. iMOOSE-An Open-Source Environment for Finite-Element Calculations. *IEEE Trans. Magn.* **2004**, *40*, 1390–1393. [CrossRef]
48. von Pfingsten, G.; Nell, M.; Hameyer, K. Hybrid Simulation Approaches for Induction Machine Calculation. *COMPEL-Int. J. Comput. Math. Electr. Electron. Eng.* **2018**, *37*, 1744–1754. [CrossRef]

energies

Article

Approach for the Model and Parameter Selection for the Calculation of Induction Machines

Martin Nell *, Alexander Kubin and Kay Hameyer

Institute of Electrical Machines (IEM), RWTH Aachen University, 52062 Aachen, Germany; alexander.kubin@rwth-aachen.de (A.K.); kay.hameyer@iem.rwth-aachen.de (K.H.)
* Correspondence: martin.nell@iem.rwth-aachen.de

Abstract: The solution of multiphysical problems in the field of electrical machines is a complex task that involves the modeling of a wide variety of coupled physical domains. Different types of models and solution methods can be used to model and solve the individual domains. In this paper a procedure for the methodical selection of the most suitable model for a given multiphysics task is presented. Furthermore, an approach for the selection of the most suitable variable machine parameters for a design optimization is presented. The model selection is presented on the basis of the electromagnetic calculation of an induction machine. For this purpose, models of different value ranges and levels of detail, such as analytical and numerical ones, are considered. The approach of the model selection is explained and applied on the basis of a coupled electromagnetic-thermal simulation of an exemplary induction machine. The results show that the model selection presented here can be used to methodically determine the most suitable model in terms of its value range, level of detail and computational effort for a given multiphysical problem.

Keywords: induction machine; electromagnetic models; model selection

Citation: Nell, M.; Kubin, A.; Hameyer, K. Approach for the Model and Parameter Selection for the Calculation of Induction Machines. *Energies* **2021**, *14*, 5623. https://doi.org/10.3390/en14185623

Academic Editor: Ryszard Palka

Received: 30 July 2021
Accepted: 2 September 2021
Published: 7 September 2021

1. Introduction

Complex tasks in the field of electrical machines usually comprise several physical domains and thus form a multiphysics problem. The aim of modeling and calculating multiphysics problems is to represent all physical effects of the individual domains that are relevant for the application. In a multiphysics problem, the individual domains can be independent or coupled. In [1], coupled problems are defined as those involving multiple domains and dependent variable sets that usually, but not necessarily, describe different physical effects, and where either no domain can be solved correctly independently of the other domains or none of the dependent variable sets can be eliminated explicitly. An example of such a coupled multiphysics problem is the thermal simulation of an operating point of an electrical machine. Here, among other things, the domains of electromagnetism and thermal are bidirectionally dependent on each other.

If electrical machines are not considered as an independent system, but as part of a system of several components from different fields of technology, the complexity of the coupled multiphysics problem increases. An example of such a system is an electric drive train. The individual components of the system, such as electric motor, gearbox or battery, can in turn be assigned to individual domains or divided into several domains.

Different models are used for modeling and calculating the individual components and their domains, depending on the required accuracy and the physical effects to be modeled. The individual models therefore differ in their value range and level of detail of the representation of the physical effects as well as in their computational or solution effort. They can be classified into the categories of empirical, analytical, lumped parameter, and numerical models. The exemplary classification in value range, level of detail and computational effort for models for the electromagnetic simulation of an Induction Machine (IM) used in this paper are shown in Figure 1. Analytical models such as the Fundamental

Wave Model (FWM) or Harmonic Wave Model (HWM) have a smaller value range of modeled physical effects and a lower level of detail of individual effects than numerical models such as the Finite Element Method (FEM), but also a lower computational effort.

Examples of models, methods, and physical effects considered for solving multiphysics problems in the field of electrical machines are explained in [2]. There, a modular computational approach for the calculation, simulation, and design of electrical machines is presented, which has been and is being applied in several scientific papers.

For the computation and simulation of the aforementioned multiphysics and coupled problems, the question arises as to which models are most suitable for which domains. Suitability refers to the fact that the particular model can represent the desired effects at an appropriate level of detail and is computationally efficient. The respective suitability of a model is strongly problem-dependent, since different problems require the modeling of various effects with different degrees of accuracy. Due to its high computational effort, the selection of the most complex model in the respective domain is not always the means of choice, especially in the design process of the machine or the overall system. Furthermore, the sensitivity with respect to the coupling variables is a necessary requirement of the used models.

In view of the modular computational approach mentioned above, the question of efficient model selection may arise in many of the calculation columns listed. For example, which electromagnetic and thermal models can be used for the thermal calculation of the electrical machine and which electromagnetic and structural dynamics models can be used for the Noise Vibration Harshness (NVH) analysis?

Design and redesign processes of electrical machines are also performed with the help of multilevel mathematical optimization algorithms. Here, the choice of the model in the individual optimization stages is an important factor influencing the quality of the result and the computational efficiency. These parameters also depend on the choice of optimization variables.

In the publications mentioned, models are mostly used without evaluating their suitability or advantages and disadvantages for solving the problem in comparison with other models. A methodical selection of the used model is not considered in detail. Therefore, such a methodological approach to model selection is addressed in this paper and presented using the electromagnetic simulation of IM as an example.

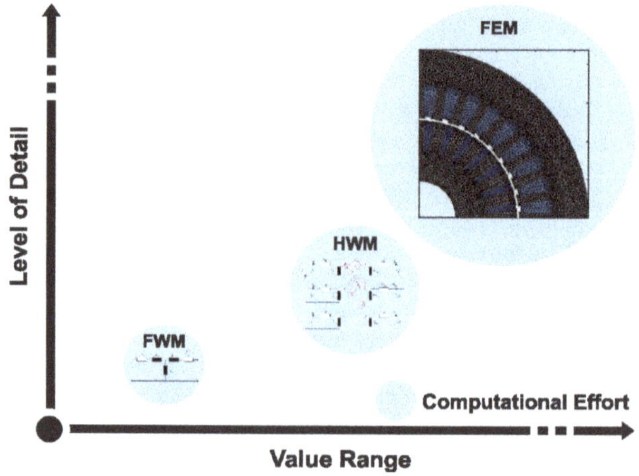

Figure 1. Classification of IM models with respect to value range, level of detail, and computational effort.

By specifying a few parameters, such as the desired effects to be modeled and their level of detail, the method offers the possibility to automatically select the model that meets the requirements and has the lowest computational cost. The selection method is based on the analysis of an exemplary IM for a given power, torque, and speed range in certain operating points and the derived ranges of values, levels of detail, and degrees of freedom of each machine model. The model selection can then be applied to problems requiring simulations of machines in a similar power, torque and speed range, and similar geometric dimensions. The proposed model selection approach can also be used in machine optimization problems in which machines of similar power range but different designs are simulated to determine the appropriate models in individual optimization stages. Such an example is given in [3]. In such optimization environments, the choice of optimization parameters is crucial and can also be done methodically. For this purpose, an approach for parameter selection is presented in this paper, which ensures an efficient design and redesign optimization.

The paper is organized as follows. First, the models of an IM considered in the model selection are introduced. These include the FWM, the HWM according to [4–6], and an Extended Harmonic Wave Model (E-HWM) developed in this work to consider saturation, three analytical models, and the Time Harmonic Finite Element Model (TH-FEM) and Transient Finite Element Model (T-FEM), two numerical models. Afterward, the approach of the model selection is presented. The input parameters of the methodology and the procedure for the analysis of the value ranges, levels of detail, and computational efforts of the models are discussed. In the following, the parameter selection, which can be applied in optimization environments, is presented. Subsequently, the approach for the model selection is applied on the basis of an exemplary problem. As an example, a weakly coupled electromagnetic-thermal simulation of an IM is used to analyze the thermal operating behavior of the machine. The parameter selection in combination with the model selection is applied in [3] in an optimization of the machine design of an IM.

2. Induction Machine Models

In order to apply the model selection approach presented later to the example of the electromagnetic simulation of an IM, different models with different ranges of values and levels of detail are necessary. In this paper, five different models of an IM are considered. These include the FWM, the HWM, and an E-HWM, three analytical machine models, and the TH-FEM and T-FEM, two numerical models. They are briefly described below. The order corresponds to an increasing value range and level of detail. At the end of this section, the iron loss model used by all models is described.

2.1. Fundamental Wave Model

Modeling of a squirrel cage IM in the FWM is performed using the single-phase Equivalent Circuit Diagram (ECD) presented in Figure 2. In this paper, the T-ECD is used, which allows the consideration of saturation in terms of a current-dependent main inductance. The use of a stator flux or rotor flux based ECD is also possible but is not applied here. The ECD is composed of the stator resistance R_{S}, the rotor resistance R'_{R} divided by the slip s, the leakage reactances $X_{\sigma,\mathrm{S}}$ and $X'_{\sigma,\mathrm{R}}$, and the main reactance X_{h}. The rotor-side quantities related to the stator by the transmission factor are thereby characterized by a$'$. For the calculation of these elements of the ECD, which can be derived exclusively from the machine geometry and constant parameters, reference is made to the literature [7]. Important physical quantities and effects which affect the range of values or the level of detail of the machine modeling are briefly explained.

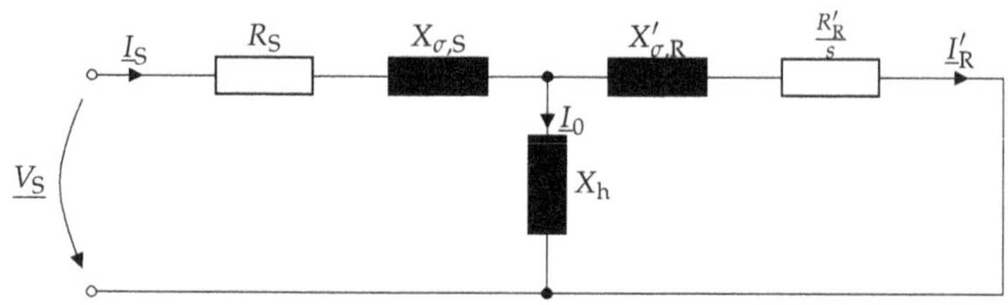

Figure 2. Representation of the single-phase ECD of the IM with a squirrel cage rotor.

2.1.1. Rotor Resistance

To consider the short circuit ring of the squirrel cage, the bar resistance R_{bar} and the ring resistance component ΔR^*_{ring} are used to calculate the two dimensional rotor resistance R_{R}. The latter results from the transformation of the resistance of a ring section ΔR_{ring} into a series equivalent resistance. The rotor resistance is thus given by:

$$R_{\text{R}} = R_{\text{bar}} + 2\Delta R^*_{\text{ring}}. \tag{1}$$

This procedure is also used for the calculation of the rotor resistance in the other machine models. A more detailed derivation of the rotor resistance for 2D modeling is given in [8].

2.1.2. Leakage Inductances

The magnetic coupling of a winding with itself is described by the leakage inductance L_σ. This coupling manifests itself in a reduction of the main flux. The essential parts of the leakage flux are the harmonic, the slot and end-face leakage flux, and the leakage flux in the tooth tip space. These effects are considered in an analytically calculated leakage flux factor $\lambda_{\text{S/R}}$ when calculating the stator and rotor inductances.

2.1.3. Main Inductance and Saturation

By calculating the main inductance via the flux linkage, it is possible to consider the material saturation of the stator and rotor laminations of the IM. With a magnetic ECD of the machine geometry, the magnetic flux densities B and magnetic voltages V in the teeth (T) and yokes (Y) of the stator (S) and rotor (R) and in the air gap (δ) are calculated. The saturated main inductance L_{h} is obtained by dividing the operating point specific main flux linkage Ψ_{h} by the magnetizing current I_0:

$$L_{\text{h}} = \frac{\Psi_{\text{h}}}{I_0}, \tag{2}$$

$$\text{with} \quad \Theta_0 = V_\delta + V_{\text{S,Y}} + V_{\text{S,T}} + V_{\text{R,Y}} + V_{\text{R,T}} \quad \text{and} \quad I_0 = \frac{\pi p}{6\sqrt{2}\chi_{\text{tot}} w_{\text{S}}} \Theta_0.$$

The factor χ_{tot} describes the total winding factor including the distribution and pitch factors, w_{S} the number of windings per phase, and p the number of pole pairs.

2.1.4. Iron Losses

The iron losses for all models in this paper are calculated using the IEM-Formula, which will be presented later in the paper. For the calculation of the iron losses, the magnetic flux density in the stator and rotor lamination is of interest. The calculation of tooth and yoke flux densities in the FWM is done by the magnetic ECD. Since a local resolution of the iron losses in the stator and rotor laminations is not possible in the FWM, the flux density in tooth and yoke is assumed to be constant and its local mean value is determined accordingly. In this model, only the fundamental component of the magnetic flux density can be calculated, which results in the neglect of the iron losses due to higher flux density harmonics.

2.2. Harmonic Wave Model

The operating behavior, in particular the electromagnetic torque, the electromagnetic forces, and the ohmic and iron losses of an IM is significantly influenced by the orders and amplitudes of the harmonics that occur. These harmonics cannot be modeled in the FWM. Therefore, the HWM considering the multiple armature reaction provides a possibility for the analytical description of these effects [4–6]. Assuming a sinusoidal stator current, the discrete distribution of the stator windings, neglecting the slot openings, results in non-sinusoidal stator Magnetomotive Force (MMF) in the form of a staircase function. The currents of different order and amplitude induced in the rotor by this stator field, in turn, each generate staircase-shaped rotor MMF, which in turn induces new voltages in the stator (*primary armature reaction*). The harmonic model according to [4–6] additionally considers the secondary, tertiary, and quaternary armature reaction [9]. In addition to the multiple armature reaction, further effects causing harmonics, such as stator and rotor slotting or rotor eccentricity, can be considered. Due to the assumption of an infinitely high iron permeability ($\mu_{\mathrm{Fe}} \rightarrow \infty$) saturation harmonics cannot be represented in this HWM. Further important physical quantities and effects affecting the value range or the level of detail of the machine modeled with the HWM are briefly explained.

2.2.1. Current Displacement

Due to the consideration of higher-order harmonics and the resulting high frequencies, current displacement must also be considered. While this effect is negligible in the stator with round wire windings due to the relatively high penetration depths at the frequencies considered [10], the influence in the rotor must be considered by a frequency-dependent reduction factor, since the bars represent solid conductors in the iron packet [11].

2.2.2. Slot Openings in the Stator and Rotor

The slot openings in the stator and rotor can be considered in the form of a permeance function Λ for a better description of the field characteristics [11,12]. Among other things, this increases the accuracy of the calculation of the iron losses and the distribution of forces in the machine. The permeance function is described as the product of the permeance functions of stator and rotor, which can be represented as infinite complex Fourier series due to the periodicity [4,11,12].

Accounting for these effects in combination with multiple armature reaction leads to a good approximation of the field behavior, which is illustrated in Figure 3 for the local variation of the air gap flux density B_{δ} compared to a linear T-FEM simulation.

2.2.3. Iron Losses

For the calculation of the iron losses in the HWM, which includes losses due to higher harmonics, the time-dependent magnetic flux density in the stator and rotor lamination is of interest. The calculation of tooth and yoke flux densities in the HWM is done by an integral approach. Since a local resolution of the iron losses in the stator and rotor laminations is also not possible in the HWM, the flux density in tooth and yoke is assumed to be constant, as in the magnetic ECD used in the FWM, and its local mean value is determined accordingly. For this purpose, it is assumed for the determination of the tooth flux density B_{T}, that the entire magnetic flux in the area of a slot pitch τ_{N} flows through the corresponding tooth.

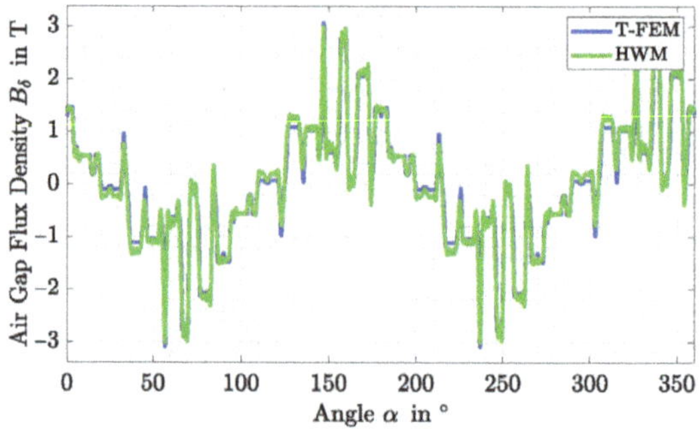

Figure 3. Simulated air gap flux density of an IM with linear stator and rotor magnetic material properties using the T-FEM and HWM.

By integrating the air gap flux density B_δ over a stator or rotor slot pitch, the magnetic flux Φ_T in the respective tooth can be determined. Since no location dependence in the tooth is considered, the area integral for calculating the tooth flux from the tooth flux density can be replaced by a multiplication with the toot width b_T and active axial length l_Fe so that the tooth flux density $\Phi_\mathrm{T}(t)$ follows from

$$\tau_\mathrm{N} = \frac{2\pi}{N}, \quad (3)$$

$$\Phi_\mathrm{T}(t) = l_\mathrm{Fe} \int_{\tau_\mathrm{N}} B_\delta(\Theta, t) r \,\mathrm{d}\Theta, \\ = l_\mathrm{Fe} b_\mathrm{T} B_\mathrm{T}(t), \quad (4)$$

with N being the number of slots and r being the middle radius of the air gap.

The assumption underlying the calculation of the yoke flux densities B_Y describes that the magnetic flux is equally distributed over both paths in the yoke via a pole pitch, which is why half the air gap flux is present in each case. By converting the area integral for the determination of the yoke flux from the yoke flux density into a multiplication with the yoke height h_Y and l_Fe the yoke flux density results in

$$\tau_\mathrm{p} = \frac{\pi}{p} \quad (5)$$

$$\Phi_\mathrm{Y}(t) = \frac{l_\mathrm{Fe}}{2} l_\mathrm{Fe} \int_{\tau_\mathrm{p}} B_\delta(\Theta, t) r \,\mathrm{d}\Theta \\ = l_\mathrm{Fe} h_\mathrm{Y} B_\mathrm{Y}(t). \quad (6)$$

2.3. *Extended Harmonic Wave Model*

The neglect of the iron saturation in the HWM according to [4–6] represents an essential limitation in the operating point calculation of the IM. Therefore, an E-HWM is introduced in this paper, which provides an approach to account for the iron saturation.

2.3.1. Approach

If the influence of the iron saturation on the harmonics is to be considered, the flattening of the hysteresis curve $B(H)$ in the nonlinear region and the resulting flattening of the air gap flux density must be modeled, which is shown for the idealized case of a sinusoidal air gap flux density in the middle plot of Figure 4. This can be realized mathematically by a circumferential location Θ dependent description of an effective air gap. Here, as a consequence of the main field saturation, the air gap is increased on average by a saturation factor $k_\mathrm{h} \geq 1$. In the region of large iron saturation, i.e., at the maximum of the air gap flux density B_δ, the air gap is increased by another saturation factor k_h1 and decreased at zero crossings. Thus, the time- and location-dependent air gap conductance function shown in the left plot of Figure 4 can be defined. This moves synchronously with the fundamental wave field and therefore results in

$$\lambda(\Theta,t) = \frac{1}{k_{\mathrm{h}}} - \frac{1}{k_{\mathrm{h1}}}\cos\left(\frac{2\pi\Theta}{\tau_{\mathrm{p}}} - 2\omega t\right), \tag{7}$$

where the factor of two in the cosine argument is a consequence of the simultaneous iron saturation by the north and south poles of the air gap field. The flattened airgap flux density $B_{\delta,\mathrm{sat}}$ follows from multiplying the original airgap flux density by the airgap conductance function given by

$$B_{\delta,\mathrm{sat}}(\Theta,t) = B_{\delta}(\Theta,t)\cdot\lambda(\Theta,t). \tag{8}$$

In the right plot of the Figure 4 exemplary flux density curves simulated by a TH-FEM, a HWM, and an E-HWM are shown. Here the effect of the flattening of the flux density by the application of the air gap conductance function can be seen.

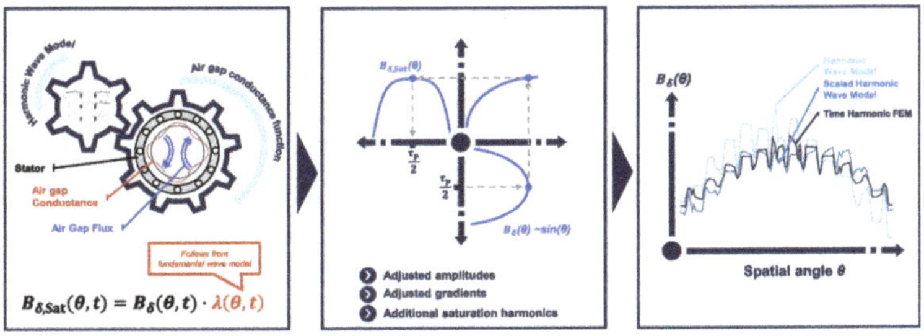

Figure 4. Approach of the consideration of saturation in the E-HWM.

2.3.2. Calculation of the Air Gap Conductance Function

The derivation of the saturation factors k_{h} and k_{h1} is done using the FWM. By dividing the amplitude of the air gap flux density $\widehat{B}_{\delta,\mathrm{FWM}}$ of the FWM by the fundamental wave $\widehat{B}_{\delta,1,\mathrm{HWM}}$ of the HWM the scaling factor

$$s_{\max} = \frac{\widehat{B}_{\delta,\mathrm{FWM}}}{\widehat{B}_{\delta,1,\mathrm{HWM}}} \tag{9}$$

at maximum saturation can be calculated. This factor thus represents the minimum of the air gap conductance function. Assuming a cosine fundamental wave, the amplitude $\frac{1}{k_{\mathrm{h1}}}$ can be expressed as

$$\frac{1}{k_{\mathrm{h1}}} = \arccos(s_{\max})\frac{4}{2\pi}s_{\max} \tag{10}$$

and thus describes the part of a period that the magnitude of the fundamental wave of the HWM is above the amplitude of the air gap flux density of the FWM, related to the scaling factor at maximum saturation. The larger this part, the greater the difference between the saturated and unsaturated regions, and thus the amplitude of the air gap conductance function. The mean value $\frac{1}{k_{\mathrm{h}}}$ follows from the addition of the minimum and amplitude. Therefore k_{h} can be expressed as

$$k_{\mathrm{h}} = \frac{1}{1+\arccos(s_{\max})\frac{4}{2\pi}}\cdot\frac{1}{s_{\max}}. \tag{11}$$

To consider the saturation effects on the curves of the flux densities in the HWM, the calculated time and local curves of the air gap flux density, as well as the time curves of the tooth and yoke flux densities, can be multiplied by the function resulting from (7). For the latter progressions, the scaling factors at maximum saturation must thereby be calculated with the amplitudes of the tooth and yoke flux densities of the FWM and HWM, respectively, resulting in different scaling functions than in the air gap.

The introduction of saturation phenomena in the iron loss calculation in the E-HWM is based on the scaling of existing harmonics as well as the consideration of additional saturation harmonics [7]. From (8), with the help of trigonometric relations, the scaling of existing harmonics can be converted to

$$\widehat{B}_{1,\mathrm{sat}} = \left(\frac{1}{k_\mathrm{h}} - \frac{1}{2k_\mathrm{h1}}\right) \cdot \widehat{B}_1 \tag{12}$$

$$\widehat{B}_{\mathrm{n,sat}} = \frac{1}{k_\mathrm{h}} \cdot \widehat{B}_\mathrm{n} \tag{13}$$

where the fundamental is scaled differently from the other harmonics. The derivation of the additional saturation harmonics is done analogously. The flattening of the air gap flux density results in particular in a dominant third harmonic, which is calculated differently from the other saturation harmonics. In general, the n-th harmonic results in two additional saturation harmonics with the order $n \pm 2$. Their amplitudes result in

$$\widehat{B}_3 = \frac{1}{1 - 2\frac{k_\mathrm{h1}}{k_\mathrm{h}}} \cdot \widehat{B}_{1,\mathrm{sat}} \tag{14}$$

$$\widehat{B}_{\mathrm{n}\pm2} = -\frac{1}{2k_\mathrm{h1}} \cdot \widehat{B}_{\mathrm{n,sat}}. \tag{15}$$

Since a change in the flux densities also results in new induced currents in the rotor, an iterative adaptation in the E-HWM is required. For this purpose, starting from the scaled flux densities, the rotor current is updated, which in turn changes the flux densities. The updated flux densities are scaled again, and so on. To adjust the induced rotor current, a constant scaling factor is applied according to

$$s_\mathrm{ind} = \overline{\lambda} = \frac{1}{k_\mathrm{h}} \tag{16}$$

by which the inductances are multiplied. Based on the scaled inductances, the rotor current can be updated. By integrating a relaxation factor, the convergence behavior of the successive substitution can also be improved. In Figure 5 the same operating point of an exemplary IM as in Figure 3 is simulated using non linear magnetic material properties. The air gap flux density compared to the linear material in Figure 3 differs strongly. Nevertheless, the use of the E-HWM makes it possible to simulate the saturation-dependent flux density approximately well.

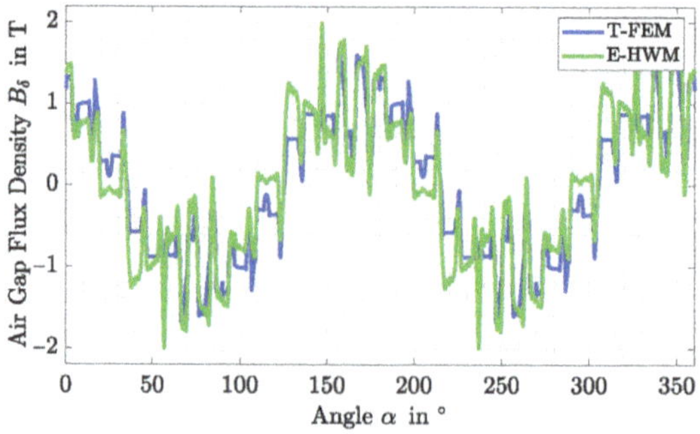

Figure 5. Simulated air gap flux density of an IM with non-linear stator and rotor magnetic material properties using the T-FEM and E-HWM.

2.4. Numerical Models Based on the Finite Element Method

The analytical models of the FWM, HWM, and E-HWM of the IM are limited in terms of the value range and level of detail of the effects that can be modeled. Effects such as ferromagnetic saturation, induced eddy currents, and a location-dependent air gap reluctance can only be described in a limited and simplified way, as can be seen in the model descriptions. Others, such as cut edge effects [13] or geometry- or motion-dependent phenomena, cannot be modeled at all [14]. An extension of the range of values and level of detail of machine modeling is offered by numerical models. However, the higher level of detail is accompanied by an increase in computational effort. In this paper, the TH-FEM and T-FEM are used and briefly explained in the following. In both cases, a 2D model is used with a current excitation in the form of a specified current density in the stator slots.

2.4.1. Mathematical Description of the FEM

In the FE models of this paper, a magnetoquasistatic formulation over the magnetic vector potential $\vec{A}$ is considered. From the Maxwell equations the parabolic partial differential equation

$$\nabla \times (\nu \nabla \times \vec{A}) + \sigma \frac{\partial \vec{A}}{\partial t} = \vec{J}_{\mathrm{s}} \tag{17}$$

the reluctance ν, the conductivity σ, and the injected stator current density $\vec{J}_{\mathrm{s}}$ can be derived [14]. A discretization of the magnetoquasistatic formulation in (17) applying the weighted-residual method, which is necessary for the FEM, leads to the weak vector potential formulation

$$\int_V (\nabla \times (\nu \nabla \times \vec{A}) + \sigma \frac{\partial \vec{A}}{\partial t}) \cdot \vec{w}_{\mathrm{i}} \, \mathrm{d}V = \int_V \vec{J}_{\mathrm{s}} \cdot \vec{w}_{\mathrm{i}} \, \mathrm{d}V \tag{18a}$$

$$\vec{A} = \sum_j a_{\mathrm{j}} \vec{v}_{\mathrm{j}}, \tag{18b}$$

with $\vec{w}_{\mathrm{i}}$ being weighting functions. The magnetic vector potential is thereby divided into a finite sum of shape functions $\vec{v}_{\mathrm{j}}$, and thus is described by its degrees of freedom a_{j} [14]. Using Galerkin's method [14], which prescribes $\vec{v}_{\mathrm{j}} = \vec{w}_{\mathrm{j}}$, by transformations based on integral theorems and boundary conditions of the solution space, the system of equations and the resulting matrix notation can be expressed as [15]

$$\sum_j (a_{\mathrm{j}} \underbrace{\int_V \nu \nabla \times \vec{w}_{\mathrm{i}} \cdot \nabla \times \vec{w}_{\mathrm{j}} \, \mathrm{d}V}_{K_{\nu,ij}} + \frac{\partial a_{\mathrm{j}}}{\partial t} \underbrace{\int_V \sigma \vec{w}_{\mathrm{i}} \cdot \vec{w}_{\mathrm{j}} \, \mathrm{d}V}_{M_{\sigma,ij}}) = \underbrace{\int_V \vec{J}_{\mathrm{s}} \cdot \vec{w}_{\mathrm{i}} \, \mathrm{d}V}_{f_{\mathrm{i}}} \tag{19a}$$

$$\mathbf{K}_{\nu} \vec{a} + \mathbf{M}_{\sigma} \frac{\partial \vec{a}}{\partial t} = \vec{f}. \tag{19b}$$

Here $\mathbf{K}_{\nu}$ is called stiffness matrix, $\mathbf{M}_{\sigma}$ is called mass matrix, and $\vec{f}$ is called load vector. From this system, the degrees of freedom of the magnetic vector potential and from it the flux density distribution in the machine can be determined.

2.4.2. Time Harmonic Finite Element Model

In the TH-FEM, the time courses of physical quantities are simplified as sinusoidal and can therefore be described by complex phasors [14]

$$y(t) = \Re(\underline{\hat{y}} e^{\mathrm{j}\omega t}). \tag{20}$$

This has the advantage that the time derivative $\frac{\partial}{\partial t}$ passes into a multiplication by $\mathrm{j}\omega$. The differential equation from (19b) thus transforms into the linear system of equations

$$(\mathbf{K}_{\nu} + \mathrm{j}\omega\mathbf{M}_{\sigma})\vec{a} = \vec{f} \tag{21}$$

which can be solved with low computational effort. The disadvantage of the time-harmonic simulation is the underlying assumption that no time harmonics with an order greater than one exist, which is why in particular the accuracy of the calculated iron losses decreases compared with the T-FEM [16].

Slip Transformation

Since the linear equation system (21) simulates the stator frequency ω_{S} in the entire solution domain, the lower frequency ω_{R} in the rotor is not considered. This can be accounted for by a slip transformation [17,18]. For this, the conductivity of the cage $\sigma_{\mathrm{R,comp}}$ is scaled with the slip s, which changes the mass matrix $\mathbf{M}_{\sigma}$ for rotor nodes such that the lower rotor frequency is considered in the induction effect.

Non Linear Material Properties

To consider the non-linear material behavior of the stator and rotor laminations, an iterative procedure is used in the field solution. For this purpose, the successive substitution approach or the Newton method can be used.

2.5. Transient Finite Element Model

The transient simulation solves the differential Equation (19b) for discrete time steps t_{k} within a time interval ΔT considering the field solution of the previous time step and the motion of the rotor. Due to the discretization, the time derivative can be described as the difference quotient $\frac{\vec{a}_{\mathrm{k}+1}-\vec{a}_{\mathrm{k}}}{\Delta t}$ with the time difference Δt between two time steps. From this the linear equation system results in

$$(\mathbf{K}_{\nu}\Delta t + \mathbf{M}_{\sigma})\vec{a}_{\mathrm{k}+1} = \vec{f}(t_{\mathrm{k}+1})\Delta t + \mathbf{M}_{\sigma}\vec{a}_{\mathrm{k}}, \tag{22}$$

which must be solved for each time step [14]. The corresponding rotor position Θ follows from the Newton equation of motion

$$J\frac{\partial^2\Theta}{\partial t^2} = T(t) \tag{23}$$

with the moment of inertia J of the rotating body. This equation is also discretized in time, with the torque T updated for each time step based on the calculated field solution. By solving the differential equation in the time domain, temporal harmonics in particular are modeled, which allows a more detailed view of the machine behavior. Since the Shannon sampling theorem prescribes a sampling frequency f_{sample} greater than or equal to $2f$ for physical consideration of a given frequency f [19], the time difference between each time step must be small enough to consider the higher order harmonics, but the total time interval large enough for consideration of the low frequencies in the rotor. This, in combination with the required transient time of the vector potential solution, leads to a high calculation effort of the transient Finite Element (FE) simulation. Faster transient is made possible by a hybrid approach via estimation of a starting solution from an open-circuit simulation [20].

In Figure 6 the simulated magnetic flux density of an exemplary IM at an saturated operating point is shown. In Figure 6a the simulation was conducted using the TH-FEM and in Figure 6b using the T-FEM. The simulation results show quite similar field characteristics.

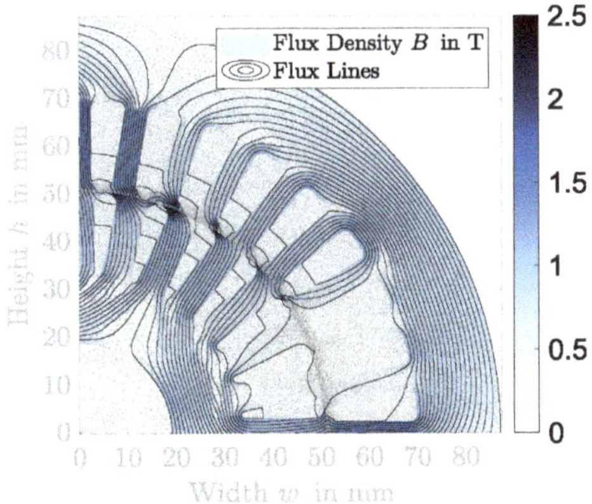

(a) TH-FEM

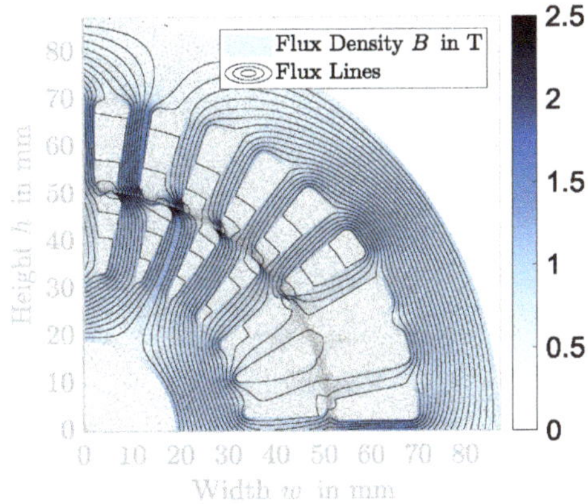

(b) T-FEM

Figure 6. Simulated magnetic flux density and isolines for an exemplary IM.

2.6. Used Iron Loss Model

The IEM-5 parameter formula is used to calculate the iron losses in each model. It is based on the Bertotti model and adds an additional term considering the non-linear material behavior [21]. The IEM-formula to calculate the iron loss density p_{Fe} is given by:

$$p_{\mathrm{Fe}} = a_1 \hat{B}^{\alpha} f + a_2 \hat{B}^2 f^2 (1 + a_3 \hat{B}^{a_4}) + a_5 \hat{B}^{1,5} f^{1,5}, \tag{24}$$

where a_1, a_2, a_5 are the hysteresis, eddy current, and excess loss factors, a_3 and a_5 are loss parameters describing the non-linear saturation losses, f the frequency, and $\hat{B}$ the amplitude of the magnetic flux density for the given frequency. In the numerical models, the iron loss densities are calculated element by element and weighted by the element area. The summation of the iron loss densities over all elements and multiplication by the iron length then results in the iron losses. In the analytical models, the iron loss densities are calculated for the stator and rotor yokes and teeth respectively with the calculated mean tooth and yoke flux densities. Multiplication by the yoke and teeth mass then also results in the value of the iron losses.

3. Model Selection Approach

The four-step generic procedure of the approach for model selection is shown in Figure 7. The problem definition is based on four input variables. These describe the searched output quantities, the effects to be investigated are whose influence on the output quantities is to be modeled, and the respective required precision, i.e., the problem-specific level of detail. Since the precision of the models depends on the operating point, a selection of the operating points to be considered is also necessary. The problem-specific output quantities to be considered define the electromagnetic coupling quantities for a coupled model. Analogously, the effects to be studied describe the external coupling variables. By means of this problem definition, suitable models can be derived on the basis of the value ranges and levels of detail of the available models and simulation methods, respectively. From the consideration of the value range the possible models follow, which represent all output quantities and effects and can model an influence of the output quantities by the effects. For the possible models, the required level of detail is examined in the following step and the suitable models that meet all requirements are derived from this. From these, the one with the lowest computational effort is selected.

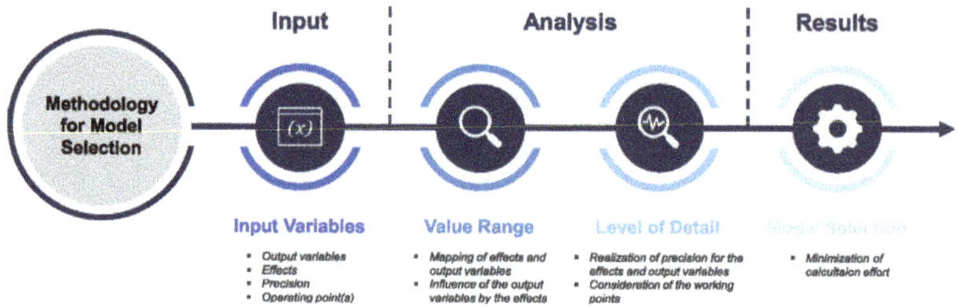

Figure 7. Process of the approach for the model selection.

One possibility to characterize the range of values and the level of detail of the models are tables, which contain entries for possible output quantities and effects, separated by the available models. The value range table contains a reference to the general ability of the models to describe the output quantities and effects, whereas the level of detail table contains the respective precision. The precision can be given in relation to a reference model, if quantifiable. Alternatively, qualitative scales can be used to describe precision. It is important to have a stringent procedure for the definition of these precisions in order to allow a future consideration of further effects and to make them comparable. Another possibility to describe the range of values and the level of detail, e.g. to consider geometry and material effects, is given by sensitivity analyses, where the parameters to be investigated are slightly varied and it is checked whether the models represent an influence of the parameters on the searched output quantities. Precision is again defined by a reference model. The characterization of the solution effort of the models is done via the number of the respective degrees of freedom since the computation time depends on computer-specific parameters such as the clock speed, the number of cores and the main memory and thus cannot be generalized. In this paper, the described approach for the model selection is presented using the example of an IM. An automated implementation allows finding the most suitable problem-specific model based on the four problem-defining input variables. In the following, the sensitivity analysis used in the selection process and the four steps of the model selection process are presented.

3.1. Sensitivity Analysis

The influence of changes in geometry and material parameters on physical quantities as well as temporal and spatial characteristics is investigated and quantified by means of sensitivity analyses. In the case of coupling one of the presented electromagnetic machine models with another, for example structural dynamic or thermal model, they also provide information about the required accuracy of the coupled model, since the latter does not have to have an influence on machine parameters to which the electromagnetic model is not sensitive.

The starting point of the sensitivity analysis for the IM is a reference geometry. Based on this, a second, slightly modified machine geometry is generated, depending on the geometry or material parameter to be investigated. For this, the parameter is changed by a user defined factor r_{sens}. In this case, too small a change leads to numerical instabilities, and too large changes possibly lead to inconsistent machine geometries. If the sensitivity to a geometry parameter is to be investigated, it is also necessary to consider the influence of this parameter on other geometry variables. For example, if the air gap width is changed, the stator or rotor diameter must also be adjusted. These influences are described via a correlation matrix, which describes for each possible variable geometry parameter which other geometry parameters must also be changed. Subsequently, the machine models are simulated for both geometries and the output quantities are compared. In this work, a T-FEM simulation is not performed in the context of the sensitivity analysis due to the high computational effort associated with it. Instead, the same results are assigned to the T-FEM as to the TH-FEM, since the value ranges and levels of detail are very similar due to analogous procedures in meshing. The following findings can be derived from the comparison of the output quantities of both geometries:

1. Sensitivity: For each of the models considered in the sensitivity analysis, it can be stated separately for all physical quantities whether they are sensitive to the geometry or material parameter under investigation.
2. Influence on Harmonics: The influence of the parameter to be examined on the temporal and spatial courses can be described on the basis of the change of the respective harmonic orders of the FFT spectrum. This influence can be recorded qualitatively, but can also be normalized in relation to the largest change of the occurring orders and thus quantified. In this way, the harmonics whose relative change is the greatest can be recorded in particular.
3. Elasticity: The elasticity ϵ describes the ratio of the change of an output quantity, related to the change of the input quantity and therefore results in the case of the sensitivity analysis for a physical quantity to

$$\epsilon_{\text{phys}} = \frac{r_{\text{phys}}}{r_{\text{sens}}}, \quad (25)$$

where r_{phys} describes the relative change of the physical quantity between the two simulated machine models. This allows us to capture, for each model, how much a certain quantity changes when the parameter under investigation is modified.
4. Level of Detail: The description of the level of detail of the detection of a geometry or material change, related to a physical quantity can be done by means of a reference model. For this purpose, the TH-FEM is selected in this work, since it has the highest average precision after the T-FEM. The degree of detail of a physical quantity, therefore, follows from the relative deviation of the elasticity of the model under consideration from the elasticity of the TH-FEM.

3.2. Input Variables

Possible output quantities of the IM to be investigated are physical quantities such as torque and losses, but also temporal and spatial flux density and torque characteristics. The effects considered in this work include saturation phenomena, harmonics, and other physical effects, but also geometry and material effects as well as dynamic processes. A geometric effect is the change in a geometric quantity, such as the air gap width or tooth width. A material effect is the change of a material property due to external influences such as temperature.

3.3. Analysis of the Value Range and Degrees of Freedom

In the first step, the range of values of the described models is considered. Here, it is investigated for all output quantities and effects whether these can be represented by the value range of the individual models. When considering geometry or material effects, each of the selected output quantities must be influenced by at least one of these effects. The characterization of the value range of the IM is done by means of a table for the output quantities and physical effects and by means of the sensitivity analysis for geometry or material effects. The value range is analyzed in terms of spatial and temporal resolution, the number of degrees of freedom, physical effects, dynamic processes, and physical quantities for the five models presented. Here, for the FE simulations, the degrees of freedom are composed of the number of operating points, the nodes in the mesh, the iterations to account for saturation, and the transient steps. For the HWM, the degrees of freedom are also described by the number of operating points and, in addition, by the stator and rotor orders considered and the Fourier coefficients of the stator and rotor permeance functions. In the E-HWM, the iterations for updating the rotor current are added. The number of degrees of freedom in the FWM is composed only of the simulated stator flux linkages and the operating points.

3.4. Analysis of the Level of Detail

The procedure for characterizing the level of detail is analogous to that for characterizing the range of values via tables and sensitivity analysis. The tables contain precision values of the different physical output parameters and physical effects for linear and non-linear operating points of an exemplary machine in the given power range. The quantification of the precision values of the physical output parameters and effects is performed on the basis of individual categories. These are subdivided as follows:

- Physical Values: For the physical output quantities, such as ohmic losses, and iron losses, the simulation results of the T-FEM is used as a reference. The relative deviations from the results of the T-FEM are determined for exemplary operating points in the linear and non-linear range.
- Temporal and Spatial Characteristic: The precision of temporal and spatial characteristic, such as the magnetic flux density in the stator tooth over time or the magnetic air gap flux over time or position, is determined with reference to the simulation results of the T-FEM. For each characteristic, an exemplary operating point in the linear range and one in saturation are analyzed. The analysis is performed on the basis of the frequency spectrum of the considered quantity. Spatial characteristics, such as the spatial characteristic of the air gap flux density, are decomposed into their spectral components using the Fast Fourier Transformation (FFT) with a rectangular window. The level of detail is then separated into three subcategories. The precision of the fundamental wave, the precision of the harmonic amplitudes, and the precision of the occurring frequency orders. The slip dependence of an operating point during the operation of the IM leads to the fact that frequency components occur in the time variables which do not correspond to an integer multiple of the fundamental frequency. The use of a rectangular window in the FFT therefore leads to large spectral leakage. Therefore, a window function is used for the temporal quantities, which reduces the spectral leakage. A suitable window function is the Hann window. Figure 8a shows an example of the temporal course of the air gap flux density simulated by means of two different models at a particular position in the air gap. In Figure 8b the frequency spectra are plotted by means of a FFT with a rectangular window and by means of a FFT with a Hann window. To determine the level of detail, the output signal of the FFT with the Hann window of the reference signal as well as that of the comparison signal are analyzed for their local maxima. From the local maxima which are larger than a certain threshold value, the level of detail of the fundamental, the amplitudes of the harmonics and the occurring frequencies are then determined, as in the case of the location-dependent characteristics. In Figure 8b the maxima used to analyze the level of detail are marked with numbers.
- Physical Effects and Dynamic Processes: Since the level of detail of physical effects and dynamic processes, such as saturation, skin effect, or cutting edge effects, cannot be considered separately from other quantities, these precision are defined using a subjective point scale from one to ten, where one describes a very high precision and ten a very low precision.

The level of detail is analyzed for all output quantities and effects and matched with the required level of detail. The affiliation of an operating point to the linear and non-linear operating range of the machine is thereby determined on the basis of the FWM. The influence of geometry and material effects to be considered must achieve the required precision for each individual influenced output variable.

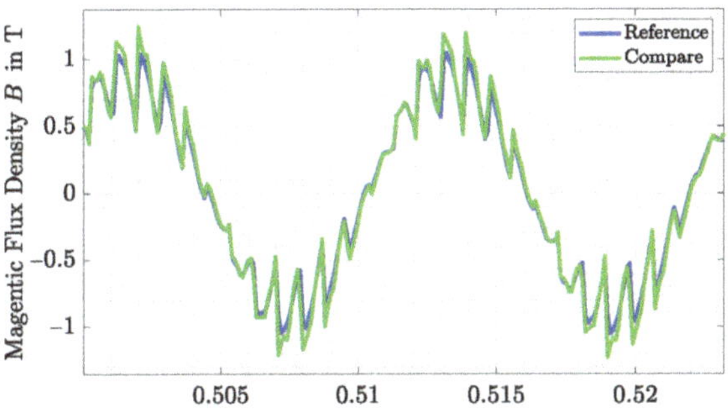

(**a**) Flux density over time.

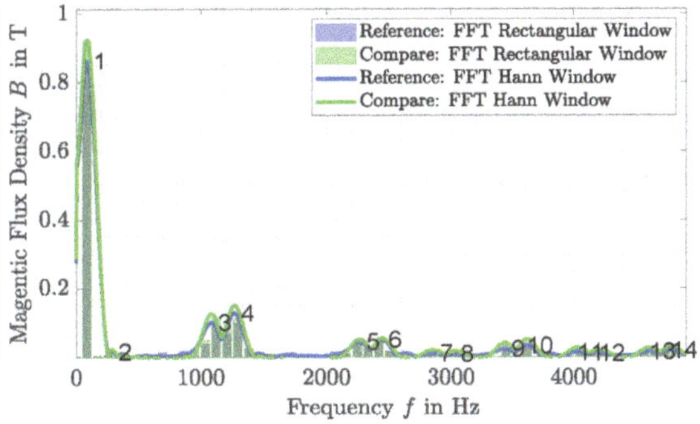

(**b**) FFT using rectangular and Hann window.

Figure 8. Magnetic air gap flux density simulated by two different models.

3.5. Model Selection

Based on the results of the value range analysis and the level of detail analysis, the model with the lowest degree of freedom is then selected from all suitable models. This promises the lowest computational effort and thus the most efficient calculation for the precision requirements. If a working point matrix is considered in the model selection process, the selection of possible models is done analogously based on the range of values. For the resulting models, the precision of the output quantities and effects to be examined are considered in each working point. The selection of the most suitable machine model can then be made using two procedures. One option is to consider those models that have the required level of detail at all operating points. From these, the one with the lowest computational effort is then derived. An alternative is a operating point specific model selection by means of a Branch and Bound optimization, on the basis of which different models can be assigned to different operating points, so that the required level of detail is achieved at each point, but the overall solution effort is minimized.

4. Approach for Parameter Selection

Mathematical optimization algorithms are gaining importance for the design, revision, and optimization of electrical machines. The optimization parameters in the context of such machine optimizations represent those geometry and material parameters that are varied during the optimization procedure in order to find a better solution. These parameters should therefore cover a high degree of freedom of the geometry. However, as the number of optimization parameters increases, so does the search space and the associated solution effort. For this reason, it is advisable to select those geometry parameters that have the greatest influence on the searched output quantities but have the lowest degree of redundancy among themselves. The parameter selection approach presented in this paper describes the selection of such parameters as a methodical procedure. The generic three-step procedure of parameter selection is shown in Figure 9. The process of the parameter selection is problem-specific and is also influenced by the selected system model, which follows from the approach for the model selection. In the context of the optimization environment, the output quantities and effects to be considered in the model selection describe those variables that influence the decision parameters of the optimization problem. The input variables for the problem definition of the parameter selection describe on the one hand the resulting model and the output quantities already described for the methodology for the model selection, but also those problem variables which come into question as optimization parameters. In addition, a selection of the number of parameters to be determined is required, which defines the degrees of freedom and thus the accuracy of the optimization environment, but also describes the required solution effort.

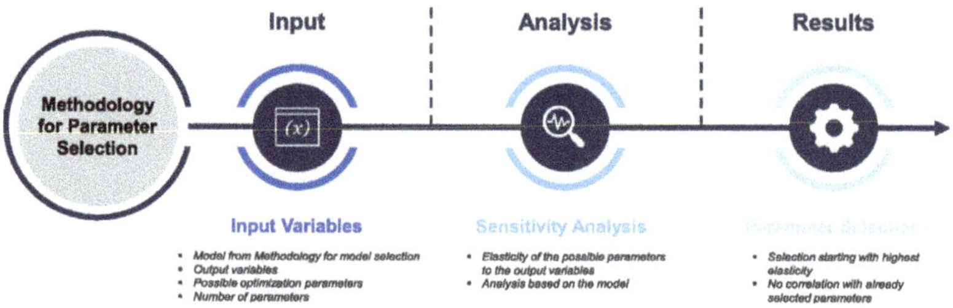

Figure 9. Process of the approach for the parameter selection.

Based on these input variables, a sensitivity analysis is performed for each possible optimization parameter using the model resulting from the model selection approach. Here, the elasticities of the output quantities relevant to the optimization are examined so that the possible optimization parameters can be sorted based on their elasticity. This identifies those parameters that have the greatest possible influence on the optimization problem, minimizing the global optimum in particular, since this correlates negatively with the elasticity of the optimization parameters. From the sorting of the optimization parameters, starting with the highest elasticity, the given number of parameters can be selected. For each new optimization parameter to be added, the correlation with already selected parameters must be checked, since the individual optimization variables must be independent of each other and must not influence each other. Otherwise, contradictory solutions may result. The optimization parameters resulting from the parameter selection procedure serve as input variables of an optimization environment. Here, a problem-specific, experience-based selection of upper and lower bounds of the parameters is important to reduce the size of the solution space and thus the computational effort. A multi-step optimization environment for the design of an IM using the model and parameter selection procedure presented here is given in [3].

5. Application Example

The model selection procedure is exemplary applied to a multiphysics weakly coupled simulation in this work. The weakly coupled simulation consists of an electromagnetic and a thermal model of an exemplary IM. The electromagnetic model determines the actual temperature-dependent ohmic and iron losses of the IM at different stator winding and rotor cage temperatures. The calculated losses are the input parameters of the thermal model. The thermal model is then used to update the machine temperatures and return them to the electromagnetic model, resulting in a change of the material conductivities. The electromagnetic model to be used is selected by applying the presented approach of model selection. For the thermal model, an implementation in MATLAB® is used, which is based on the Lumped Parameter Thermal Network (LPTN) model of Motor-CAD®. The resulting simulated temperature curves are compared with measured results of the same machine at the same operating point. The use of the parameter selection in combination with the model selection is investigated in [3] using a multi-stage optimization of the design of an IM.

5.1. Exemplary Induction Machine

The exemplary IM is an aluminum die-cast squirrel cage IM. The IM is used as a traction drive in an electrical vehicle. The machine, the cross-sectional area of the stator and rotor and important geometrical and electrical parameters are given in Figure 10. The studied IM delivers a S2 power of 30 kW and is equipped with a forced air cooling in the housing. The temperature measurement of the IM was performed on a test bench. The test bench setup is described in [22]. Seven temperature sensors were used to measure the average temperature of the stator winding. These were placed at different positions within the winding.

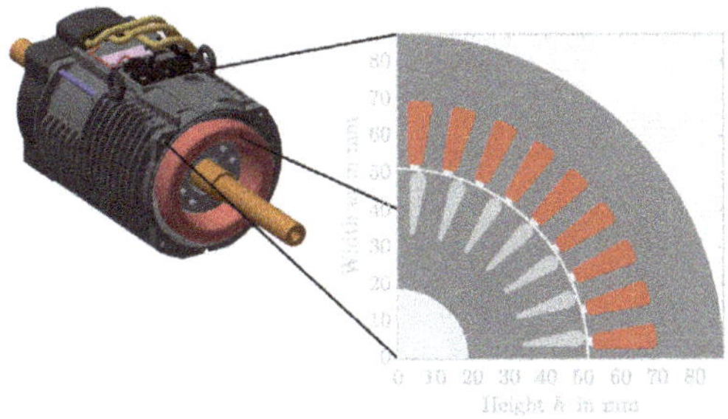

Parameter	Value	Unit
Number of pole pairs	2	-
Number of stator slots	36	-
Number of rotor slots	28	-
Outer stator diameter	170	mm
Outer rotor diameter	102	mm
Air gap width	0.5	mm
Active iron length	200	mm
DC Link Voltage	130	V
Maximum Phase Current	350	A

Figure 10. Cross section and parameter of the exemplary IM.

The rotor temperature measurement was performed using a non-contact infrared (IR) sensor. The temperature was measured at the short-circuit ring of the squirrel-cage. To achieve a high emissivity of the short-circuit ring, it was painted black [23]. The validation of the IR temperature measurement was performed according to the procedure described in [23].

5.2. Thermal Model

An LPTN model of the IM was used as the thermal model. This was built in Motor CAD using the geometry and winding data of the exemplary machine. The LPTN created by Motor-CAD® was extracted and imported into MATLAB®. The thermal parameters and the LPTN of the machine are explained in [24]. The LPTN is solved in Matlab using the modified nodal potential analysis and the Backward Euler method.

Two options can be used to calculate the electromagnetic losses used as input for the thermal simulation. Both options are equivalent in their results. First, in each step where the losses are updated, the electromagnetic model can be calculated with the actual temperatures of the stator winding and the rotor cage. This implies a high computational effort, especially for the numerical models, since the FE simulation must be performed at each step. On the other hand, for a given stator winding and rotor cage temperature, the machine can first be calculated with the selected model and the influences of the temperature changes on the losses can be determined using the scaling laws of IM described in [8,22,25,26]. This reduces the computational effort while maintaining the quality of the calculated losses, as shown in [8,22,25,26].

5.3. Model Selection Approach

The selection of the electromagnetic simulation model of the IM to be used for the described field of application of a coupled electromagnetic-thermal operating point simulation is performed automatically with the presented approach of the model selection. The output quantities and effects to be investigated to describe the problem and the required levels of detail are shown in Table 1. A required level of detail of, for example, 25% means a maximum permissible deviation of 25% from the T-FEM. The required level of detail of the ohmic and stator and rotor losses is obtained assuming a measurement deviation of 5% and a deviation of the electromagnetic reference model, the transient FEM, of 10%. The level of detail of the iron loss components was chosen to be 25% because they have lower losses compared to the ohmic losses at the considered operating point. The required level of detail of the saturation was chosen to be 5 on the subjective point scale, which corresponds to the middle value of the scale. A saturated point in the efficiency maximum of the machine is selected as the operating point to be investigated. With a speed of 3700 rpm and a torque of 60 Nm, this lies in the limit between the base speed and the field weakening range of the IM.

Table 1. Required level of detail of the output quantities and effects of the electromagnetic model.

Output Quantities and Effects		Level of Detail
Output Quantities	Ohmic Stator Losses	14.5%
	Ohmic Rotor Losses	14.5%
	Hysteresis Losses	25%
	Eddy Current Losses	25%
	Excess Losses	25%
	Saturation Losses	25%
Effects	Saturation	5
	Winding Temperature	14.5%
	Rotor Temperature	14.5%

The model resulting from the procedure, which can represent both the range of values and the required levels of detail, is the E-HWM. The TH-FEM and T-FEM also meet the desired levels of detail, but were not chosen due to the higher degrees of freedom and associated higher computational effort.

5.4. Simulation Results

Using the determined electromagnetic E-HWM, the thermal behavior of the machine is simulated by means of the coupled simulation at the given operating point of 3700 rpm and 60 Nm. The simulated and measured temperature of the stator winding and the rotor cage are shown in Figure 11. The stator winding temperature has a maximum deviation of 7.8 °C and 5.7%, respectively, with respect to the measured temperatures during the heat-up phase. The rotor cage temperature shows a deviation of 13.9 °C and 8.2% respectively. These values are therefore within the required level of detail of the losses. For comparison, the simulation is also performed with T-FEM, which has a higher level of detail than the E-HWM, and the FWM, which has a lower level of detail. The results are also plotted in Figure 11. The maximum deviation related to the measurements of the stator winding temperature for the simulation using the T-FEM is 3.7 °C and 3.4%, respectively, and that of the rotor cage temperature is 4.3 °C and 2.5%, respectively. The deviations of the simulations using the FWM are above 25% in both temperatures.

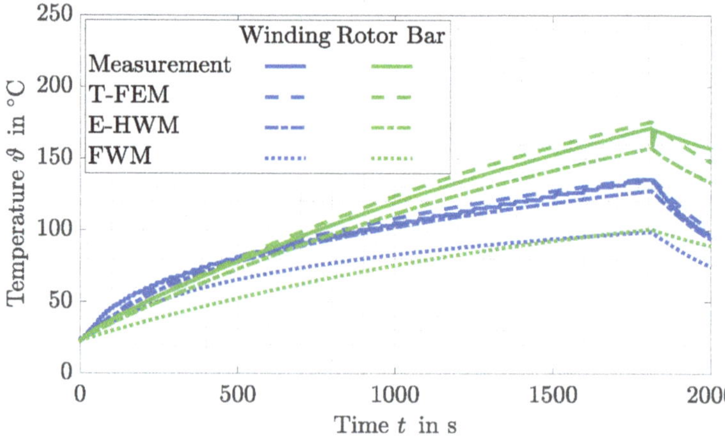

Figure 11. Simulated and measured stator winding and rotor bar temperature of the exemplary IM using the T-FEM, the E-HWM, and FWM.

6. Discussion

The model and parameter selection approaches presented in this paper are based on the analysis of different analytical and numerical models of an exemplary IM for a given power, torque, and speed range. The analysis is performed at certain operating points and the models are evaluated in terms of their value ranges and levels of detail. The model and parameter selection approaches can then be applied to problems requiring simulations of IM in a similar power, torque, and speed range and with similar geometric dimensions.

In an example simulation of the thermal behavior of an IM the model selection approach is used to determine the most appropriate electromagnetic machine model. The results of this application example show that the model selection approach is suitable for methodically determining the most suitable and computationally efficient model for given requirements in terms of value range and level of detail. Thus, it is an efficient tool for the characterization and methodological analysis of models, especially when selecting one of several models with many different output quantities, effects, and parameters.

The parameter selection procedure provides an approach to identify machine parameters that have a high impact on the defined target variable. This procedure lends itself to the selection of optimization parameters in machine design optimizations as presented in [3].

Author Contributions: Conceptualization, M.N. and A.K.; methodology, M.N. and A.K.; software, M.N. and A.K.; validation, M.N.; formal analysis, M.N.; investigation, M.N. and A.K.; writing—original draft preparation, M.N.; writing—review and editing, M.N., A.K. and K.H.; visualization, M.N. and A.K.; supervision, K.H. All authors have read and agreed to the published version of the manuscript.

Funding: This research received no external funding.

Institutional Review Board Statement: Not applicable.

Informed Consent Statement: Not applicable.

Conflicts of Interest: The authors declare no conflict of interest.

Abbreviations

The following abbreviations are used in this manuscript:

ECD	Equivalent Circuit Diagram
E-HWM	Extended Harmonic Wave Model
FE	Finite Element
FWM	Finite Element Model
FFT	Fast Fourier Transformation
FWM	Fundamental Wave Model
HWM	Harmonic Wave Model
IM	Induction Machine
IR	Infrared
LPTN	Lumped Parameter Thermal Network
MMF	Magnetomotive Force
NVH	Noise Vibration Harshness
T-FEM	Transient Finite Element Model
TH-FEM	Time Harmonic Finite Element Model

References

1. Zienkiewicz, O.C.; Chan, A.H.C. Coupled Problems and Their Numerical Solution. In *Advances in Computational Nonlinear Mechanics*; Springer: Vienna, Austria, 1989; Volume 300, pp. 139–176. [CrossRef]
2. Nell, M.; Leuning, N.; Mönninghoff, S.; Groschup, B.; Müller, F.; Karthaus, J.; Jaeger, M.; Schröder, M.; Hameyer, K. Complete and Accurate Modular Numerical Computation Scheme for Multi-Coupled Electric Drive Systems. *IET Sci. Meas. Technol.* **2020**, *14*, 259–271. [CrossRef]
3. Nell, M.; Kubin, A.; Hameyer, K. Multi-Stage Optimization of Induction Machines Using Methods for Model and Parameter Selection. *Energies* **2021**, *14*, 5537. [CrossRef]
4. Oberretl, K. Allgemeine Oberfeldtheorie für ein- und dreiphasige Asynchron- und Linearmotoren mit Käfig unter Berücksichtigung der mehrfachen Ankerrückwirkung und der Nutöffnungen. Teil I: Theorie und Berechnungsverfahren. *Arch. Elektrotechnik* **1993**, *76*, 111–120. [CrossRef]

5. Oberretl, K. Allgemeine Oberfeldtheorie für ein- und dreiphasige Asynchron- und Linearmotoren mit Käfig unter Berücksichtigung der mehrfachen Ankerrückwirkung und der Nutöffnungen. Teil II: Resultate, Vergleich mit Messungen. *Arch. Elektrotechnik* **1993**, *76*, 201–212. [CrossRef]
6. Oberretl, K. Losses, Torques and Magnetic Noise in Induction Motors with Static Converter Supply, Taking Multiple Armature Reaction and Slot Openings into Account. *IET Electr. Power Appl.* **2007**, *1*, 517–531. [CrossRef]
7. Binder, A. *Elektrische Maschinen und Antriebe: Grundlagen, Betriebsverhalten*; Springer: Berlin/Heidelberg, Germany, 2012. [CrossRef]
8. Nell, M.; Lenz, J.; Hameyer, K. Scaling Laws for the FE Solutions of Induction Machines. *Arch. Electr. Eng.* **2019**, *68*, 677–695. [CrossRef]
9. Oberretl, K. Die Oberfeldtheorie des Käfigmotors unter Berücksichtigung der durch die Ankerrückwirkung verursachten Statoroberströme und der parallelen Wicklungszweige. *Arch. Elektrotechnik* **1965**, *49*, 343–364. [CrossRef]
10. Stoyanov, L.; Lazarov, V.; Zarkov, Z.; Popov, E. Influence of Skin Effect on Stator Windings Resistance of AC Machines for Electric Drives. In Proceedings of the 16th Conference on Electrical Machines, Drives and Power Systems (ELMA), Varna, Bulgaria, 6–8 June 2019; pp. 1–6. [CrossRef]
11. Oberretl, K. Einseitiger Linearmotor mit Käfig im Sekundärteil. *Arch. Elektrotechnik* **1974**, *56*, 305–319. [CrossRef]
12. Zhu, Z.Q.; Howe, D. Instantaneous Magnetic Field Distribution in Brushless Permanent Magnet DC Motors. III. Effect of Stator Slotting. *IEEE Trans. Magn.* **1993**, *29*, 143–151. [CrossRef]
13. Elfgen, S.; Steentjes, S.; Bohmer, S.; Franck, D.; Hameyer, K. Influences of Material Degradation Due to Laser Cutting on the Operating Behavior of PMSM Using a Continuous Local Material Model. *IEEE Trans. Ind. Appl.* **2017**, *53*, 1978–1984. [CrossRef]
14. Hameyer, K.; Belmans, R. Numerical Modelling and Design of Electrical Machines and Devices. In *Advances in Electrical and Electronic Engineering*; WIT Press: Southampton, UK, 1999; Volume 1.
15. Schöps, S.; De Gersem, H.; Weiland, T. Winding Functions in Transient Magnetoquasistatic Field-Circuit Coupled Simulations. *COMPEL Int. J. Comput. Math. Electr. Electron. Eng.* **2013**, *32*, 2063–2083. [CrossRef]
16. Vassent, E.; Meunier, G.; Foggia, A. Simulation of Induction Machines Using Complex Magnetodynamic Finite Element Method Coupled with the Circuit Equations. *IEEE Trans. Magn.* **1991**, *27*, 4246–4249. [CrossRef]
17. De Gersem, H.; Hameyer, K. Air-Gap Flux Splitting for the Time-Harmonic Finite-Element Simulation of Single-Phase Induction Machines. *IEEE Trans. Magn.* **2002**, *38*, 1221–1224. [CrossRef]
18. De Geserm, H.; Mertens, R.; Kay, H. Comparison of time-harmonic and transient finite element models for asynchronous machines. In Proceedings of the Internationel Conference on Electrical Machines (ICEM00), Espoo, Finland, 28–30 August 2000; Volume 1, pp. 66–70.
19. Shannon, C.E. A Mathematical Theory of Communication. *Bell Syst. Tech. J.* **1948**, *27*, 379–423. [CrossRef]
20. von Pfingsten, G.; Nell, M.; Hameyer, K. Hybrid Simulation Approaches for Induction Machine Calculation. *COMPEL—Int. J. Comput. Math. Electr. Electron. Eng.* **2018**, *37*, 1744–1754. [CrossRef]
21. Steentjes, S.; von Pfingsten, G.; Hombitzer, M.; Hameyer, K. Iron-Loss Model With Consideration of Minor Loops Applied to FE-Simulations of Electrical Machines. *IEEE Trans. Magn.* **2013**, *49*, 3945–3948. [CrossRef]
22. Nell, M.; Groschup, B.; Hameyer, K. Using Scaled Fe Solutions for an Efficient Coupled Electromagnetic—Thermal Induction Machine Model. *COMPEL—Int. J. Comput. Math. Electr. Electron. Eng.* **2021**, *40*, 267–279. [CrossRef]
23. von Pfingsten, G.; Steentjes, S.; Hameyer, K. Operating Point Resolved Loss Calculation Approach in Saturated Induction Machines. *IEEE Trans. Ind. Electron.* **2017**, *64*, 2538–2546. [CrossRef]
24. Groschup, B.; Nell, M.; Pauli, F.; Hameyer, K. Characteristic Thermal Parameters in Electric Motors: Comparison between Induction- and Permanent Magnet Excited Machine. *IEEE Trans. Energy Convers.* **2021**, *36*, 2239–2248. [CrossRef]
25. Nell, M.; Lenz, J.; Hameyer, K. Efficient Numerical Optimization of Induction Machines by Scaled FE Simulations. In Proceedings of the XIII International Conference on Electrical Machines (ICEM), Alexandroupoli, Greece, 3–6 September 2018; pp. 198–204. [CrossRef]
26. Nell, M.; Groschup, B.; Hameyer, K. Efficient Coupled Electromagnetic-Thermal Induction Machine Model Using Scaled FE-Solutions. In Proceedings of the 19th International Symposium on Electromagnetic Fields in Mechatronics, Electrical and Electronic Engineering (ISEF), Nancy, France, 29–31 August 2019; pp. 1–2. [CrossRef]

energies

MDPI

Article

Multiple Fault Detection in Induction Motors through Homogeneity and Kurtosis Computation

Ana L. Martinez-Herrera, Edna R. Ferrucho-Alvarez, Luis M. Ledesma-Carrillo, Ruth I. Mata-Chavez, Misael Lopez-Ramirez and Eduardo Cabal-Yepez *

Multidisciplinary Studies Department, Engineering Division, Campus Irapuato-Salamanca, University of Guanajuato, Guanajuato 38944, Mexico; martinez.al@ugto.mx (A.L.M.-H.); er.ferruchoalvaez@ugto.mx (E.R.F.-A.); lm.ledesma@ugto.mx (L.M.L.-C.); ruth@ugto.mx (R.I.M.-C.); lopez.misael@ugto.mx (M.L.-R.)
* Correspondence: educabal@ugto.mx; Tel.: +52-445-458-9040

Abstract: In the last few years, induction motor fault detection has provoked great interest among researchers because it is a fundamental element of the electric-power industry, manufacturing enterprise, and services. Hence, considerable efforts have been carried out on developing reliable, low-cost procedures for fault diagnosis in induction motors (IM) since the early detection of any failure may prevent the machine from suffering a catastrophic damage. Therefore, many methodologies based on the IM startup transient current analysis have been proposed whose major disadvantages are the high mathematical complexity and demanding computational cost for their development. In this study, a straightforward procedure was introduced for identifying and classifying faults in IM. The proposed approach is based on the analysis of the startup transient current signal through the current signal homogeneity and the fourth central moment (kurtosis) analysis. These features are used for training a feed-forward, backpropagation artificial neural network used as a classifier. From experimentally obtained results, it was demonstrated that the brought-in scheme attained high certainty in recognizing and discriminating among five induction motor conditions, i.e., a motor in good physical condition (HLT), a motor with one broken rotor bar (1BRB), a motor with two broken rotor bars (2BRB), a motor with damage on the bearing outer race (BRN), and a motor with an unbalanced mechanical load (UNB).

Keywords: artificial neural network; fourth central moment; homogeneity analysis; induction motors; mechanical unbalance; one broken rotor bar; outer-race bearing fault; startup transient current; two broken rotor bars

Citation: Martinez-Herrera, A.L.; Ferrucho-Alvarez, E.R.; Ledesma-Carrillo, L.M.; Mata-Chavez, R.I.; Lopez-Ramirez, M.; Cabal-Yepez, E. Multiple Fault Detection in Induction Motors through Homogeneity and Kurtosis Computation. *Energies* **2022**, *15*, 1541. https://doi.org/10.3390/en15041541

Academic Editors: Mario Marchesoni and Ryszard Palka

Received: 4 January 2022
Accepted: 16 February 2022
Published: 19 February 2022

1. Introduction

Induction motors have stayed for many years as essential components of every electrical and manufacturing process. Because of their low cost, stiffness, and quality of performing consistently well, they are extensively used around the planet. The squirrel cage induction motor (SCIM) provides the most common type of electromechanical drive for commercial, domestic, and, the most important, industrial applications, corresponding to around 85% of the electrical energy utilization in this area [1]. In the past decades, profound efforts have been devoted to induction motor (IM) fault diagnosis due to the economic and technical consequences of an unexpected downtime caused by a failure. As any other electrical device, an IM is vulnerable to numerous kinds of failures that can be classified as bearings' faults, with 50% of incidence, rotor faults, with the 10%, and stator-related faults, with the 40% [2]. Some of the symptoms produced by these failures are excessive vibrations, unbalanced line currents and/or voltages, torque pulsations and decreased average torque, and excessive heating, among others, aggravating efficiency losses on any process. Stator faults have been lessened at present by improving the SCIM design and its building quality. On the other hand, broken rotor bars (BRB), bearing faults (BRN), and rotor unbalance (UNB) constitute very common problems, particularly in heavy duty systems [3]. Many efforts have been made to prevent catastrophic failures to occur with the application of several techniques for fault detection; unfortunately, most of them focus on detecting a single, specific fault separately, such as BRB [4], BRN [5], or UNB [6]. These techniques are usually based on monitoring and analyzing current and vibration signals [7]. Motor current signature analysis (MCSA) is one of the most popular and effective techniques for induction motor fault detection. It has the advantages of being noninvasive and simple to carry out, coming out with good results during faulty condition identification [8]. MCSA processes either the startup transient or the steady-state electric current signal fed to the SCIM stator, which is gathered with a current clamp probe to carry out early detection of these types of faults, trying to avoid unscheduled maintenance and interruption of production lines, which yield to critical outcomes in produced merchandise conditions, manufacturing prices, and security.

One of the most difficult faults to detect is broken rotor bars (BRB) because they usually do not lead to an abrupt total motor failure but to a progressive deterioration that may not be detected until the motor is severely damaged, causing a shutdown in the production line. In [9], a new technique was developed using the fast Fourier transform (FFT), and an index-based classifier was introduced for BRB diagnosis. Nevertheless, several studies have shown that analyzing the steady-state signal might not be an effective approach for identifying certain operational conditions such as voltage fluctuations, bearing failures, noise, and mechanical load changes [10]. Therefore, non-stationary signal analysis for SCIM fault detection has generated great interest in recent years. Some advantages of analyzing and monitoring the induction motor (IM) current signals during its transient state are related to the close relation between the signal noise and the rotor fast slip [7], making it easier to identify BRB and other kinds of faults during this regime. However, the greatest obstacle for this type of monitoring comes from the startup span, which is very short; besides that, non-stationary signals represent a challenging task as they cannot be analyzed separately in time or frequency domains. The short-time Fourier transform (STFT) [11] and the Hilbert transform [4] are well-known techniques for identifying faulty conditions in SCIM. On the other hand, the wavelet transform (WT) has attained great interest among researchers for BRB fault detection, as well as the high-resolution technique known as multiple signal classification (MUSIC) [12]. On the other hand, most of the techniques utilized for identifying bearing faults (BRN) rely on the analysis of vibration signals [5]. However, the STFT and the WT and its variations directly depend on the correct selection of a suitable window size and a mother wavelet function, respectively, to perform an effective signal analysis. Other works based on quadratic distributions (QD) [13–15] provide a time–frequency representation for non-stationary signals. However, QD may generate spurious frequencies called cross terms that compromise the correct identification of the fault-related frequencies. In an equivalent way, mechanical unbalance fault (UNB) diagnosis has been traditionally addressed by analyzing vibration signals [16–18], too.

Although the methods and techniques mentioned before are suitable for detecting and diagnosing independent induction motor faults, most of them rely on the combination of complex mathematical bases that demand specialized hardware and software for their implementation in order to take the time-domain signals into the frequency domain and back to the time domain. This requires a long execution time and computational resources for the signal processing. Furthermore, some of these techniques involve the analysis of the electric current from the three phases along with the multi-axis vibration signals from the SCIM. Therefore, in this study, an approach based on the examination of just one phase from the electrical current fed to the IM during its startup transient through homogeneity and kurtosis computations was presented for detecting and classifying distinct induction motor faults, i.e., one broken rotor bar (1BRB), two broken rotor bars (2BRB), a motor with damage on the bearing outer race (BRN), and a motor with an unbalanced mechanical load (UNB). The introduced methodology has a low computational complexity compared to other methods in related literature for signal examination aimed to IM fault detection;

hence, the proposed approach required a short processing time, making it feasible for being utilized in online-processing applications.

The paper is organized as follows. Section 2 provides a theoretical background for fault and indexes' description. Section 3 describes the experimental setup. Section 4 presents the obtained results, and, finally, some conclusions are provided in Section 5.

2. Theoretical Framework

This section provides a mathematical background about the induction motor faults treated in this work, as well as the signal processing techniques utilized for analyzing the startup electric current signal and the artificial neural network utilized for classifying the IM operational condition.

2.1. Broken Rotor Bar Fault (BRB)

The BRB fault is the most common rotor-related failure that affects SCIM, and it is very difficult to detect because, under this state, the motor operates apparently under normal condition. BRB is mainly caused by overload and thermal imbalances, electromagnetic forces and noise, vibrations, environmental damage, or by manufacturing processes.

An induction motor operating with BRB defects generates an opposing succession of rotor currents caused by the asymmetries, which bring on a distinctive element in the frequency spectrum of the stator current. The fault-related frequencies (f_{BRB}) indicating the presence of BRB are given by:

$$f_{BRBs} = (1 - 2ks)f_s,\ k = 1,2,3,\ldots \tag{1}$$

where k is an integer number, f_s is the main frequency component of the electric power supply, and the motor slip is represented by s, which takes values in the range from 0 to 1 [19].

2.2. Bearing Fault (BRN)

Characteristic frequency components are generated on the stator current when an SCIM has a faulty bearing. These specific frequency components can be predicted since they are related to both the power supply frequency and the mechanical system frequency. When there is a fault in any component of a bearing, for instance, its inner raceway, outer raceway, or rolling elements, specific components are induced in the vibration and current signals of the machine. These characteristic frequencies associated to the bearing faults depend directly on the bearing geometry and the machine rotating speed. A defect on the outer race will cause an impulse every time the rolling elements make contact with the defect. The outer raceway-related frequency can be theoretically determined by

$$f_o = \frac{n}{2} f_r \left[1 - \left(\frac{BD}{PD}\right)\cos(\beta)\right] \tag{2}$$

where n is the number of balls (rolling elements), f_r is the shaft rotating frequency, BD is the diameter of the balls, PD is the bearing race diameter, and β is the angle between the ball in the race [20]

2.3. Mechanical Unbalance Fault (UNB)

When the mechanical load of the induction motor is not evenly disseminated, displacing the center of mass out of the motor rotating axis, there is an unbalance fault. A manufacturing defect is the principal cause of rotor unbalance; also, heating dilation has an effect on the internal misalignment or the shaft deviation that produces an unbalanced rotor. The unbalance state occurs when there is an uneven distribution of weight around the rotor center of rotation, generating an unbalance force U, which is given by

$$U = m \times r \tag{3}$$

where m is the mass and r is its eccentricity, which is the separation between the rotor center of mass and its rotating axis. Unbalance force fluctuates with the rotating speed and drags the rotor off from the stator center bore to a distinct position, which is known as the heavy spot. Mutual inductances between stator and rotor loops get to be uneven because of the rotor unbalance, which results in frequency components induced in the stator current that are given by

$$f_{unb} = f_s[1 \pm k(1-s)/p],\ k = 1,2,3,\ldots \quad (4)$$

where f_s is the fundamental frequency of the electric current supply, k is an integer number, s is the motor slip, and p is the number of pole pairs in the SCIM [21].

2.4. Homogeneity

In image classification, homogeneity is a textural attribute that estimates the variability of the gray level in the pixels from an image. It is derived from the gray level co-occurrence matrix (GLCM) [22], and it measures the closeness of the element distribution in the GLCM regarding its diagonal. The GLCM shows how many times each gray level comes about at a pixel situated at a predetermined geometric position regarding any other pixel through a function of their gray levels. Homogeneity ranges from 0 to 1 and reaches its maximum value when the diagonal elements have a value of 1. Homogeneity can be computed by

$$H = \sum_i \sum_j \frac{1}{1+(i-j)^2} p(i,j) \quad (5)$$

where $p(i, j)$ is the (i, j)th element of the normalized GLCM. Homogeneity can be used as an index for fault detection and classification in SCIM since distinct fault-associated frequency elements are induced in the electrical current signal, changing its uniformity [23].

2.5. Kurtosis

Kurtosis has the capability of measuring the deviation, i.e., tailedness, of a probability distribution, and discriminating between distributions with different shapes; therefore, it can be used as an efficient indicator for SCIM fault detection. Kurtosis is the fourth-order moment that describes the shape of a probability distribution from a signal. If there is a high impulsive component, with a sharp signal intensity distribution, then there is a high kurtosis value. Kurtosis of a random event X is computed as

$$Kurt[X] = \frac{\frac{1}{N}\sum_{i=1}^{N}(x_i-\mu)^4}{\left(\frac{1}{N}\sum_{i=1}^{N}(x_i-\mu)^2\right)^2} = \frac{\mu_4}{\sigma^4} \quad (6)$$

where N is the number of samples, x_i is the time raw-signal sample for $i = 1, 2, \ldots, N$, and μ is the mean of the random event $X = [x_1, x_2, x_3, \ldots, x_N]$.

2.6. Artificial Neural Networks

Artificial neural networks (ANN) provide a powerful and speedy tool for classification problems. Figure 1 shows the basic design of ANN with two layers where each neuron produces a unique number. Inputs are multiplied by corresponding weights and summed up. The corresponding output is derived after adding a bias term to the in-between results. The sum of the weighted inputs is transformed through a nonlinear activation function to get the outcome of each corresponding neuron. Several triggering functions are viable; however, in this work, from an empirical analysis, a hyperbolic tangent function was used. The multilayer perceptron architecture is used in the ANN during experimentation. It is made up of a feed-forward architecture composed of an input layer, one or more inside layers, and one output layer. The number of inside layers and neurons on each layer is determined by the dealt issue; in this case, the inputs correspond to the signal-obtained features, i.e., homogeneity and kurtosis. The hidden layer is set heuristically by experimentation, whereas the number of output neurons is related to the distinct categories being recognized. The ANN employed in this experimentation was trained using a Levenberg–Marquardt backpropagation scheme [24].

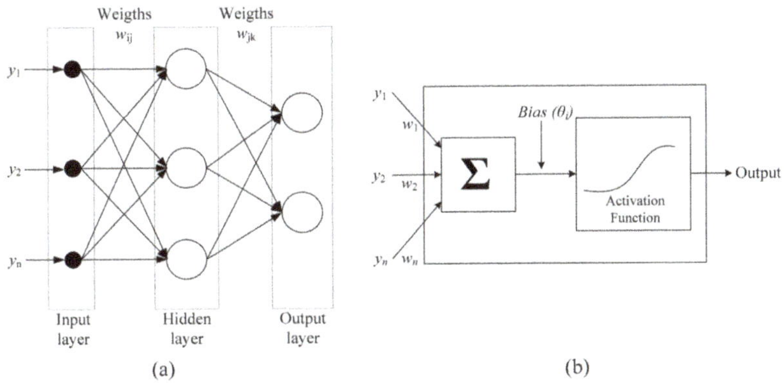

Figure 1. (**a**) Typical scheme of an ANN, (**b**) operation of an artificial neuron in a layer.

3. Experimentation

The electrical current signal from the startup transient of a SCIM is used for identifying and classifying a healthy motor (HLT) or a motor with a faulty condition from those treated in this study, i.e., one broken rotor bar (1BRB), two broken rotor bars (2BRB), outer-race faulty bearing (BRN), and mechanical unbalance (UNB). Figure 2 shows the testbench configuration, which employs distinct 1-HP SCIM (model WEG 00136APE48T) for assessing the feasibility of using the introduced procedure to identify and classify distinct operational states.

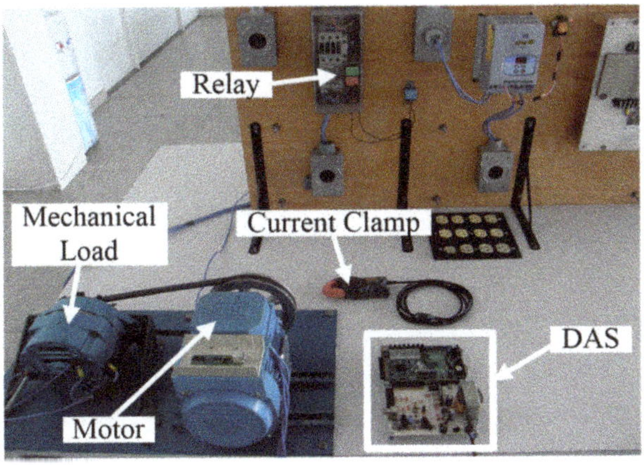

Figure 2. Testbench used for assessing the proposed method for detecting and classifying distinct faults in SCIM.

The motors under test received an electric power supply of 220 V ac at 60 Hz. They had 28 bars in the rotor, two poles, and the used mechanical load was an ordinary alternator that was equivalent to one-quarter of the SCIM nominal load. One phase of the three-phase electric current supply signal was collected through an i200s, ac current clamp from Fluke. The data acquisition system (DAS) used a 16-bit analog-to-digital converter (ADS7809). The instrumentation system used a sampling frequency f_0 of 1.5 kHz, obtaining 4096 samples in 2.7 s during the induction motor startup transient.

For this study, one motor was kept in healthy condition, i.e., in good physical condition, to be used as benchmark. The 1BRB and 2BRB conditions were generated in a synthetic way by drilling one hole and two holes, respectively, with a diameter of 7.938 mm, without damaging the rotor shaft, as shown in Figure 3a,b, respectively. On the other hand, the bearing was synthetically harmed on its outer race by boring a 1.191-mm hole utilizing a tungsten drill bit. Figure 3c displays the bearing model 6203-2ZNR that was synthetically damaged to carry out the experimentation. The mechanical unbalance fault was generated by adding a mass in one of the pulley arms. In a drilled hole with 8 mm of diameter, a two-sided screw was placed and secured using female screws on both sides of the pulley arm, as depicted in Figure 3d. A total of 100 trials were executed for each motor condition.

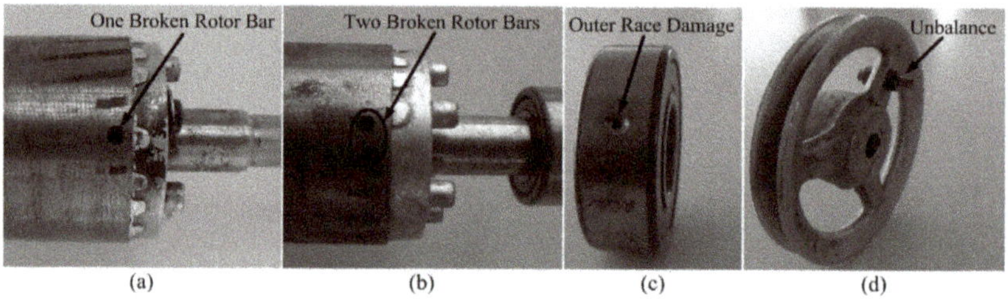

Figure 3. Artificially generated faults. (**a**) One broken rotor bar (1BRB); (**b**) two broken rotor bars (2BRB); (**c**) bearing with outer race damaged (BRN), and (**d**) mechanical unbalance.

Figure 4 depicts the proposed methodology for multiple fault diagnosis and classification. The electrical current signal of the startup transient was obtained by the current clamp; then, it was adjusted and analog-to-digital transformed in the DAS. The resulting discrete current signal was treated for obtaining the desired features, homogeneity and kurtosis, which were used as entries to the artificial neural network, a multilayer perceptron with a feed-forward architecture.

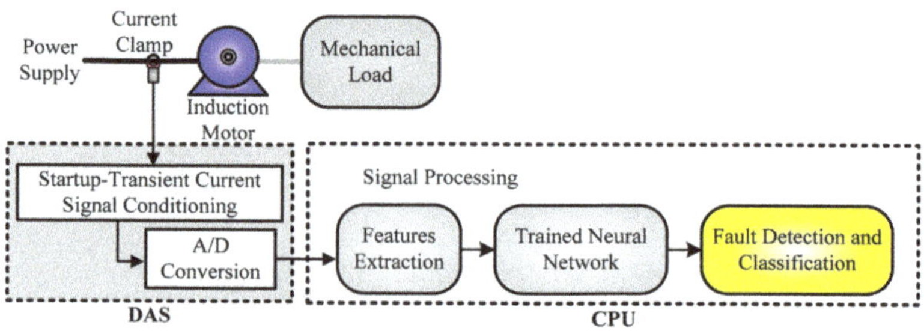

Figure 4. Proposed approach for distinct fault detection and classification.

Homogeneity and kurtosis values were obtained and normalized for each motor condition by the definitions (5) and (6), respectively. For each motor condition, a statistical analysis was performed. The mean (μ) and standard deviation (σ) of the homogeneity and kurtosis values for a motor without harm (HLT), a motor with one separated rotor bar (1BRB), a motor with two shattered rotor bars (2BRB), a motor with outer-race bearing damage (BRN), and a motor with unbalance (UNB) show that the respective probability density functions (PDF) partially cover each other in some degree, impeding a direct classification. Figure 5 shows the PDF of homogeneity and kurtosis features where the overlap among some of the treated conditions is evident. Therefore, a neural network classification was used for refining the proper operation that allowed a precise identification of multiple operational conditions.

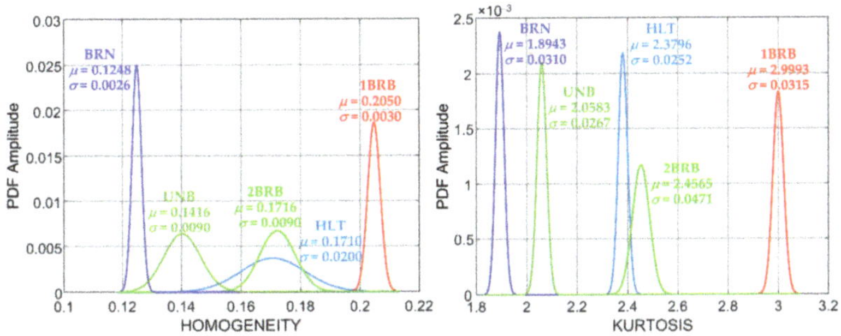

Figure 5. Homogeneity and kurtosis PDF values for the different motor states: HLT, healthy; 1BRB, one broken rotor bar; 2BRB, two broken rotor bars; BRN, outer-race bearing damage; and UNB, unbalance.

The classification was carried out utilizing the homogeneity and kurtosis data as entries to the ANN. The artificial neural network was a multilayer perceptron with a feed-forward architecture with two inputs (kurtosis values and homogeneity values) only, one hidden layer, and one output. A Levenberg–Marquardt backpropagation algorithm was used for training the ANN and the mean-square-error index was used for performance assessment. Inputs to the ANN were the feature vectors composed of the homogeneity and kurtosis data, where each motor condition had a total of 20 values in its validation data set, given as a result of feature vectors of length 100.

4. Obtained Results

The proposed technique efficacy was validated through 100 different trials under each treated motor condition. A holdout-type data set was employed, where the first 70 experiments were employed for teaching the network and the remaining ones were used for checking the proposed methodology efficacy. Table 1 shows the performance results, through a confusion matrix, of the introduced technique during the identification and classification of the IM operational condition as HLT, 1BRB, 2BRB, BRN, and UNB. The right-most column in Table 1 displays the average success rate of the suggested procedure.

The results obtained experimentally are depicted in Table 1 and they show that the introduced methodology reached 100% of effectiveness on identifying and discriminating among a healthy motor (HLT), a motor with one broken rotor bar (1BRB), a motor with two broken rotor bars (2BRB), a motor with outer-race bearing damage (BRN), and a motor with an unbalanced mechanical load (UNB). The proposed technique was executed in a 2.20-GHz Intel Core i7-8750 processor, making use of the software MATLAB 2020a.

Table 1. Confusion matrix and overall effectiveness of using homogeneity and kurtosis features for fault detection and classification in SCIM.

IM Condition	HLT	1BRB	2BRB	BRN	UNB	Overall Effectiveness
HLT	20	0	0	0	0	100%
1BRB	0	20	0	0	0	100%
2BRB	0	0	20	0	0	100%
BRN	0	0	0	20	0	100%
UNB	0	0	0	0	20	100%

Discussion

The proposed methodology is compared against previous approaches used for detecting SCIM faults in Table 2. The obtained results from real experimentation demonstrated the usefulness of homogeneity and kurtosis as indexes for IM diagnosis and their high reliability as indicators for multiple fault identification and classification [25,26]. The proposed approach can recognize and classify the operational condition of an induction motor as in a good state (HLT), a motor with one split rotor bar (1BRB), a motor with two damaged rotor bars (2BRB), a motor with outer race damage in the bearing (BRN), and a motor with an unbalanced mechanical load (UNB) with high certainty, attaining up to 100% of effectiveness, utilizing just two features of a single phase from the three-phase electrical current supply to the SCIM, as inputs to a multilayer perceptron ANN. This is different from other approaches reported in the reviewed literature that even require the signal transformation from the time domain into the frequency domain and back to the time domain to carry out the signal processing in order to extract up to 29 features of the current signals from the three phases and the multi-axis vibration signals, in conjunction, in order to be capable of performing the fault detection. In the proposed procedure, the electrical current signal from the SCIM startup transient was analyzed just in time domain without any pre-treatment, which is an evident convenience compared to the previous works in Table 2, which require the combination of two or more processing techniques to carry out a qualitative diagnosis or to perform it in a quantitative style by analyzing current and vibration signals in time, frequency, and even time-frequency domains. Therefore, the proposed methodology is a reliable tool that ensures high certainty during different fault detections and classifications in induction motors through the analysis of just one phase from startup electric current supply, outperforming previous approaches in the state of the art.

Table 2. Comparison chart of the proposed methodology against the state of the art in related literature for fault detection in IM.

Method	Accuracy Rate	Applied Techniques	Detected Fault
Garcia-Bracamonte et al. [8]	From 90% to 99%	Autocorrelation; FFT; independent component analysis; region-of-interest segmentation; and 1-D, 2-D, and 3-D vector extraction.	BRB only
Yang and Shi [27]	Qualitative	Wavelet packet, threshold optimization, Shannon entropy, wavelet packet reconstruction, and FFT computation.	
Haroun et al. [28]	From 81.4% to 100%	Zero-crossing time, envelope extraction of the three phase currents, frequency domain characterization, ReliefF algorithm, and self-organizing map.	
Li et al. [29]	Qualitative	Fourier transform, power spectral density estimation, local characteristic frequency bands' synchronization, spectrum transformation, central point computation, and kurtosis energy-based spectrum.	
Gong et al. [30]	Qualitative	Wavelet packet transform and spectral kurtosis.	BRN only
Gao and Xiang [31]	Qualitative	Ensemble empirical mode decomposition, L-Kurtosis value, FFT, clustering-based segmentation, inverse FFT, and Hilbert envelope spectrum computation.	
Navasari et al. [32]	From 98% to 100%	Wavelet decomposition, sampling of the decomposition streams, energy computation, and ANN.	
Ben Abid et al. [33]	100%	Stationary wavelet packet transform, root mean square (RMS), aiNet clustering algorithm, and directed acyclic graph support vector machine.	
Rahman and Uddin [16]	Qualitative	Standard deviation, crest factor, and kurtosis computation; FFT; DWT-based frequency domain analysis; Hilbert transform; and envelope detection.	UNB only
Tahir et al. [17]	100%	Multi-axis RMS value, variance, skewness, kurtosis, impulse factor, and range computation; signed distance computation; and SVM.	
Guo et al. [18]	86.87%	DC part removal, signal resampling, continuous wavelet transform scalogram (CWTS), cropping, and convolutional neural network.	
Cunha Palacios et al. [34]	From 99.7% to 100%	Signal segmentation, peak value, module computation, crossover detection, normalization, input selection, classification through different intelligent algorithms.	BRN, BRB, and Stator Faults
Jigyasu et al. [35]	From 99.7% to 100%	RMS, variance, kurtosis, peak value, skewness, median, crest factor, margin factor, impulse factor, shape, and median range extraction; different neural network structures.	BRN, BRB, and Eccentricity
Proposed Approach	**100%**	**Homogeneity, kurtosis, and an ANN.**	**1BRB, 2BRB, BRN, and UNB**

5. Conclusions

Recently proposed techniques can detect one single induction motor fault with an adequate certainty; however, most of them rely on the combination of complex mathematical operations that require specific hardware and software for their implementation. Furthermore, they involve the monitoring and processing of different signals as electric current supply and multi-axis vibration signals to obtain time, frequency, and even time–frequency features that allow them to attain high certainty on the induction motor diagnosis. Therefore, in this work, a straightforward technique for multiple IM fault detection, which just requires the computation of homogeneity and kurtosis from a single phase of the supplied electrical current signal during the SCIM startup transient, was introduced. The obtained results from experimental studies demonstrated that the proposed methodology provides highly reliable results on detecting and classifying distinct induction motor faults by computing the homogeneity and kurtosis on the time domain, allowing the identification of five different operational conditions, a motor in healthy state (HLT), a motor with one broken bar (1BRB), a motor with two broken rotor bars (2BRB), a motor with outer-race damage on its bearing (BRN), and a motor with an unbalanced mechanical load (UNB), with remarkable certainty. A thorough comparison against the state of the art in the subject of induction motor fault detection showed that the proposed method outperformed previous approaches in the reviewed literature, which usually just detect one single type of fault, by implementing a low-cost computational technique suitable for online applications.

Future work will focus on assessing the proposed technique for multiple IM fault detection under different scenarios. It will contemplate incorporating additional faulty conditions and signal examination techniques to recover other signal characteristics, as well as assessing distinct kinds of classifications to carry out the fault identification and sorting with high precision.

Author Contributions: A.L.M.-H. and E.R.F.-A. performed the methodology implementation and helped during experimentation for data acquisition. L.M.L.-C. helped during data analysis and result interpretation. R.I.M.-C. helped during the manuscript writing—review and editing. M.L.-R. helped during experimentation and data acquisition. E.C.-Y. supervised the project and the experimentation implementation, assessed the obtained results, and helped in the document preparation. All authors have read and agreed to the published version of the manuscript.

Funding: This research received no external funding.

Institutional Review Board Statement: Not applicable.

Informed Consent Statement: Not applicable.

Acknowledgments: This work was supported in part by the National Council on Science and Technology (CONACYT), Mexico, under Grants 443689 and 710888, and in part by DAIP—U. de Gto. under the Convocatoria Institucional de Inves-tigacion Cientifica 2022.

Conflicts of Interest: The authors declare no conflict of interest.

References

1. Hassan, O.E.; Amer, M.; Abdelsalam, A.K.; Williams, B.W. Induction Motor Broken Rotor Bar Fault Detection Techniques Based on Fault Signature Analysis—A Review. *IET Electr. Power Appl.* **2018**, *12*, 895–907. [CrossRef]
2. Thomson, W.T.; Fenger, M. Current Signature Analysis to Detect Induction Motor Faults. *IEEE Ind. Appl. Mag.* **2001**, *7*, 26–34. [CrossRef]
3. Cubert, J.M. Use of Electronic Controllers in Order to Increase the Service Life on Asynchronous Motors. In Proceedings of the European Seminar on Electro-Technologies for Industry, Bilbao, Spain, 20–22 May 1992; pp. 393–404.
4. Abd-el-Malek, M.B.; Abdelsalam, A.K.; Hassan, O.E. Novel Approach Using Hilbert Transform for Multiple Broken Rotor Bars Fault Location Detection for Three Phase Induction Motor. *ISA Trans.* **2018**, *80*, 439–457. [CrossRef] [PubMed]
5. Deekshit Kompella, K.C.; Venu Gopala Rao, M.; Srinivasa Rao, R. Bearing Fault Detection in a 3 Phase Induction Motor Using Stator Current Frequency Spectral Subtraction with Various Wavelet Decomposition Techniques. *Ain Shams Eng. J.* **2018**, *9*, 2427–2439. [CrossRef]

6. Rahman, M.M.; Uddin, M.N. Online Unbalanced Rotor Fault Detection of an IM Drive Based on Both Time and Frequency Domain Analyses. In Proceedings of the 2015 Industry Applications Society Annual Meeting, Addison, TX, USA, 18–22 October 2015; pp. 1–8. [CrossRef]
7. Liu, Y.; Bazzi, A. A Review and Comparison of Fault Detection and Diagnosis Methods for Squirrel-Cage Induction Motors: State of the Art. *ISA Trans.* **2017**, *70*, 400–409. [CrossRef] [PubMed]
8. Garcia-Bracamonte, J.E.; Ramirez-Cortes, J.M.; de Rangel-Magdaleno, J.J.; Gomez-Gil, P.; Peregrina-Barreto, H.; Alarcon-Aquino, V. An Approach on MCSA-Based Fault Detection Using Independent Component Analysis and Neural Networks. *IEEE Trans. Instrum. Meas.* **2019**, *68*, 1353–1361. [CrossRef]
9. Rivera-Guillen, J.R.; De Santiago-Perez, J.J.; Amezquita-Sanchez, J.P.; Valtierra-Rodriguez, M.; Romero-Troncoso, R.J. Enhanced FFT-Based Method for Incipient Broken Rotor Bar Detection in Induction Motors during the Startup Transient. *Measurement* **2018**, *124*, 277–285. [CrossRef]
10. Zajac, M.; Sulowicz, M. Wavelet Detectors for Extraction of Characteristic Features of Induction Motor Rotor Faults. In Proceedings of the 2016 International Conference on Signals and Electronic Systems (ICSES), Krakow, Poland, 5–7 September 2016; pp. 212–218. [CrossRef]
11. Ayon-Sicaeros, R.A.; Cabal-Yepez, E.; Ledesma-Carrillo, L.M.; Hernandez-Gomez, G. Broken-Rotor-Bar Detection Through STFT and Windowing Functions. In Proceedings of the 2019 IEEE Sensors Applications Symposium (SAS), Sophia Antipolis, France, 11–13 March 2019; pp. 1–5. [CrossRef]
12. Bensaoucha, S.; Bessedik, S.A.; Ameur, A.; Moreau, S.; Teta, A. A Comparative Study for Broken Rotor Bars Fault Detection in Induction Machine Using DWT and MUSIC Techniques. In Proceedings of the 2020 1st International Conference on Communications, Control Systems and Signal Processing (CCSSP), El Oued, Algeria, 16–17 May 2020; pp. 523–528. [CrossRef]
13. Climente-Alarcon, V.; Antonino-Daviu, J.A.; Riera-Guasp, M.; Vlcek, M. Induction Motor Diagnosis by Advanced Notch FIR Filters and the Wigner–Ville Distribution. *IEEE Trans. Ind. Electron.* **2014**, *61*, 4217–4227. [CrossRef]
14. Perez-Ramirez, C.A.; Amezquita-Sanchez, J.P.; Dominguez-Gonzalez, A.; Valtierra-Rodriguez, M.; Camarena-Martinez, D. Compact Kernel Distribution-Based Approach for Broken Bars Detection on Induction Motors. In Proceedings of the 2015 IEEE International Autumn Meeting on Power, Electronics and Computing (ROPEC), Ixtapa, Mexico, 4–6 November 2015; pp. 1–6. [CrossRef]
15. Quinde, I.R.; Sumba, J.C.; Ochoa, L.E.; Guevara, A.V., Jr.; Morales-Menendez, R. Bearing Fault Diagnosis Based on Optimal Time-Frequency Representation Method. *IFAC-PapersOnLine* **2019**, *52*, 194–199. [CrossRef]
16. Rahman, M.M.; Uddin, M.N. Online Unbalanced Rotor Fault Detection of an IM Drive Based on Both Time and Frequency Domain Analyses. *IEEE Trans. Ind. Appl.* **2017**, *53*, 4087–4096. [CrossRef]
17. Tahir, M.M.; Hussain, A.; Badshah, S.; Khan, A.Q.; Iqbal, N. Classification of Unbalance and Misalignment Faults in Rotor Using Multi-Axis Time Domain Features. In Proceedings of the 2016 International Conference on Emerging Technologies (ICET), Islamabad, Pakistan, 18–19 October 2016; pp. 1–4. [CrossRef]
18. Guo, S.; Yang, T.; Gao, W.; Zhang, C. A Novel Fault Diagnosis Method for Rotating Machinery Based on a Convolutional Neural Network. *Sensors* **2018**, *18*, 1429. [CrossRef] [PubMed]
19. Burnett, R.; Watson, J.F.; Elder, S. The Application of Modern Signal Processing Techniques for Use in Rotor Fault Detection and Location within Three-Phase Induction Motors. *Signal Process.* **1996**, *49*, 57–70. [CrossRef]
20. Frosini, L.; Bassi, E. Stator Current and Motor Efficiency as Indicators for Different Types of Bearing Faults in Induction Motors. *IEEE Trans. Ind. Electron.* **2010**, *57*, 244–251. [CrossRef]
21. Benbouzid, M.E.H.; Vieira, M.; Theys, C. Induction Motors' Faults Detection and Localization Using Stator Current Advanced Signal Processing Techniques. *IEEE Trans. Power Electron.* **1999**, *14*, 14–22. [CrossRef]
22. Haralick, R.M.; Shanmugam, K.; Dinstein, I. Textural Features for Image Classification. *IEEE Trans. Syst. Man Cybern.* **1973**, *SMC-3*, 610–621. [CrossRef]
23. Sapena-Bano, A.; Pineda-Sanchez, M.; Puche-Panadero, R.; Martinez-Roman, J.; Matic, D. Fault Diagnosis of Rotating Electrical Machines in Transient Regime Using a Single Stator Current's FFT. *IEEE Trans. Instrum. Meas.* **2015**, *64*, 3137–3146. [CrossRef]
24. Kowalski, C.T.; Orlowska-Kowalska, T. Neural Networks Application for Induction Motor Faults Diagnosis. *Math. Comput. Simul.* **2003**, *63*, 435–448. [CrossRef]
25. Yang, R.; Li, H.-K.; Tang, D.-L.; Hou, M.-F. Based on the Optimal Frequency Band of Maximum Correlation Kurtosis De-Convolution for Bearing Weak Fault Diagnosis. In Proceedings of the 2017 Prognostics and System Health Management Conference (PHM-Harbin), Harbin, China, 9–12 July 2017; pp. 1–5. [CrossRef]
26. Lizarraga-Morales, R.A.; Rodriguez-Donate, C.; Cabal-Yepez, E.; Lopez-Ramirez, M.; Ledesma-Carrillo, L.M.; Ferrucho-Alvarez, E.R. Novel FPGA-Based Methodology for Early Broken Rotor Bar Detection and Classification Through Homogeneity Estimation. *IEEE Trans. Instrum. Meas.* **2017**, *66*, 1760–1769. [CrossRef]
27. Yang, M.; Shi, W. Research on Feature Extraction Method for Motor Rotor Broken Bar Based on Optimized Wavelet Packet Analysis. In Proceedings of the 2018 IEEE 4th Information Technology and Mechatronics Engineering Conference (ITOEC), Chongqing, China, 14–16 December 2018; pp. 24–28. [CrossRef]
28. Haroun, S.; Nait Seghir, A.; Touati, S. Self-Organizing Map and Feature Selection for IM Broken Rotor Bars Faults Detection and Diagnosis. In Proceedings of the 2018 International Conference on Electrical Sciences and Technologies in Maghreb (CISTEM), Algiers, Algeria, 28–31 October 2018; pp. 1–6. [CrossRef]

29. Li, D.D.; Wang, W.; Ismail, F. An Energy Spectral Technique for Induction Motor Fault Detection. In Proceedings of the 2017 International Conference on Sensing, Diagnostics, Prognostics, and Control (SDPC), Shanghai, China, 16–18 August 2017; pp. 704–709. [CrossRef]
30. Gong, X.; Zhang, W.; Jing, Y.; Du, W.; Jing, Y. Bearing Fault Diagnosis for Coupling Faults of Rotor System Using Spectral Kurtosis and Wavelet Packet. In Proceedings of the 2019 International Conference on Sensing, Diagnostics, Prognostics, and Control (SDPC), Beijing, China, 15–17 August 2019; pp. 479–483. [CrossRef]
31. Gao, Q.; Xiang, J. A Method Using EEMD and L-Kurtosis to Detect Faults in Roller Bearings. In Proceedings of the 2018 Prognostics and System Health Management Conference (PHM-Chongqing), Chongqing, China, 26–28 October 2018; pp. 71–76. [CrossRef]
32. Navasari, E.; Asfani, D.A.; Negara, M.Y. Detection of Induction Motor Bearing Damage with Starting Current Analysis Using Wavelet Discrete Transform and Artificial Neural Network. In Proceedings of the 2018 10th International Conference on Information Technology and Electrical Engineering (ICITEE), Bali, Indonesia, 24–26 July 2018; pp. 316–319. [CrossRef]
33. Ben Abid, F.; Zgarni, S.; Braham, A. Distinct Bearing Faults Detection in Induction Motor by a Hybrid Optimized SWPT and AiNet-DAG SVM. *IEEE Trans. Energy Convers.* **2018**, *33*, 1692–1699. [CrossRef]
34. Cunha Palácios, R.H.; da Silva, I.N.; Goedtel, A.; Godoy, W.F. A Comprehensive Evaluation of Intelligent Classifiers for Fault Identification in Three-Phase Induction Motors. *Electr. Power Syst. Res.* **2015**, *127*, 249–258. [CrossRef]
35. Jigyasu, R.; Mathew, L.; Sharma, A. Multiple Faults Diagnosis of Induction Motor Using Artificial Neural Network. In Proceedings of the 2018 Second International Conference (ICAICR), Shimla, India, 14–15 July 2018; pp. 701–710. [CrossRef]

energies

Article

Model-Based Predictive Rotor Field-Oriented Angle Compensation for Induction Machine Drives

Yang Liu *, Jin Zhao and Quan Yin

School of Artificial Intelligence and Automation, Huazhong University of Science and Technology, Wuhan 430074, China; jinzhao617@hust.edu.cn (J.Z.); yinquans@hust.edu.cn (Q.Y.)
* Correspondence: yangliu30@hust.edu.cn; Tel.: +86-138-7110-9968

Abstract: In this paper, a model-based predictive rotor field-oriented angle compensation approach is proposed for induction machine drives. Indirect rotor field-oriented control is widely used in induction machine drives for its simple implementation and low cost. However, the accuracy of the rotor field-oriented angle is affected by variable parameters such as the rotor resistance and inductance. An inaccurate rotor field-oriented angle leads to a degradation of the torque and dynamic performance, especially in the high-speed flux-weakening region. Therefore, the d-axis and q-axis currents in the rotation reference frame are predicted based on the model and compared with the feedback current to correct the rotor field-oriented angle. To improve the stability and robustness, the proposed predictive algorithm is based on the storage current, voltage, and velocity data. The algorithm can be easily realized in real-time. Finally, the simulated and experimental results verify the algorithm's effectiveness on a 7.5 kW induction machine setup.

Keywords: rotor field-oriented angle error; indirect rotor field-oriented control; induction machine drives; model-based prediction

Citation: Liu, Y.; Zhao, J.; Yin, Q. Model-Based Predictive Rotor Field-Oriented Angle Compensation for Induction Machine Drives. *Energies* **2021**, *14*, 2049. https://doi.org/10.3390/en14082049

Academic Editor: Ryszard Palka

Received: 14 March 2021
Accepted: 5 April 2021
Published: 7 April 2021

1. Introduction

Indirect rotor field-oriented control (IRFOC) is widely used in induction machine drives because of its high performance in the base speed and field-weakening region. The control scheme of IRFOC is shown in Figure 1. Currently, the flux level and torque control in IRFOC are the research highlights in the field-weakening region [1–6]. The solutions are based on the accuracy of the rotor field-oriented angle. However, the rotor field-oriented method based on the integration of the rotor angular velocity and rotor slip angular velocity in IRFOC is affected by variations in parameters such as the rotor resistance. The rotor resistance varies with temperature and can be more than twice that of the normal resistance at 25 °C. The well-known solution to rotor field-oriented inaccuracy is parameter identification [5–10] and observers [11–15]. In [5,6], a magnetizing curve of induction in the field-weakening region and saturated region is proposed. Off-line parameter identification methods are proposed in [7,8]. A simple calculation based on the specification of an induction machine is introduced in [9]. These methods are useful and easy to apply in industry. Online parameter identification schemes are proposed in [10,11]. Solutions to address the parameter sensitivity problem in speed sensorless control of induction machines have been proposed, such as a sliding mode observer [12–15], a low-pass filter [16], square-wave voltage injection [17], and model reference adaptive control [18]. These algorithms require considerable computational resources and mainly aim to reduce the risk of instability phenomena. However, in the IRFOC of induction machines, the inaccurate field orientation caused by variable parameters is due to not only the instability but also the load capacity and dynamic performance.

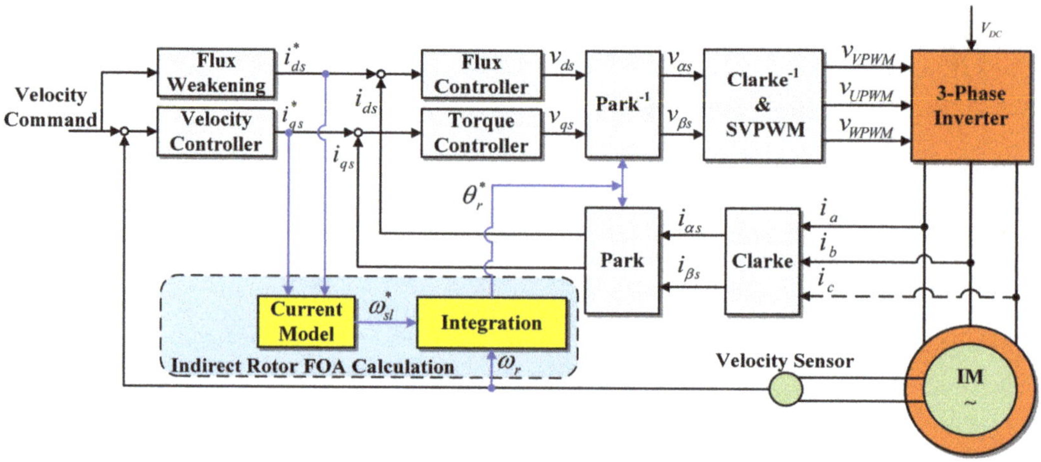

Figure 1. Control diagram of induction machine IRFOC.

Model-based predictive control (MPC) for machine drives and power electronics is an alternative control strategy that has gained attention in recent years. This approach can be used to address multivariable system constraints and nonlinearities in a very intuitive way [19]. Therefore, MPC has been successfully used for different applications, such as power converters connected to resistor–inductor (RL) loads [20], power electronics fault tolerance [21–23], energy management of electric vehicles [24,25], autonomous vehicle control [26,27], and high-performance drives for AC machines [28–32]. In [28], an MPC-based vector control method named GTV-MPTC for induction machines is proposed to cause the instantaneous torque to reach its reference value at the end of the next control period. The weighting factors in MPC are eliminated by investigating the relationship between the torque and stator flux to avoid tedious tuning work in [29]. However, the impact of variable parameters such as stator and rotor resistors is not given. In [30–32], MPC is used to improve the dynamic performance and reduce torque ripples in permanent-magnet synchronous motors (PMSM) drives. Compared to the vector control of induction machines, the control of PMSMs does not require a calculation of the slip velocity. Therefore, the rotor flux orientation for PMSMs is easy and accurate using a speed sensor.

In this paper, a compensation approach based on a model predictive algorithm of the rotor field-oriented angle error is proposed for IRFOC of induction machines. The d-axis and q-axis currents in the rotation reference frame are predicted and compared with the currents by current sensors to correct the rotor flux oriented angle. To improve the stability and robustness, the proposed predictive algorithm is based on the current, voltage, and velocity data stored in the memory. The algorithm can be realized easily in real-time. After compensation of the rotor field-oriented angle error, the output torque and current control performance can be improved. In Section 2, the mathematical model of IRFOC is introduced. Section 3 demonstrates the proposed model-based predictive algorithm and its implementation based on the stored data. The simulated results are shown in Section 4. Finally, the experimental results are presented to verify the proposed method in Section 5.

2. Induction Machine Control Model

2.1. Induction Machine Model

The dynamic model of the induction machine is important for the study of transient analysis on computers. If the currents in the rotating reference frame are selected as the main variables, then the state space stator voltage equations in the rotating reference frame can be obtained as

$$\begin{bmatrix} v_{qs} \\ v_{ds} \end{bmatrix} = \begin{bmatrix} R_s + L_s p & \omega_e L_s & L_m p & \omega_e L_m \\ -\omega_e L_s & R_s + L_s p & \omega_e L_m & L_m p \end{bmatrix} \begin{bmatrix} i_{qs} \\ i_{ds} \\ i_{qr} \\ i_{dr} \end{bmatrix} \tag{1}$$

where v_{qs}, v_{ds}, i_{qs}, and i_{ds} are the stator q-axis and d-axis voltages and currents, respectively; i_{qr} and i_{dr} are the rotor q-axis and d-axis currents, respectively; R_s and L_s the stator resistance and inductance, respectively; L_m is the magnetizing inductance; ω_e is the synchronous rotating angular velocity; and p is the differential factor.

The rotor flux linkage expressions in terms of the currents can be written as

$$\begin{bmatrix} \psi_{qr} \\ \psi_{dr} \end{bmatrix} = \begin{bmatrix} L_m & 0 & L_r + L_m & 0 \\ 0 & L_m & 0 & L_r + L_m \end{bmatrix} \begin{bmatrix} i_{qs} \\ i_{ds} \\ i_{qr} \\ i_{dr} \end{bmatrix}, \tag{2}$$

where ψ_{qr} and ψ_{dr} are the rotor q-axis and d-axis flux linkages, respectively. L_r is the inductance of the rotor.

The d-axis is located on the rotor flux linkage in the IRFOC. Therefore, $\psi_{qr} = 0$ and $\psi_r = \psi_{dr}$. By Equation (2), the rotor current in Equation (1) can be substituted as

$$v_{qs} = R_s i_{qs} + \delta L_s \frac{di_{qs}}{dt} + \omega_e L_s i_{ds} + \frac{\omega_e L_m}{L_r}(\psi_r - L_m i_{ds}), \tag{3}$$

$$v_{ds} = R_s i_{ds} + L_s \frac{di_{ds}}{dt} - \omega_e \delta L_s i_{qs} + \frac{L_m}{L_r}\frac{d(\psi_r - L_m i_{ds})}{dt}, \tag{4}$$

where $\delta = 1 - \frac{L_m^2}{L_s L_r}$.

The rotor flux in the IRFOC can be expressed as

$$\frac{L_r}{R_r}\frac{d\psi_r}{dt} + \psi_r = L_m i_{ds}, \tag{5}$$

Then, Equation (5) is substituted into Equation (3) to obtain

$$v_{qs} = R_s i_{qs} + \delta L_s \frac{di_{qs}}{dt} + \omega_e L_s i_{ds} - \frac{\omega_e L_m}{R_r}\frac{d\psi_r}{dt} \tag{6}$$

And Equation (4) can be given as

$$v_{ds} = R_s i_{ds} + \delta L_s \frac{di_{ds}}{dt} - \omega_e \delta L_s i_{qs} + \frac{L_m}{L_r}\frac{d\psi_r}{dt} \tag{7}$$

2.2. Discrete-Time Model

The first-order approximation, as shown in Equation (8), is usually used to transfer the continuous-time model to the discrete-time model.

$$\frac{dx}{dt} = \frac{x(k) - x(k-1)}{T_s} \tag{8}$$

where T_s is the sample period. By substituting Equation (8) into Equations (6) and (7), the discrete-time model of the induction machine control system can be obtained as

$$\begin{aligned} v_{qs}(k) = (R_s + \tfrac{\delta L_s}{T_s}) i_{qs}(k) + \omega_e L_s i_{ds}(k) - \\ \tfrac{\delta L_s}{T_s} i_{qs}(k-1) + \tfrac{\omega_e L_m}{R_r T_s}(\psi_r(k) - \psi_r(k-1)) \end{aligned} \tag{9}$$

$$v_{ds}(k) = (R_s + \frac{\delta L_s}{T_s})i_{ds}(k) - \frac{\delta L_s}{T_s} i_{ds}(k-1) - \omega_e \delta L_s i_{qs}(k) + \frac{L_m}{L_r T_s}(\psi_r(k) - \psi_r(k-1)) \tag{10}$$

The rotor flux varies slowly compared to the variation in the current and voltage. Therefore, Equations (9) and (10) can be simplified as

$$v_{qs}(k) = (R_s + \frac{\delta L_s}{T_s})i_{qs}(k) - \frac{\delta L_s}{T_s} i_{qs}(k-1) + \omega_e L_s i_{ds}(k) \tag{11}$$

$$v_{ds}(k) = (R_s + \frac{\delta L_s}{T_s})i_{ds}(k) - \frac{\delta L_s}{T_s} i_{ds}(k-1) - \omega_e \delta L_s i_{qs}(k) \tag{12}$$

3. Model-Based Predictive Algorithm and Implementation

3.1. Model-Based Predictive Algorithm

In the discrete model, based on Equations (11) and (12), the d-axis and q-axis currents at the $k+1$ instant are predicted by

$$\begin{cases} (R_s + \frac{\delta L_s}{T_s})i_{qs}(k+1) + \omega_e L_s i_{ds}(k+1) = \frac{\delta L_s}{T_s} i_{qs}(k) + v_{qs}(k+1) \\ -\omega_e \delta L_s i_{qs}(k+1) + (R_s + \frac{\delta L_s}{T_s})i_{ds}(k+1) = \frac{\delta L_s}{T_s} i_{ds}(k) + v_{ds}(k+1) \end{cases} \tag{13}$$

Then,

$$\begin{cases} i_{qs}(k+1) = \frac{C \times A - D \times E}{A^2 + F \times E} \\ i_{ds}(k+1) = \frac{C \times F + D \times A}{A^2 + F \times E} \end{cases} \tag{14}$$

where $A = R_s + \frac{\delta L_s}{T_s}$, $C = \frac{\delta L_s}{T_s} i_{qs}(k) + v_{qs}(k+1)$, $D = \frac{\delta L_s}{T_s} i_{ds}(k) + v_{ds}(k+1)$, $E = \omega_e(k) L_s$, and $F = \omega_e(k) \delta L_s$. The voltages $v_{ds}(k+1)$ and $v_{qs}(k+1)$ are the d-axis and q-axis voltages at $k+1$ instant, respectively. When the IRFOC is realized in a digital signal processor (DSP) or microcontroller unit (MCU), $v_{ds}(k+1)$ and $v_{qs}(k+1)$ can be obtained by the pulse width modulation (PWM) duty cycle at the k instant because the PWM duty cycle calculated at the k instant will be active at the $k+1$ instant. Therefore, $v_{ds}(k+1)$ and $v_{qs}(k+1)$ do not need to be predicted.

Based on (14), the predicted currents, $i_{qs}(k+1)$ and $i_{ds}(k+1)$, at the $k+1$ instant can be obtained without the rotor field-oriented angle. According to Equations (13) and (14), the stator resistance R_s is needed and varies with temperature. However, the variation in the stator resistance can be neglected because it is very small compared with the other parts of Equation (14). Therefore, the stator resistance on the motor plate can be used. The predictive model control diagram is shown in Figure 2. θ_{com} in Figure 2 is the compensated angle of the rotor field-oriented error; θ_{we} is the rotor field-oriented angle with compensation; $i_{an}(k)$, $i_{bn}(k)$, and $i_{cn}(k)$ are the three-phase currents at the k instant; $v_{an}(k+1)$, $v_{bn}(k+1)$, and $v_{cn}(k+1)$ are the phase voltages at the $k+1$ instant; $i_{dsfed}(k+1)$ and $i_{qsfed}(k+1)$ are the d-axis and q-axis currents based on the feedback current at the $k+1$ instant.

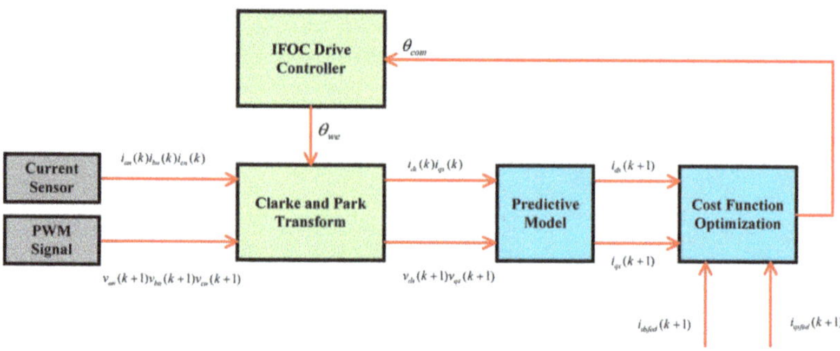

Figure 2. Control diagram of model-based predictive rotor field-oriented angle compensation.

The cost function is $g(k+1)$.

$$g(k+1) = p_1 g_1(k+1) + p_2 g_2(k+1) + \ldots + p_n g_n(k+1) \tag{15}$$

where $p_1, p_2 \ldots p_n$ are the weighting coefficients and $p_1 + p_2 + \ldots + p_n = 1$. The functions $g_1(k+1), g_2(k+1) \ldots g_n(k+1)$ are the different optimization objective cost functions. Here,

$$g_1(k+1) = i_{ds}^*(k+1) - i_{ds}(k+1) \tag{16}$$

$$g_2(k+1) = i_{qs}^*(k+1) - i_{qs}(k+1) \tag{17}$$

If the rotor field-oriented angle is accurate, then $g_1(k+1) = 0$ and $g_2(k+1) = 0$. Therefore, $g(k+1) = 0$. If the $g(k+1) \neq 0$, then the angle θ_{we} will be compensated by θ_{com}. The proportional integral (PI) regulator shown in Figure 3 is used to linearly tune the θ_{com}. Zero is used as the input of the PI regulator. It means that the rotor field-oriented angle error is none.

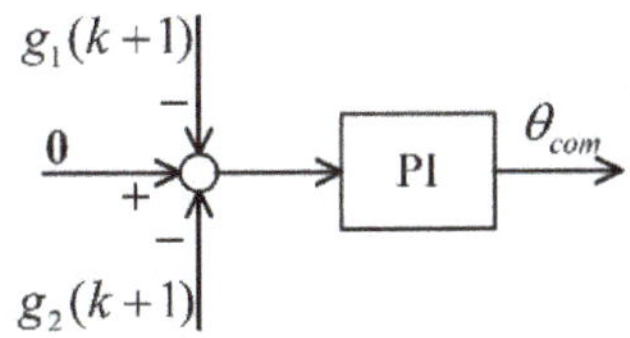

Figure 3. PI regulator used to tune the θ_{com}.

3.2. Implemented Algorithm

The currents in the IRFOC are always in a dynamic state. The current at the $k+1$ instant cannot be sensed at present. Therefore, an approach algorithm that takes advantage of the historic current and voltage is proposed.

The historic data of the current, voltage, and field-oriented angle, e.g., $i(k-n)$, $i(k-n+1) \ldots i(k-1), i(k)$, as shown in Figure 4, can be recorded in the random-access memory (RAM) of the DSP or MCU. According to (14), the current at the instant $k-n+1$ can be predicted with the current $i(k-n)$ and $v(k-n+1)$. That is

$$\begin{cases} i_{qs}^*(k-n+1) = \frac{C' \times A - D' \times E'}{A^2 + F' \times E'} \\ i_{ds}^*(k-n+1) = \frac{C' \times F' + D' \times A}{A^2 + F' \times E'} \end{cases} \tag{18}$$

where $C' = \frac{\delta L_s}{T_s} i_{qs}(k-n) + v_{qs}(k-n+1)$, $D' = \frac{L_s}{T_s} i_{ds}(k-n) + v_{ds}(k-n+1)$, $E' = \omega_e(k-n)L_s$, $F' = \omega_e(k-n)\delta L_s$, $i_{qs}^*(k-n+1)$ and $i_{ds}^*(k-n+1)$ are the predicted currents at the instant $k-n$.

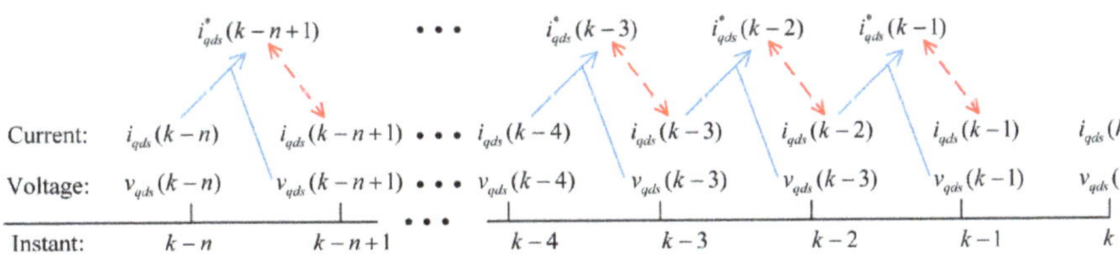

Figure 4. Proposed implemented algorithm diagram.

The accuracy of the predicted current at the next instant can be improved with the data at the previous instants. The current and voltage are always in a dynamic state because the period of the current and voltage sample is very short and the closed-loop control period is usually the same as the sample period. Therefore, the predicted current is not stable enough to tune θ_{com}. By using the saved data, a filter for the predicted current can be used to improve the stability of the predicted and feedback current as

$$i^*_{qds} = \frac{i^*_{qds}(k-4) + i^*_{qds}(k-3) + i^*_{qds}(k-2) + i^*_{qds}(k-1)}{4} \quad (19)$$

$$i_{qds} = \frac{i_{qds}(k-4) + i_{qds}(k-3) + i_{qds}(k-2) + i_{qds}(k-1)}{4} \quad (20)$$

where $i^*_{qds} = i^*_{qs} - ji^*_{ds}$ and $i_{qds} = i_{qs} - ji_{ds}$ are the predicted current and the actual feedback current, respectively. Then, according to Equations (16) and (17), the cost function can be calculated. Finally, θ_{com} can be tuned using the PI regulator in Figure 3.

4. Simulation Results

4.1. System Description

Simulations were performed in the below-based speed region and the field-weakening region based on the IRFOC with a speed sensor. The rotor field angle is usually calculated by (21) and (22) in IRFOC.

$$\theta_r = \int \omega_r dt + \int \omega_{sl} dt \quad (21)$$

$$\omega_{sl} = \frac{1}{t_r} \times \frac{i^*_{qs}}{i^*_{ds}} \quad (22)$$

where θ_r is the rotor field angle, ω_r is the actual electric angular velocity of the rotor, which can usually be obtained with a photoelectric encoder or rotating transformer, ω_{sl} is the slip angular velocity, t_r is the rotor time constant and $t_r = L_r/R_r$. R_r is the resistance of the rotor, i^*_{qs} is the torque current command, and i^*_{ds} is the magnetic current command. To reflect the inaccuracy of the rotor field orientation, the slip angular velocity is calculated with different rotor resistances, namely,$0.5R_r$, $0.8R_r$, and $1.5R_r$. The specifications of the simulated and experimental induction machines are shown in Table 1.

Table 1. Specification of the simulated and experimental induction machines.

Machine Type	3 ph IM	Stator leakage inductance	3.3 mH
Rated power	7.5 kW	Rotor leakage inductance	5.6 mH
Rated speed	1450 rpm	Magnetizing inductance	56.4 mH
Maximum speed	12,000 rpm	Inertia	0.029 kgm2
Rated frequency	50 Hz	Number of pole pairs	2
Rated torque at rated speed	48.8 Nm	Rated DC-line voltage	540 V
Stator resistance	0.374Ω	Rated rotor flux level	0.73 Wb
Rotor resistance	0.267Ω	Rated current	18.8 A

4.2. Simulated Results

Simulations were performed with the Saber simulator. Figure 5 shows the simulated results at 1200 rpm. In the IRFOC, the d-axis current, referred to as the magnetic current, is kept constant in the below-based speed region. According to Equation (5), the rotor flux should be kept constant. A step load is set at 2.0 s with 30 Nm and at 4.0 s with 60 Nm. Figure 5a–c depicts the simulated results when the rotor resistance R^*_r is used in the slip angular velocity calculation is set as $0.5R_r$, $0.8R_r$ and $1.5R_r$, respectively. The rotor field-oriented angle is compensated from 1.0 s. The rated flux level is set to 0.73 Wb based on the motor parameters. The flux level varies because of the inaccurate rotor field-oriented angle without compensation. This effect can lead to the degradation of dynamic and stable performance. After compensation, we observe that the flux intensity under the three simulation conditions could reach 0.73 Wb at 2.0 s. The flux level could be kept almost constant during the load step.

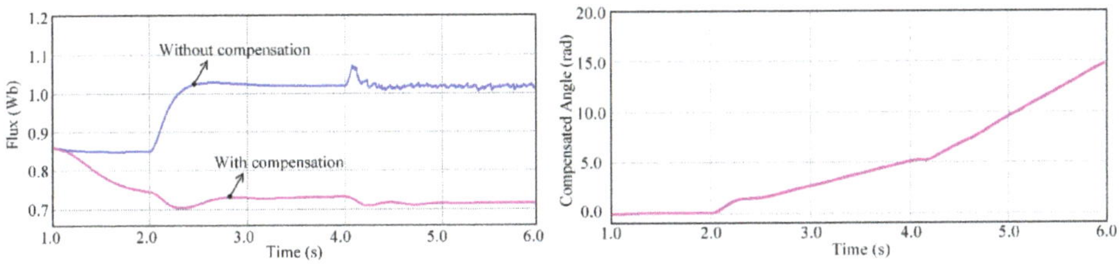

(a)

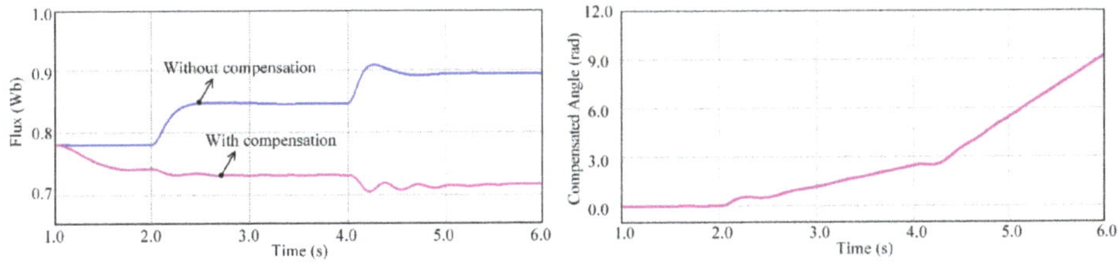

(b)

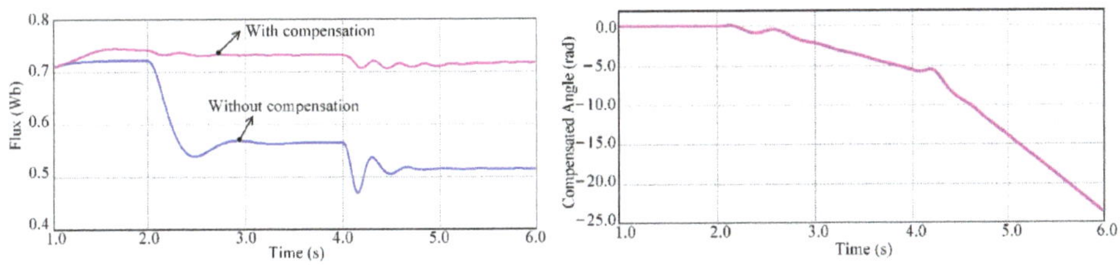

(c)

Figure 5. Flux and compensated angle at 1200 rpm with 30 Nm and 60 Nm step loads at 2.0 s and 4.0 s, respectively. **(a)** $R_r^* = 0.5R_r$; **(b)** $R_r^* = 0.8R_r$; **(c)** $R_r^* = 1.5R_r$.

According to Equation (18), the predictive model needs stator resistance and inductance. Inductance is almost constant. The variation in the stator resistance is neglected in the proposed algorithm. To verify that the neglect is accepted, a simulation was performed. In the simulation, the stator resistance was changed from 0.374 Ω to 0.748 Ω linearly, and the rotor resistance was set to $0.5R_r$. A step load is also set at 2.0 s with 30 Nm and at 4.0 s with 60 Nm. Compared with the simulated results in Figure 5a, the rotor flux and compensated angle indicate little difference, as shown in Figure 6. The flux level with angle error compensation, regardless of whether the stator resistance is changed, is much better than that without compensation. Although the maximum deviation of the compensated angle is almost 2.5 rad when the stator resistance changed to twice the nominal resistance, the deviation of the rotor flux level is only 0.03 Wb. In the application, this small difference of the rotor flux level could be neglected, and the algorithm proposed here is almost not affected by the variation in the stator resistance.

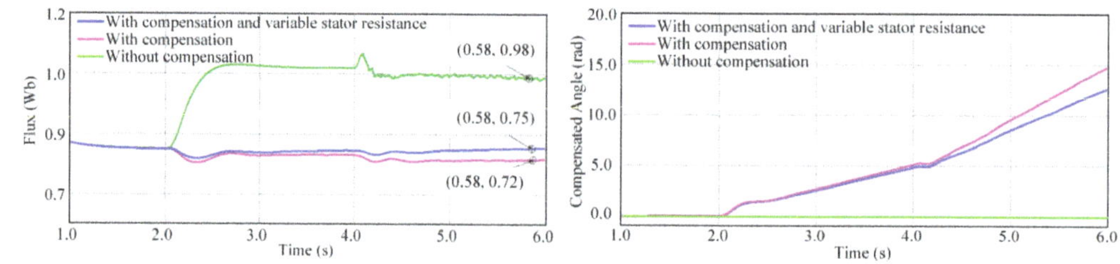

Figure 6. Rotor flux and compensated angle at 1200 rpm with the variable stator resistance.

In the IRFOC, the q-axis current reflects the developed torque when the motor runs at a constant speed. Therefore, the q-axis current should be proportional to the torque. That is, the q-axis current with a 60 Nm load should be twice the value with a 30 Nm load at 1200 rpm. The q-axis current with different loads is presented in Figure 7. After compensation, the q-axis current is changed from approximately 15 A to 30 A when the load rises from 30 Nm to 60 Nm regardless of the R_r^* set. However, with no compensation, the q-axis current is different from the same load and is not proportional to the torque. The simulated current is compared in Table 2.

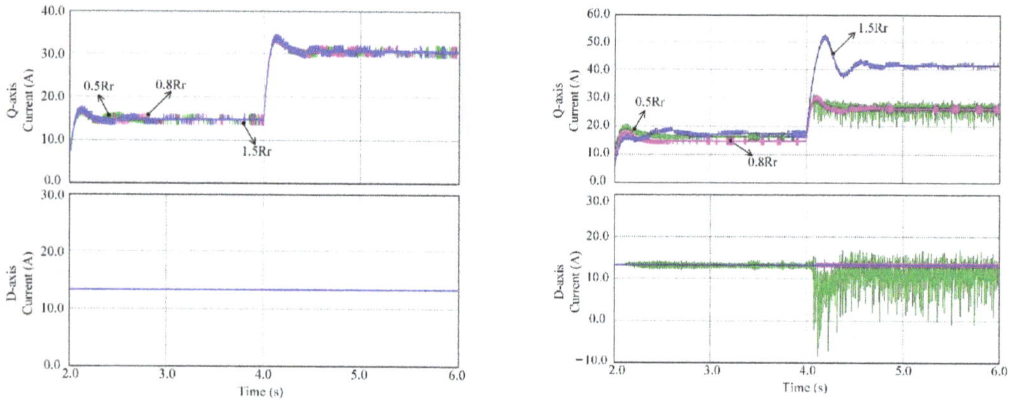

(**a**) (**b**)

Figure 7. The q-axis current comparative waveforms at 1200 rpm with a 30 Nm load at 2.0 s and 60 Nm load at 4.0 s. (**a**) Current with compensation; (**b**) Current without compensation.

Table 2. Q-axis current comparison.

	Without Compensation			With Compensation		
	R_r^*=0.5R_r	R_r^*=0.8R_r	R_r^*=1.5R_r	R_r^*=0.5R_r	R_r^*=0.8R_r	R_r^*=1.5R_r
i_{qs1}(A)	17.1	15.01	17.9	14.9	14.92	14.95
i_{qs2} (A)	26.1	26.5	42.2	30.1	30	30.15
ratio (i_{qs2}/i_{qs1})	1.52	1.77	2.35	2.02	2.01	2.02

In Table 2, R_r^* is the rotor resistance, which is used to calculate the slip angular velocity in the program; R_r is the actual resistance of the induction machine; i_{qs1} and i_{qs2} are the values of the q-axis current when the loads are 30 Nm and 60 Nm, respectively. If the rotor field orientation is accurate, then the ratio of i_{qs2} and i_{qs1} should be 2. It can be seen that the q-axis current is not proportional to the load when the rotor resistance is not the actual value. After compensation, the q-axis current is almost proportional to the load.

Figure 8 shows the simulation results at 3000 rpm. A step load at 20 Nm was set at 2.0 s. When $R_r^* = 0.5R_r$, the speed could no longer be kept at 3000 rpm without compensation. This is different from the simulation result at 1200 rpm shown in Figure 7 because the voltage is limited to the supply torque current in the field-weakening region if the rotor flux level is not sufficiently reduced, as shown in Figure 8b. The actual flux without compensation when $R_r^* = 0.5R_r$ was much higher than the normal level. Therefore, the voltage could not supply enough q-axis current when the speed was 3000 rpm, and then the speed was decreased. This finding means that an inaccurate field-oriented angle can affect the maximum output torque of induction machines. This results of q-axis current can also be seen in the following experimental results at 1200 rpm. Although the flux levels with compensation are not constant, the variable in Figure 8a is much smaller than that without compensation, as shown in Figure 8b.

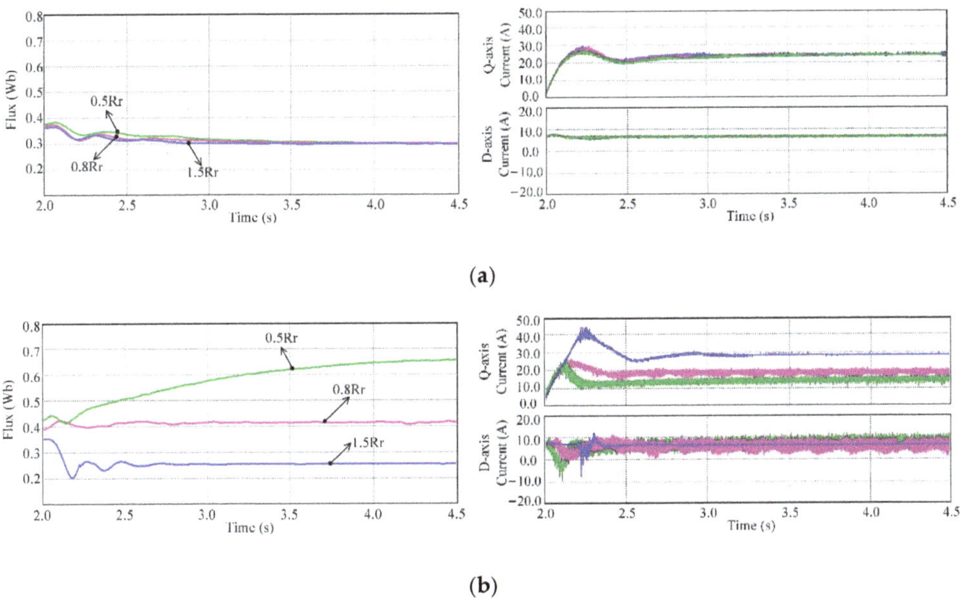

Figure 8. Rotor flux, q-axis current, and d-axis current comparative waveforms at 3000 rpm with a 20 Nm load at 2.0 s. (**a**) Rotor flux and d/q current with compensation; (**b**) Rotor flux and d/q current without compensation.

Figure 9 compares the waveform with and without the proposed compensation. In the simulation, the rotor resistance began to change from 0.268 Ω to 0.536 Ω linearly at the instant of 2.0 s during the following 2 s interval. Then, the rotor resistance was changed back to 0.268 Ω linearly at the instant of 4.0 s during the next 2 s interval. After compensation, the torque, rotor flux, and q-axis current are almost the same. There are some fluctuations in the flux, especially when the resistance was changed instantly. This change is mainly because the compensation of the proposed algorithm requires some time to realize. In practice, the rotor resistance cannot be changed so fast. The proposed algorithm has enough time to regulate the slip coefficient. The simulation at 3000 rpm yielded similar results, as shown in Figure 10.

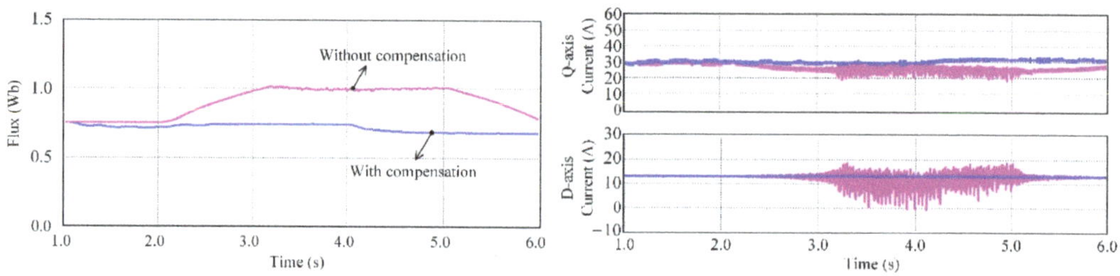

Figure 9. Rotor flux, d-axis, and q-axis current comparative waveforms at 1200 rpm with a 60 Nm load.

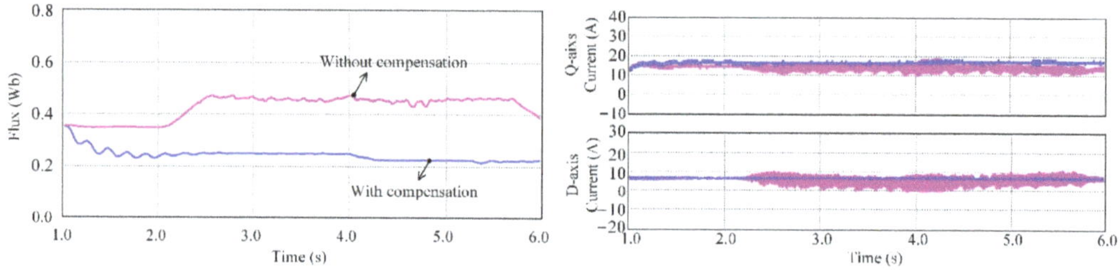

Figure 10. Rotor flux, d-axis, and q-axis current comparative waveforms at 3000 rpm with a 15 Nm load.

5. Experimental Results

Experiments were performed on the setup, as shown in Figure 11. A 7.5 kW spindle motor was used, and its parameters are shown in Table 1. The DC generator was used as a load in the high-speed region. The output torque can be obtained by the torque transducer, which was equipped between the spindle motor and the DC generator. A DSP TMS320F28377D (Texas Instruments, Texas, USA), which is a two core MCU was used to realize the proposed algorithm and IRFOC algorithm. A CPLD EPM240T100C5N (Altera, California, USA) was used to realize PWM. In the experiment, the PWM frequency was 10 kHz. The current sample and control frequency was 20 kHz. The phase currents were obtained by the two current sensors on the power converter board.

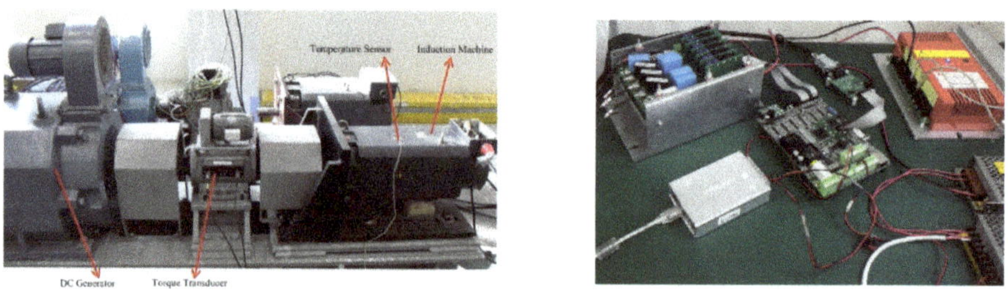

(a) (b)

Figure 11. Experiment setup. (**a**) is the induction machine and load. and (**b**) is the control and power board.

In the experiment, the maximum q-axis current was set to $3\sqrt{2}i_{rated}$. As the simulation shows, the q-axis current was not proportional to the load at the same flux level when the field-oriented angle is inaccurate. The q-axis current comparative experiment results are given in Figure 12 at 1200 rpm with different loads. The torque was applied by a DC generator. The q-axis current was calculated by the DSP and stored in RAM. Therefore, the q-axis current of the experiment was read by the DSP and transferred to a computer.

The experimental results are similar to the simulated results shown in Table 2. After compensating for the rotor flux oriented angle error, the q-axis current is proportional to the load torque and is almost the same regardless of which rotor resistance was used to calculate the slip angular velocity.

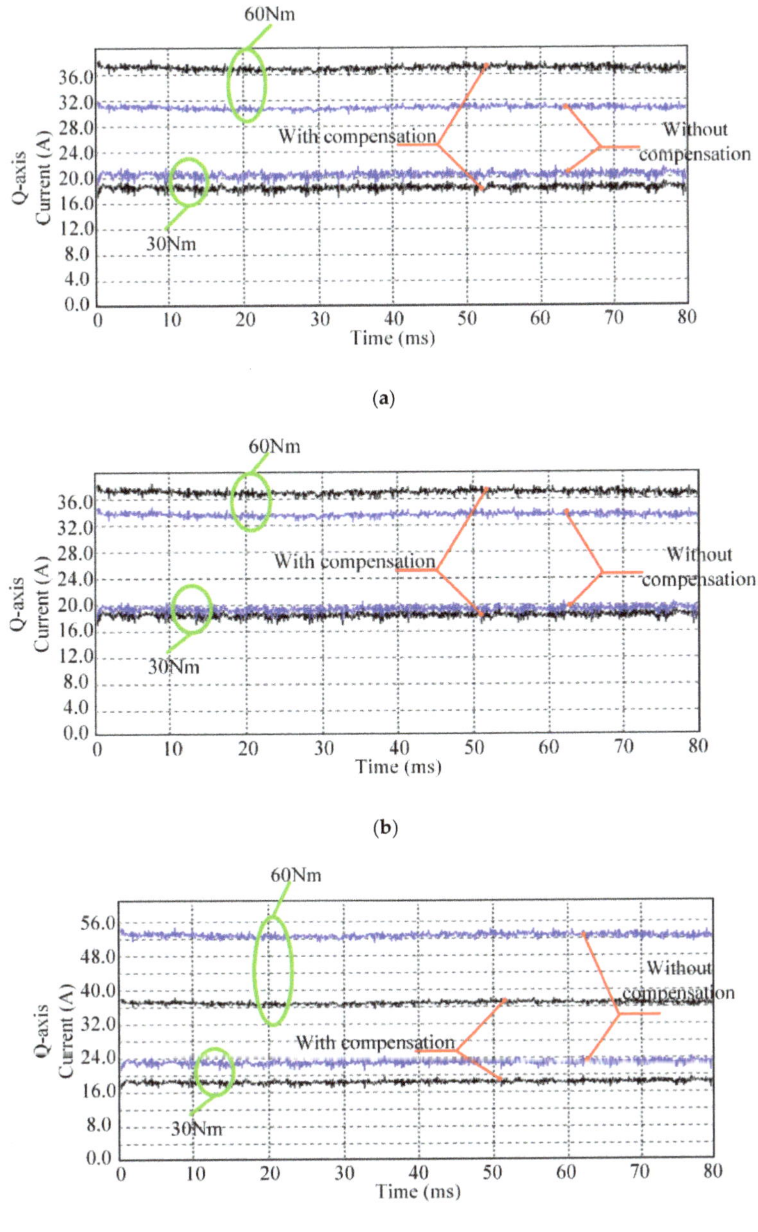

(a) (b) (c)

Figure 12. Q-axis current at 1200 rpm with 30 Nm, and 60 Nm loads. (**a**) $R_r^* = 0.5R_r$; (**b**) $R_r^* = 0.8R_r$; (**c**) $R_r^* = 1.5R_r$.

If the field-oriented angle is not accurate, the maximum torque will decrease, especially in the flux weakening region. The maximum output torque at different speeds was recorded, as shown in Figure 13. The maximum torque decreased with increasing speed when the rotor resistance used in the slip angular velocity calculation was not accurate. After compensation, the output torque was increased.

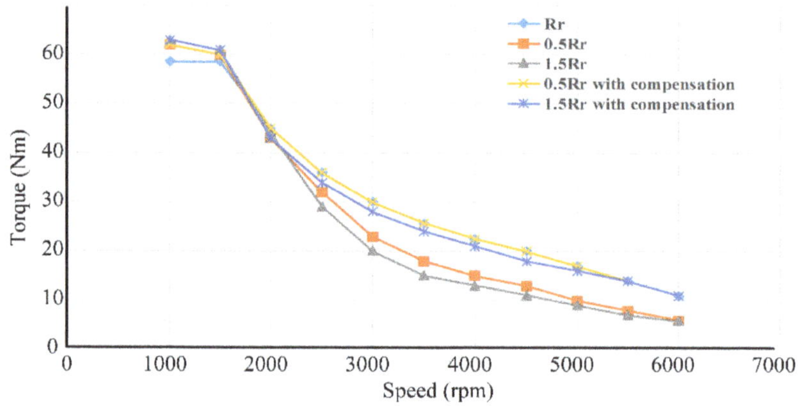

Figure 13. Maximum output torque from 1000 rpm to 6000 rpm.

6. Conclusions

This paper focuses on the correction of the field-oriented inaccuracy by a model predictive method in induction machine drives, including the base speed region and field-weakening region. The inaccuracy of the field-oriented control will lead to the actual flux variation, which could be larger or smaller than the reference value. Furthermore, the flux level variation will decrease the output torque and degrade the dynamic performance and current control regulator, especially in the field-weakening region. Therefore, a q-axis and d-axis current predictive method-based correction method for the field-oriented angle is proposed in this paper. To easily realize the proposed method in real-time, the data, such as the current, voltage and velocity used in MPC, are stored in the RAM of the DSP. The effectiveness of this approach is verified by simulations and experiments on a 7.5 kW induction machine setup.

Author Contributions: Y.L. is the corresponding author. J.Z. was in charge of the experiments. Q.Y. was in charge of the experimental and simulated data analysis. All authors have read and agreed to the published version of the manuscript.

Funding: This paper received no external funding.

Institutional Review Board Statement: Not applicable.

Data Availability Statement: Data available in a publicly accessible repository.

Conflicts of Interest: The authors declare no conflict of interest.

References

1. Lin, F.J.; Liao, Y.H.; Lin, J.R.; Lin, W.T. Interior Permanent Magnet Synchronous Motor Drive System with Machine Learning-Based Maximum Torque per Ampere and Flux-Weakening Control. *Energies* **2021**, *14*, 346. [CrossRef]
2. Wu, C.; Yang, J.; Li, Q. GPIO-Based Nonlinear Predictive Control for Flux-Weakening Current Control of the IPMSM Servo System. *Energies* **2020**, *13*, 1716. [CrossRef]
3. Zhou, K.; Ai, M.; Sun, D.; Jin, N.; Wu, X. Field Weakening Operation Control Strategies of PMSM Based on Feedback Linearization. *Energies* **2019**, *12*, 4526. [CrossRef]
4. Marques, G.D.; Iacchetti, M.F. Field-Weakening Control for Efficiency Optimization in a DFIG Connected to a DC-Link. *IEEE Trans. Ind. Electron.* **2016**, *63*, 3409–3419. [CrossRef]

5. Wang, K.; Yao, W.; Chen, B.; Shen, G.; Lee, K.; Lu, Z. Magnetizing Curve Identification for Induction Motors at Standstill Without Assumption of Analytical Curve Functions. *IEEE Trans. Ind. Electron.* **2014**, *62*, 2144–2155. [CrossRef]
6. Ruan, J.Y.; Wang, S.M. Magnetizing Curve Estimation of Induction Motors in Single-phaseMagnetization Mode Considering Differential Inductance Effect. *IEEE Trans. Power Electron.* **2015**, *31*, 497–506. [CrossRef]
7. Al-Jufout, S.A.; Al-Rousan, W.H.; Wang, C. Optimization of Induction Motor Equivalent Circuit Parameter Estimation Based on Manufacturer's Data. *Energies* **2018**, *11*, 1792. [CrossRef]
8. Alonge, F.; Cirrincione, M.; D'Ippolito, F.; Pucci, M.; Sferlazza, A. Parameter Identification of Linear Induction Motor Model in Extended Range of Operation by Means of Input-Output Data. *IEEE Trans. Ind. Appl.* **2013**, *50*, 959–972. [CrossRef]
9. Guimaraes, J.; Bernardes, J.; Hermeto, A.; Bortoni, E. Parameter determination of asynchronous machines from manufacturer data sheet. *IEEE Trans. Energy Convers.* **2014**, *29*, 689–697. [CrossRef]
10. Liu, Y.; Fang, J.; Tan, K.; Huang, B.; He, W. Sliding Mode Observer with Adaptive Parameter Estimation for Sensorless Control of IPMSM. *Energies* **2020**, *13*, 5991. [CrossRef]
11. Wang, K.; Chen, B.; Shen, G.; Yao, W.; Lee, K.; Lu, Z. Online Updating of Rotor Time Constant Based on Combined Voltage and Current Mode Flux Observer for Speed-Sensorless AC Drives. *IEEE Trans. Ind. Electron.* **2013**, *61*, 4583–4593. [CrossRef]
12. Bao, D.; Wu, H.; Wang, R.; Zhao, F.; Pan, X. Full-Order Sliding Mode Observer Based on Synchronous Frequency Tracking Filter for High-Speed Interior PMSM Sensorless Drives. *Energies* **2020**, *13*, 6511. [CrossRef]
13. Wang, C.; Cao, D. New Sensorless Speed Control of a Hybrid Stepper Motor Based on Fuzzy Sliding Mode Observer. *Energies* **2020**, *13*, 4939. [CrossRef]
14. Comanescu, M. Design and Implementation of a Highly Robust Sensorless Sliding Mode Observer for the Flux Magnitude of the Induction Motor. *IEEE Trans. Energy Convers.* **2016**, *31*, 649–657. [CrossRef]
15. Dominguez, J.R. Discrete-Time Modeling and Control of Induction Motors by Means of Variational Integrators and Sliding Modes—Part II: Control Design. *IEEE Trans. Ind. Electron.* **2015**, *62*, 6183–6193. [CrossRef]
16. Stojic, D.; Milinkovic, M.; Veinovic, S.; Klasnic, I. Improved Stator Flux Estimator for Speed Sensorless Induction Motor Drives. *IEEE Trans. Power Electron.* **2015**, *30*, 2363–2371. [CrossRef]
17. Yoon, Y.D.; Sul, S.K. Sensorless Control for Induction Machines Based on Square-Wave Voltage Injection. *IEEE Trans. Power Electron.* **2014**, *29*, 3637–3645. [CrossRef]
18. Smith, A.N.; Gadoue, S.M.; Finch, J.W. Improved Rotor Flux Estimation at Low Speeds for Torque MRAS-Based Sensorless Induction Motor Drives. *IEEE Trans. Energy Convers.* **2015**, *31*, 270–282. [CrossRef]
19. Vazquez, S.; Leon, J.I.; Franquelo, L.G.; Rodriguez, J.; Young, H.; Marquez, A.; Zanchetta, P. Model Predictive Control: A Review of Its Applications in Power Electronics. *IEEE Ind. Electron. Mag.* **2014**, *8*, 16–31. [CrossRef]
20. Yaramasu, V.; Rivera, M.; Wu, B.; Rodriguez, J. Model Predictive Current Control of Two-Level Four-Leg Inverters—Part I: Concept, Algorithm, and Simulation Analysis. *IEEE Trans. Power Electron.* **2012**, *28*, 3459–3468. [CrossRef]
21. Barrero, F.; Bermudez, M.; Duran, M.J.; Salas, P.; Gonzalez-Prieto, I. Assessment of a Universal Reconfiguration-less Control Approach in Open-Phase Fault Operation for Multiphase Drives. *Energies* **2019**, *12*, 4698. [CrossRef]
22. Guzman, H.; Duran, M.J.; Barrero, F.; Zarri, L.; Bogado, B.; Prieto, I.G.; Arahal, M.R. Comparative Study of Predictive and Resonant Controllers in Fault-Tolerant Five-Phase Induction Motor Drives. *IEEE Trans. Ind. Electron.* **2016**, *63*, 606–617. [CrossRef]
23. Druant, J.; Vyncke, T.; Melkebeek, J.A. Adding inverter fault detection to model-based predictive control for flying-capacitor inverters. *IEEE Trans. Ind. Electric.* **2015**, *62*, 2054–2063. [CrossRef]
24. Uebel, S.; Murgovski, N.; Baker, B.; Sjoberg, J. A Two-Level MPC for Energy Management Including Velocity Control of Hybrid Electric Vehicles. *IEEE Trans. Veh. Technol.* **2019**, *68*, 5494–5505. [CrossRef]
25. Buerger, J.; East, S.; Cannon, M. Fast Dual-Loop Nonlinear Receding Horizon Control for Energy Management in Hybrid Electric Vehicles. *IEEE Trans. Control. Syst. Technol.* **2018**, *27*, 1060–1070. [CrossRef]
26. Alcala, E.; Puig, V.; Quevedo, J.; Cayuela, V.P.; Casin, J.Q. TS-MPC for Autonomous Vehicles Including a TS-MHE-UIO Estimator. *IEEE Trans. Veh. Technol.* **2019**, *68*, 6403–6413. [CrossRef]
27. Liniger, A.; Lygeros, J. Real-Time Control for Autonomous Racing Based on Viability Theory. *IEEE Trans. Control. Syst. Technol.* **2019**, *27*, 464–478. [CrossRef]
28. Zhang, Y.; Yang, H. Generalized Two-Vector-Based Model-Predictive Torque Control of Induction Motor Drives. *IEEE Trans. Power Electron.* **2015**, *30*, 3818–3829. [CrossRef]
29. Zhang, Y.; Yang, H. Two-Vector-Based Model Predictive Torque Control Without Weighting Factors for Induction Motor Drives. *IEEE Trans. Power Electron.* **2016**, *31*, 1381–1390. [CrossRef]
30. Akpunar, A.; Iplikci, S. Runge-Kutta Model Predictive Speed Control for Permanent Magnet Synchronous Motors. *Energies* **2020**, *13*, 1216. [CrossRef]
31. Bao, G.; Qi, W.; He, T. Direct Torque Control of PMSM with Modified Finite Set Model Predictive Control. *Energies* **2020**, *13*, 234. [CrossRef]
32. Richter, J.; Doppelbauer, M. Predictive Trajectory Control of Permanent-Magnet Synchronous Machines With Nonlinear Magnetics. *IEEE Trans. Ind. Electron.* **2016**, *63*, 3915–3924. [CrossRef]

energies

Article

Accounting for Slot Harmonics and Nonsinusoidal Unbalanced Voltage Supply in High-Speed Solid-Rotor Induction Motor Using Complex Multi-Harmonic Finite Element Analysis

Tomasz Garbiec * and Mariusz Jagiela

Faculty of Electrical Engineering, Automatic Control and Informatics, Opole University of Technology, ul. Proszkowska 76, 45-758 Opole, Poland; m.jagiela@po.edu.pl
* Correspondence: t.garbiec@po.edu.pl

Abstract: Solid rotor induction machines are still used in high-speed systems. A two-dimensional field-circuit model based on the finite element method and the complex magnetic vector potential has been shown as a very time-effective tool in the analysis of their steady states compared to time-domain models. This continuation work presents a validated computational algorithm that enables the inclusion of the nonsinusoidal and/or asymmetrical voltage supply in the multi-harmonic field-circuit model of these machines that was presented in the previous works by the authors. The extended model accounts for both spatial harmonics due to slotting and/or winding distribution and the time-harmonics due to voltage waveform. The applicability range of the model therefore increases to cases when the machine is supplied with a nonsinusoidal three-phase system of voltages with symmetry or asymmetry that can be decomposed into three symmetrical components. Its short execution time characteristic allows for much more insightful design studies of the contribution of voltage supply- and slotting-related harmonics to the overall efficiency of the machine than is possible with the time-consuming time-domain models. The proposed computational framework has never been presented in the literature. The model is verified positively by the comprehensive time-domain model. It is especially useful in design studies on solid rotor induction motors related to the optimisation of the efficiency of induction motor-based drive systems.

Keywords: induction motor; solid rotor; effective parameters; finite element method

Citation: Garbiec, T.; Jagiela, M. Accounting for Slot Harmonics and Nonsinusoidal Unbalanced Voltage Supply in High-Speed Solid-Rotor Induction Motor Using Complex Multi-Harmonic Finite Element Analysis. *Energies* **2021**, *14*, 5404. https://doi.org/10.3390/en14175404

Academic Editor: Armando Pires

Received: 19 July 2021
Accepted: 27 August 2021
Published: 30 August 2021

1. Introduction

Despite the growing popularity of permanent magnet electric machines, induction motors still remain the main components of industrial electric drives. This is due to the simplicity of construction, reliability and the development of static converters that allow the shaping of static characteristics and dynamic parameters of the drive system. Among the various designs, the induction machine with solid rotor deserves special attention, being one of the simplest and oldest AC electric machines, although requiring entirely different design routines that have been in development for nearly a century [1–5]. The mechanical properties of the solid rotor, as well as their resistance to aggressive chemical compounds, are incomparable to any other structure, which makes them a suitable source of drive-in applications requiring high or very high rotational speed such as those shown in [6–13].

The analysis of the properties and design of the discussed machine is still considered a difficult task, mainly because of the very complex electromagnetic phenomena occurring in the solid rotor. The only method that allows for the overall effect of these phenomena to be taken into account is to utilize a non-linear three-dimensional numerical model formulated in the time domain. Unfortunately, as the calculations would take a long time, this method is difficult to apply in practice. For this reason, alternative approaches have been developed which, in principle, can be divided into two groups: methods based on the analytical solution of the field problem in the solid rotor presented in [14–18], and methods utilizing the two-dimensional numerical models formulated as time-domain or frequency-domain models used in works [7,9–11,19–21].

The basic problem related to the analysis of high-speed induction machines with solid rotors using two-dimensional numerical models formulated in the time domain is a very long computation time resulting from a significant ratio of the electromagnetic time constant of the machine to the supply voltage period. In the case where the subjects of consideration are the steady states of the machine, the computation time can be shortened by applying special modifications of the model formulated in the time-domain, such as: elimination of the DC flux linkage in [22], elimination of the DC current at each step in [23], transient magnetic solution initialized with proper initial values [24] or initial current extraction in [19].

A very promising alternative to the models formulated in the time domain is the use of modified time-harmonic models, such as, for example, the multi-harmonic field model with a strong coupling, which takes into account the influence of the spatial harmonics due to the slotting or winding distribution on the losses in the solid rotor and the value of the developed electromagnetic torque [25–31]. In these models these effects can be accounted for using the modified slip

$$s' = 1 - \nu(1 - s) \tag{1}$$

where ν is the ordinal number of spatial harmonic ($\nu \in \mathbb{Z}$) and s is the fundamental slip.

As shown in the works [30,31], with the appropriate formulation of the stator–rotor coupling scheme and the method of modelling materials non-linearity, it is possible to obtain high accuracy of the results of steady-state calculations in a very short time of model execution. This is because relatively coarse mesh densities are sufficient for the representation of the most important air gap magnetic field harmonics.

In this moment, the application of this type of model was limited to the case of the symmetrical sinusoidal voltage supply of the stator winding. In the present work, the concept of multi-harmonic effective magnetic permeability proposed in [31] was used to develop an original approach that allows for taking into account the nonsinusoidal voltage supply waveforms (that appear, for example, when supplying the machine through a quasi-square voltage inverter) in a multi-harmonic model of a solid rotor induction machine. It is shown that the steady-state time-harmonic model with only a few dominating time-harmonics of the magnetic field attributed to the voltage supply provides solutions very similar to ones obtained from a time-demanding time-domain solution. This is in terms of the magnitude and phase relationships of the waveforms as well as the computed torques and power losses. The execution time of the former is however only a fraction of that of the latter. Moreover, the model is demonstrated to work correctly in cases where the voltage supply exposes noticeable asymmetry that appears during inverter fault conditions. The above means that in the mathematical sense the model is capable of accounting for the space-harmonics and the time-harmonics that come from the power supply, but not from the motional effects which are more important in the squirrel cage machines. It is thus most suitable to analyze induction motors with solid rotors. To the best of the authors' knowledge such a model has not been proposed until this moment.

The basic operating characteristics of the machine were calculated for four different supply voltage waveforms, and then compared with the results of calculations obtained via the comprehensive model formulated in the time domain which in the steady-state condition considers the whole spectrum of magnetic field harmonics. Original contributions of this work include:

- development, implementation and validation of a cost-effective computational algorithm for nonlinear steady-state analysis of solid-rotor induction motors with an accuracy characteristic comparable to that of the comprehensive time-domain model
- determination of the torque and power dissipation in the rotor as sets of separated components associated with the time- and space-harmonics of the magnetic field

allowing for a more detailed analysis of the impact of stator slotting and power supply harmonics on machine efficiency.

2. Mathematical Model

2.1. Analysed Machine

The subject of research presented in the present work is a high-speed machine with a homogeneous solid rotor, already analysed in the previous works by the authors [30,31]. The basic data of the converter and the materials' magnetization characteristics of the stator and rotor area adopted for the calculations are shown in Table 1 and Figures 1 and 2.

Table 1. Basic specifications of tested induction motor [31].

Parameter	Value
Nominal power	125 W
Operation frequency (ω)	2000π rad/s
Number of pole pairs (p)	2
RMS phase voltage	50 V
Phase resistance	0.62 Ω
End-winding leakage inductance	98 μH
Rotor conductivity	5.2 MS/m
Machine length (l_z)	32 mm
Number of stator slots	24

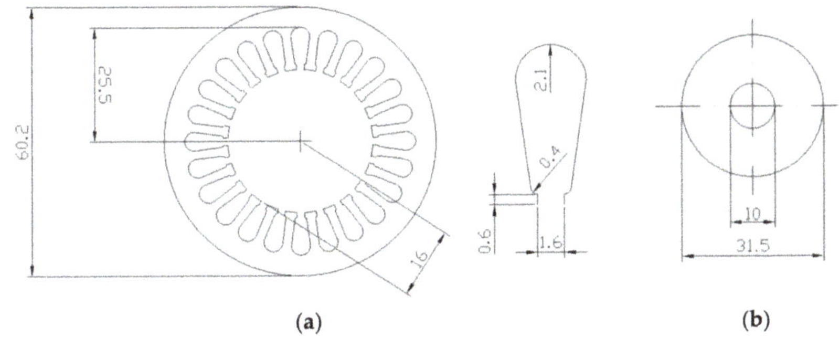

Figure 1. Solid rotor induction machine taking into consideration: (**a**) dimensions of the stator package (cross-section area), (**b**) dimensions of the rotor [31].

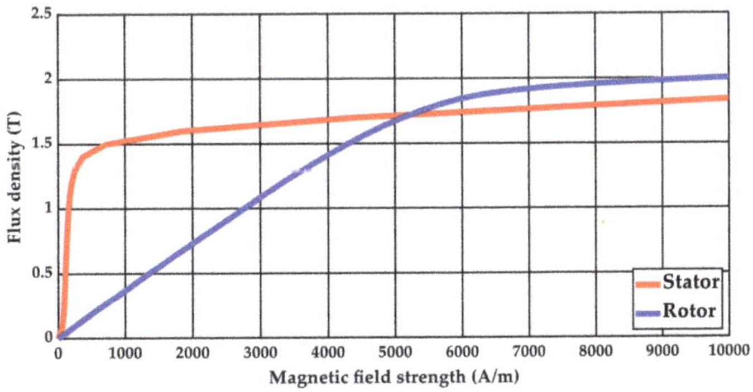

Figure 2. DC magnetization curves used in computations.

For the sake of simplicity, the influence of the rotor end-effects and losses in the stator package, and those due to the hysteresis in the rotor area, have been neglected. The influence of the rotor end-effects can be taken into account by applying the so-called rotor end-effect coefficient, that can be calculated analytically or numerically as shown in [21]. Taking into account the above losses, however, will require separate research, which is the authors' aim for the future. It should be noticed that the above assumptions are necessary from the point of view of the possibility to validate the proposed approach by the comprehensive time-domain model. It is also worth noticing that due to these assumptions the obtained results will be very hard or even impossible to verify on the experimental test-stand.

2.2. Reference Time-Stepping Model

All the results of the calculations presented in this work using the proposed approach were compared with the results of calculations using the complete model formulated in the time domain. The equations for such a model after discretisation via the Galerkin procedure and the implicit Euler method take the form of [31]:

$$\begin{bmatrix} \mathbf{S}(\mu_{DC}) + \mathbf{G}\Delta t^{-1} & -\mathbf{D}^{\mathrm{T}}\mathbf{K}^{\mathrm{T}} \\ l_z\mathbf{K}\mathbf{D}\Delta t^{-1} & \mathbf{K}(\mathbf{R} + \mathbf{L}\Delta t^{-1})\mathbf{K}^{\mathrm{T}} \end{bmatrix}^{n} \begin{bmatrix} \boldsymbol{\varphi} \\ \mathbf{i_S} \end{bmatrix}^{n} = \begin{bmatrix} \mathbf{G}\Delta t^{-1} & \mathbf{0} \\ l_z\mathbf{K}\mathbf{D}\Delta t^{-1} & \mathbf{K}\mathbf{L}\Delta t^{-1}\mathbf{K}^{\mathrm{T}} \end{bmatrix}^{n-1} \begin{bmatrix} \boldsymbol{\varphi} \\ \mathbf{i_S} \end{bmatrix}^{n-1} + \\ + \begin{bmatrix} \mathbf{0} \\ \mathbf{K}\mathbf{e_S} \end{bmatrix}^{n-1}, \quad (2)$$

where: $\mathbf{S}$—reluctivity matrix, $\mathbf{G}$—conductivity matrix, $\mathbf{D}$—matrix describing the winding, $\mathbf{K}$—matrix describing the winding connection method, $\mathbf{R}$—winding resistance matrix, $\mathbf{L}$—winding leakage inductivity matrix, Δt—time-integration step, $\boldsymbol{\varphi}$—vector of nodal values of the vector magnetic potential, $\mathbf{i_S}$—vector of instantaneous values of the stator loop currents, $\mathbf{e_S}$—vector of the instantaneous supply voltages in the stator winding, μ_{DC}—DC magnetic permeability. The rotational movement was modelled using a simple and reliable moving band technique which is presented in detail in [32,33].

2.3. Idea of the Polyharmonic Field-Circuit Model Accounting for Nonsinusoidal Supply

The analysis of the influence of higher harmonics on the operation of the squirrel cage induction motor was the subject of very intensive research that has been carried out by various authors over the last few decades [34–38]. In the context of high-speed machines with solid rotors, this issue is of particular importance due to the significant influence of the magnetic field higher harmonics on the machine operation, both resulting from the power supply and core slotting. As shown in previous works by the authors [30,31], the influence of the permeance (slot) harmonics of the magnetic field can be successfully taken into account by applying a non-linear multi-harmonic field-circuit model. A detailed analysis performed in those works proved that through the appropriate formulation of the so-called effective magnetic permeability, it is possible to use a superposition of several field rotor models associated with the appropriate harmonics of the magnetic field distribution in the air gap of the machine that are strongly coupled with the stator model. The results of this analysis prompted the authors to extend their deliberation over the case of the nonsinusoidal power supply using the same main concepts of the model. As an extension of the models presented in [30,31], the construction of a non-linear multi-harmonic model of an induction machine with a solid rotor, taking into account both the higher harmonics of the voltage supply and higher permeance harmonics, can be presented in the following form:

(I) Perform the fast Fourier transform (FFT) analysis for the adopted nonsinusoidal symmetric supply waveforms. Extract the amplitudes E_{hn} and phase angles ψ_{hn} for N the most significant harmonics of the supply voltage of order $\{h1, h2, \ldots, hN\}$.

(II) Discretise the model calculation area using standard first-order triangular elements. Re-number the mesh elements to separate the grids associated with the stator and rotor areas.

(III) Set the null magnetic field strength in all ferromagnetic areas.

(IV) Calculate the effective magnetic permeability distribution μ_{eff} for the stator and rotor core using the DC magnetization characteristics and the formula [31]:

$$\mu_{eff}(H_{h1}, H_{h2}, \ldots, H_{hN}) = \frac{\sqrt{B_{h1}^2 + B_{h2}^2 + \ldots + B_{hN}^2}}{\sqrt{H_{h1}^2 + H_{h2}^2 + \ldots + H_{hN}^2}}, \quad (3)$$

where: H_{hn}—amplitude of the magnetic field strength related to hn supply voltage harmonic, B_{hn}—magnetic flux density amplitude related to hn supply voltage harmonic:

$$B_{hn} = \frac{2}{\pi}\int_0^{\pi} \mu_{DC}(H_{h1}\sin h1\alpha + H_{h2}\sin h2\alpha + \ldots + H_{h1}\sin hN\alpha)(H_{h1}\sin h1\alpha + H_{h2}\sin h2\alpha + \ldots + H_{hN}\sin hN\alpha)\sin hn\alpha d\alpha \quad (4)$$

(V) Create and solve N of independent multi-harmonic linear field-circuit models, each including M of spatial harmonics of the magnetic field strength [30,31]:

$$\begin{bmatrix} \mathbf{M}_{11}\left(\mu_{eff}\right) & \mathbf{M}_{12} & \mathbf{M}_{13} \\ \mathbf{M}_{21} & \mathbf{M}_{22} & 0 \\ \mathbf{M}_{31} & 0 & 0 \end{bmatrix} \begin{bmatrix} \underline{\boldsymbol{\varphi}}_{hn} \\ \underline{\mathbf{I}}_{Shn} \\ \underline{\boldsymbol{\lambda}}_{hn} \end{bmatrix} = \begin{bmatrix} 0 \\ \underline{\mathbf{E}}_{Shn} \\ 0 \end{bmatrix}, \quad (5)$$

where: $\mathbf{M}_{11}$—matrix describing the magnetic and electrical properties of materials, $\mathbf{M}_{12} = -\mathbf{M}_{21}^{\mathrm{T}}/(\mathrm{j}\omega l_z)$ —matrix describing the distribution and connection method of the stator winding, $\mathbf{M}_{13} = \mathbf{M}_{31}^{\mathrm{T}}$—matrices describing coupling between the rotor models and the stator model [23], $\mathbf{M}_{22}$—stator winding impedance matrix, $\underline{\boldsymbol{\varphi}}_{hn}$ —vector of the nodal values of the complex magnetic vector potential for the model associated with hn harmonic of the supplying voltage, $\underline{\mathbf{I}}_{Shn}$ —vector of the amplitudes of the loop currents in the stator winding due to hn harmonic of the supplying voltage, $\underline{\boldsymbol{\lambda}}_{hn}$ —vector of complex circulations of the magnetic field strength vector for the model associated with hn harmonic of the supplying voltage, $\underline{\mathbf{E}}_{Shn}$ —vector of the complex voltage amplitudes in the loops in the stator winding circuit associated with hn harmonic of the supplying voltage.

(VI) Based on the calculated magnetic field distributions, $\underline{\boldsymbol{\varphi}}_{hn}$ calculate new values of the magnetic field strength H_{hn}. Update the effective magnetic permeability distribution according to (3) and the matrix parameters $\mathbf{M}_{11}$ in (5).

(VII) Repeat steps IV–VI until convergence criterion based on relative change of the norm (equal to 0.1%) of the solution vector in (5) is reached.

The above procedure requires some additional comments. Firstly, it is assumed here that the nonlinearity of the stator winding currents must be considered as a superposition of nonlinear effects on the individual harmonic waveforms multiplied or divided by a time-invariant function (impedance related with magnetic permeability dependent only on magnitudes of magnetic quantities). The above means that the proposed method does not account for the saturation harmonics of magnetic flux and thus of current. In a solid rotor induction motor the saturation harmonics are, however, not significant due to the large value of inductance of the stator winding, and as shown in [30,31], the calculation results obtained in this way are very close to the ones coming from comprehensive time-domain computations.

It is still assumed that to determine the effective magnetic permeability, the higher harmonics of the magnetic field strength are to be used, and not the higher harmonics of flux density. Due to the fact that each model associated with a given voltage harmonic is a multi-harmonic model that includes M of spatial field harmonics, it is assumed that the magnetic field strength calculated in step (VI) is the RMS value of the magnetic field strength from M considered spatial harmonics, multiplied by $\sqrt{2}$. In addition, each model related to the corresponding supply voltage harmonic must be formulated for the corresponding slip reference value related to the given voltage harmonic and the symmetrical voltage system it creates (positive, negative or zero). Theoretically, the proposed approach allows for taking into account any number of higher harmonics, because individual models are solved independently and their coupling is established only through their common value of the effective magnetic permeability, calculated after solving all particular models. Thus, the solution can be executed simultaneously on distributed or parallel systems. In [31], the effective magnetic permeability was determined as a multi-dimensional look-up table. This type of approach provides accurate results with a small (less than or equal to three) number of harmonics included. For a larger number of considered harmonics, in order to avoid large sizes of data files or interpolation errors, it is necessary to elaborate more suitable functions that calculate the value of effective magnetic permeability with the use of numerical integration procedures.

3. Calculation Results

The above approach was utilised to calculate the basic operating characteristics (phase current RMS and electromagnetic torque) of the machine described in Section 2.1 when supplied by three different waveforms presented in Figure 3 accompanied by their total harmonic distortion (THD) coefficients. The root mean square (RMS) value of the stator phase current I and the electromagnetic torque T_e are determined on the basis of the following formula:

$$I = \sqrt{\sum\nolimits_{h1}^{hN} I_{hn}^2}, \tag{6}$$

$$T_e = \sum_{h1}^{hN} T_{ehn} \tag{7}$$

where I_{hn} and T_{ehn} are, respectively, the RMS value and the electromagnetic torque, obtained as a result of solving the model related to the hn harmonic of the supply voltage.

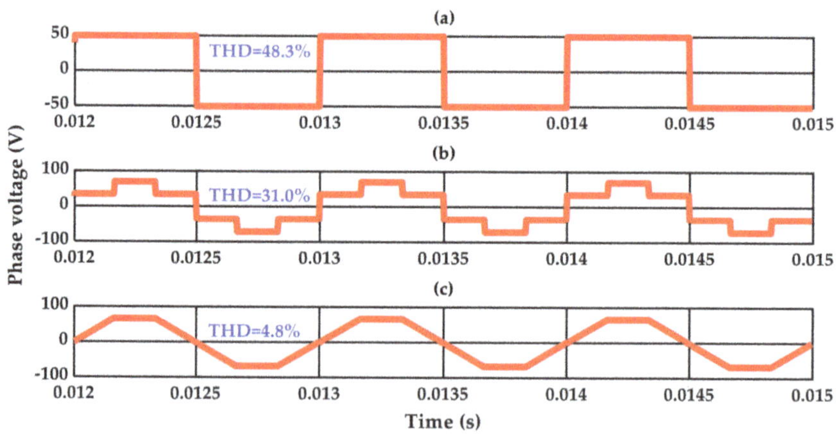

Figure 3. Three different types of supply waveforms considered in computations: square (**a**), six-step (**b**) and trapezoidal (**c**).

The calculated characteristics were compared with the results obtained with the time-domain model and by adopting only one fundamental voltage harmonic with the RMS value equal to the RMS value of the original supply waveform (note that all the waveforms shown have the same 50 V RMS value). All the considered models were implemented by the authors in the Matlab scripting language (Mathworks, Natick, MA, USA) [38]. To discretise the computational area an open-source generator (GMSH) was used [39]. In each case, the five most significant harmonics of the supply waveforms and two main slot harmonics were used (−11, +13). Due to the star-connected winding without neutral wire, harmonics that are multiplicities of the third harmonic were not taken into account. As a result, in addition to the fundamental harmonics, harmonics 5, 7, 11 and 13 were also included. The voltage harmonics 1, 7 and 13 formed positive sequence of voltages, whereas the 5 and 11 formed a negative sequence of voltages. The obtained results of the calculations are shown in Figure 4.

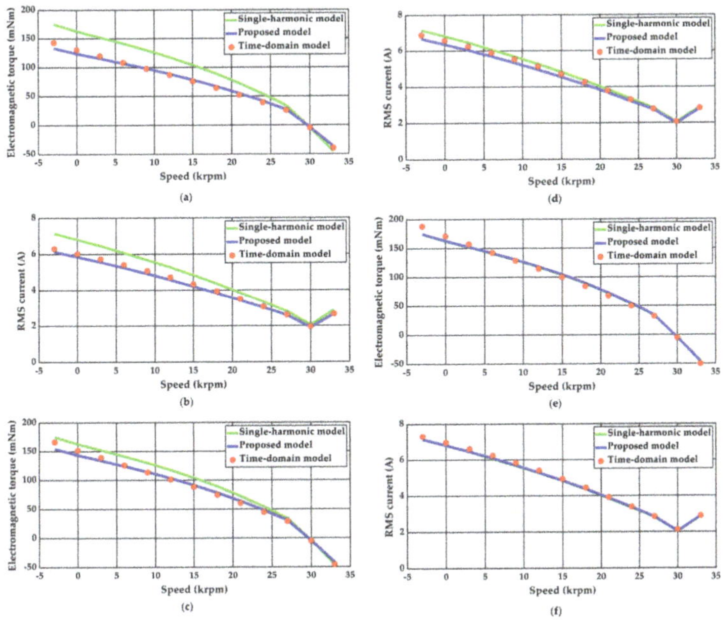

Figure 4. Results of computations: electromagnetic torque (**a**) and RMS stator current (**b**) for square supply waveforms (see Figure 3a); electromagnetic torque (**c**) and RMS stator current (**d**) for six-step supply waveforms (see Figure 3b); electromagnetic torque (**e**) and RMS stator current (**f**) for trapezoidal supply waveforms (see Figure 3c). Results obtained using the single-harmonic model and the proposed model are practically the same.

When analysing the results of the calculations one can notice a very high consistency between the characteristics calculated using the proposed approach in relation to the results of the calculations using the standard model formulated in the time domain. At the same time, one can see that by providing equality of the real and modelled waveforms the use of the RMS value is sufficient when the supply waveform distortion is small (THD < 10%). The developed methodology is also characterized by a relatively short calculation time. When the models related to individual harmonics are solved sequentially using a standard PC (Intel Core i7-5820K CPU @ 3.30 GHz, 16 GB RAM), the solution time for the speed 24 krpm with the square waveform supply is 2.5 min. For the sake of comparison, the time-domain model solution time is 4 h 16 min.

In addition to determining the static characteristics of the analysed machine, the developed method also allows for the recovering of steady-state current waveforms and making an assessment, for example, of electromagnetic torque resulting from the interaction of individual harmonics, both related to the supply waveforms and the magnetic circuit grooves.

The steady-state current waveform can be determined by solving individual component models related to the individual harmonics of the supply waveforms (see Figure 5) as

$$i(t) = \sum_{h1}^{hN} \sqrt{2}|\underline{I}_{hn}|\cos(hn\omega + \psi_{hn}) \tag{8}$$

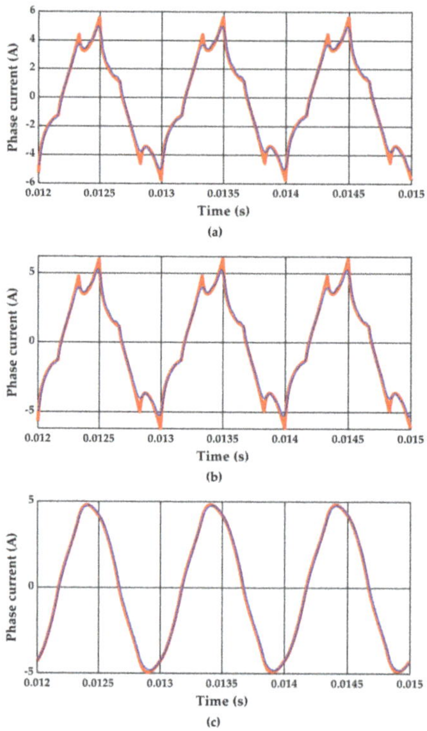

Figure 5. Comparison of the phase current waveforms calculated with the use of the time-domain model (red line) with the waveforms calculated with the use of the developed method (blue line) for a rotational speed of 24,000 rpm: square supply waveform (**a**), six-step supply waveform (**b**), trapezoidal supply waveform (**c**).

Since the electromagnetic torque is calculated as the sum of the components coming from individual slot harmonics (within the model related to the considered harmonic of the supply wave), it is possible to perform a detailed analysis of the influence of these harmonics on its value and sign. An example of such an analysis for synchronous speed with a square wave is presented in Table 2. Because the torque components correspond with different harmonic slips, these are not directly proportional to power dissipation in the rotor. The results of the rotor harmonic power dissipation computation are presented in Table 3. These clearly show that the disadvantageous effect of distorted voltage wave (first row in Table 3) is comparable to the disadvantageous effect of slotting (first column in Table 3). These results best demonstrate the machine design areas where the proposed modelling framework can be especially useful.

Table 2. The results of the electromagnetic torque computations for the synchronous speed (30,000 rpm) in mNm produced by the interaction of the harmonics of the magnetic field in the gap (n—time harmonic number of voltage supply, m—spatial harmonic ordinal number). Value of total electromagnetic torque obtained from the time-domain analysis is −4.256 mNm.

m \ n	+1	−5	+7	−11	+13	
+1	0(S)	−0.249(B)	0.065(M)	−0.013(B)	0.007(M)	Σ = −0.190
−11	−2.714(B)	−0.008(G)	−0.004(B)	0(S)	−0.005(B)	Σ = −2.731
+13	−0.560(G)	−0.002(B)	−0.003(G)	−0.002(B)	0(S)	Σ = −0.567
	Σ = −3.274	Σ = −0.259	Σ = 0.058	Σ = −0.015	Σ = 0.002	**Σ = −3.488**

(S): rotor in sync, (B): braking mode operation, (G): generating mode operation, (M): motoring mode operation, (+) means positive phase sequence, (−) negative phase sequence relative to the stator.

Table 3. The results of the rotor power dissipation computations for the synchronous speed (30,000 rpm) in Watt (n—time harmonic number of voltage supply, m—spatial harmonic ordinal number). Value of total rotor power dissipation obtained from the time-domain analysis is 19.97 W.

m \ n	+1	−5	+7	−11	+13	
+1	0(S)	4.467(B)	1.171(M)	0.455(B)	0.234(M)	Σ = 6.327
−11	8.520(B)	0.013(G)	0.017(B)	0(S)	0.003(B)	Σ = 8.553
+13	1.488(G)	0.008(B)	0.001(G)	0.001(B)	0(S)	Σ = 1.498
	Σ = 10.008	Σ = 4.488	Σ = 1.189	Σ = 0.456	Σ = 0.237	**Σ = 16.378**

(S): rotor in sync, (B): braking mode operation, (G): generating mode operation, green: motoring mode operation, (+) means positive phase sequence, (−) negative phase sequence relative to the stator.

The above analysis was carried out at synchronous speed because only in such a case can the computed value of loss torque be compared with predictions obtained from a comprehensive time-domain model. It should be, however, noticed that the proposed model can be used at any speed allowing for a more detailed investigation on power dissipation in the rotor.

4. Including the Unbalance of Nonsinusoidal Voltage Waveforms

The developed approach can be generalized to cases where asymmetry is observed in voltage supply waveforms, including the nonsinusoidal ones. The idea of the presented method is based on an independent solution of coupled sub-models by the application of single effective magnetic permeability distribution. As shown above, despite the nonlinearity of the problem, the use of the superposition principle allows one to obtain very accurate results. Therefore, in the case of supply unbalance, the method of symmetrical components can be used, allowing the presentation of the three-phase asymmetric distribution of a given harmonic of the supply waveform as a superposition of three symmetrical systems, namely the zero, positive and negative phase sequence system. To include the unbalance of the voltage supply, the model creation algorithm presented in the SubSection 2 should be modified as follows:

(I) Perform FFT analysis for the adopted non-linear asymmetrical supply waveforms $\{e_A(t)\ , e_B(t), e_C(t)\}$. Extract the amplitudes $\{E_{Ahn}, E_{Bhn}, E_{Chn}\}$ and phase angles $\{\psi_{Ahn}, \psi_{Bhn}, \psi_{Chn}\}$ for N of the most significant harmonics of the supply voltage with orders $\{h1, h2, \ldots, hN\}$.

(II) Determine the amplitudes of three-phase symmetrical systems of zero, positive and negative sequences, respectively $\{\underline{E}_{0hn}, \underline{E}_{1hn}, \underline{E}_{2hn}\}$ for each most significant harmonics of the supply voltage $\{\underline{E}_{Ahn}, \underline{E}_{Bhn}, \underline{E}_{Chn}\}$:

$$\begin{bmatrix} \underline{E}_{0hn} \\ \underline{E}_{1hn} \\ \underline{E}_{2hn} \end{bmatrix} = \frac{1}{3} \begin{bmatrix} 1 & 1 & 1 \\ 1 & a & a^2 \\ 1 & a^2 & a \end{bmatrix} \begin{bmatrix} \underline{E}_{Ahn} \\ \underline{E}_{Bhn} \\ \underline{E}_{Chn} \end{bmatrix} \tag{9}$$

where $a = e^{j\frac{2}{3}\pi}$.

(III) Discretise the model calculation area using standard first-order triangular elements. Re-number the mesh elements to separate the grids associated with the stator and rotor areas.

(IV) Set the null magnetic field strength in all ferromagnetic areas.

(V) Calculate the effective magnetic permeability distribution according to (2).

(VI) Create and solve $3N$ of the independent multi-harmonic linear field-circuit models related to zero, positive and negative sequences, assigned to individual harmonics of the supply voltage. Each one should take into account M spatial harmonics of the magnetic field strength. Calculate the requested operational parameters of the analysed machine according to (6)–(7), adopting $I_{hn} = \sqrt{I_{0hn}^2 + I_{1hn}^2 + I_{2hn}^2}$ and $T_{ehn} = T_{0ehn} + T_{1ehn} + T_{2ehn}$. For a zero sequence of the supplying voltage, adopt pulsation close to zero.

(VII) Based on the calculated magnetic field distributions, calculate new values of the magnetic field strength $H_{hn} = H_{1hn} + H_{2hn}$, where H_{1hn} and H_{2hn} are the magnetic field strength derived from hn harmonic of the supply voltage as a result of solving the model associated with the positive and negative sequence. The influence of the zero-sequence system on the saturation of the magnetic circuit is disregarded. Update the effective magnetic permeability distribution according to (3) and the matrix parameters $\mathbf{M}_{11}$ in (5).

(VIII) Repeat steps V–VII until the convergence criterion based on the relative change of the maximum value of norm (equal to 0.1%) for all vectors of the solution of (5) is reached.

The above procedure was used to calculate the operational characteristics of the tested machine when supplied by the harmonic-rich square waveforms with the amplitude of one phase reduced by 25%. As shown in Figures 6 and 7, the presented approach allows one to obtain predictions very close to ones obtained from a comprehensive time-domain model, even in the case of a lack of supply symmetry. The reason why the assumptions hold despite the asymmetry is that the circumferential distribution of winding magnetomotive force is always a periodic function as it depends on the circumferential distribution of the winding that has nothing to do with voltage asymmetry. Moreover, the degree of voltage asymmetry does not break the validity of the assumptions.

In the considered case, the solution time using the proposed model increases to 8 min, while the execution time of the complementary time-domain model solution time is similar to the previously presented case and equal to 4 h 30 min. Connected with a very good agreement of results, this result clearly exposes the benefits of the developed approach.

Looking from the practical point of view, it should be noticed that the voltage unbalance problem is effectively reduced in modern fault-tolerant power converters; however, the circumferential unbalance of magnetomotive force due to the faults of the motor winding is a much more common case of the faulty operation of induction motor drives. The corresponding computational problem can be effectively solved by the algorithm in Section 3 with both winding distribution-related and slot-related spatial harmonics taken into account.

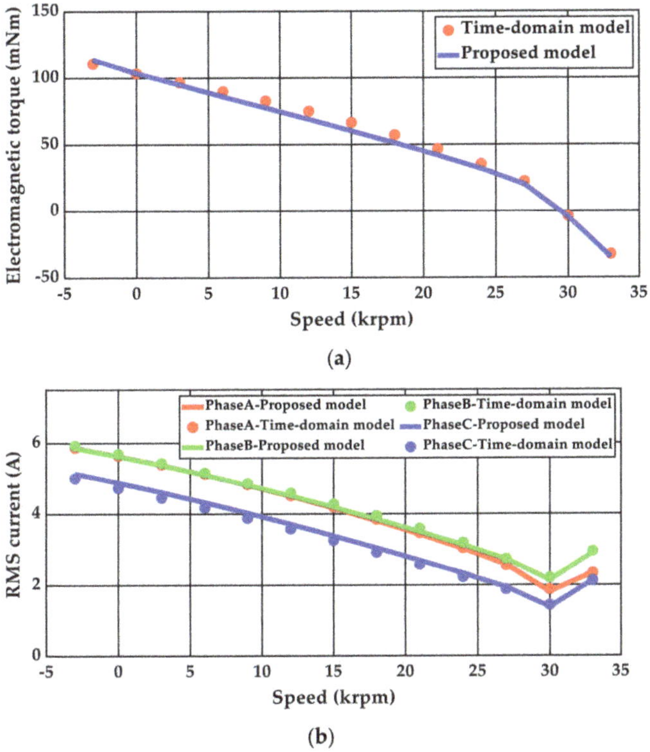

Figure 6. Results of computations for the square nonsymmetrical supplying waveforms: electromagnetic torque (**a**) RMS stator currents (**b**).

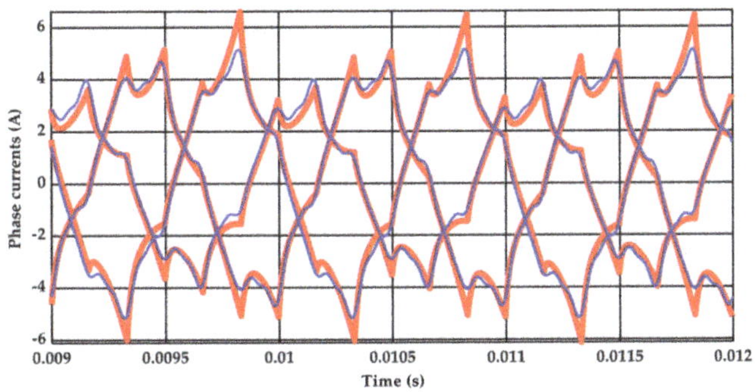

Figure 7. Comparison of the phase currents waveforms calculated with the use of the model formulated in the time domain (red line) with the waveforms calculated with the use of the developed method (blue line), as a superposition of harmonic waveforms, for the rotational speed of 24,000 rpm when supplied with square waveforms and 25% amplitude asymmetry.

5. Conclusions

The development of the strongly coupled multi-harmonic field model concept effectively accounted for the nonlinearity and asymmetry of the voltage supply in the calculation of the operating characteristics of a high-speed induction machine with a solid rotor in a steady-state complex-valued finite element modelling framework. According to the authors' opinions, in combination with the previously developed numerical method of determining the rotor end-effect coefficient [21], the multi-harmonic field-circuit model may become an effective tool in the process of designing the above-mentioned motor type. In particular, this may be an effective tool in the investigations on the reduction of losses due to higher harmonics of the magnetic field of various origins.

Of course, the present work does not cover all aspects of the issue. Further research will be undertaken to consider losses in the stator package, hysteresis losses in the solid rotor and even the influence of the power supply using pulse-width modulation (PWM) converters.

Author Contributions: Conceptualization, T.G. and M.J.; methodology, T.G. and M.J.; software, T.G.; investigation, T.G.; resources, T.G.; data curation, T.G.; writing—original draft preparation, T.G.; writing—review and editing, M.J.; visualization, T.G.; supervision, M.J.; project administration, T.G. All authors have read and agreed to the published version of the manuscript.

Funding: This research received no external funding.

Conflicts of Interest: The authors declare no conflict of interest.

References

1. O'Kelly, D. Theory and performance of solid rotor induction and hysteresis machines. *Proc. Inst. Electr. Eng.* **1976**, *123*, 421–428. [CrossRef]
2. Wood, A.J. An analysis of solid rotor machines Part I. Operational impedances and equivalent circuits. *Trans. Am. Inst. Electr. Eng.* **1959**, *78*, 1657–1665. [CrossRef]
3. Sarma, M. Current density distribution in solid-rotor induction motor. *IEEE Trans. Magn.* **1979**, *15*, 1473–1475. [CrossRef]
4. Brunelli, B.; Casadei, D.; Reggian, U.; Serra, G. Transient and steady-state behaviour of solid rotor induction machines. *IEEE Trans. Magn.* **1983**, *19*, 2650–2654. [CrossRef]
5. Zaim, M.E. Non-linear models for the design of solid rotor induction machines. *IEEE Trans. Magn.* **1999**, *35*, 1310–1313. [CrossRef]
6. Kurvinen, E.; Chong, D.; Petrov, I.; Nerg, J.; Liukkonen, O. Design And Manufacturing of A Modular Low-Voltage Multi-Megawatt High-Speed Solid-Rotor Induction Motor. *IEEE Trans. Ind. Appl.* **2021**. [CrossRef]
7. Dongdong, Z.; Ruichi, A.; Chengyuan, H.; Lixiao, B.; Wu, T. Electromagnetic Design of A Megawatt High Efficiency High Speed Solid Rotor Induction Motor. In Proceedings of the IEEE International Electric Machines and Drives Conference (IEMDC), Miami, FL, USA, 21–24 May 2017. [CrossRef]
8. Haisen, Z.; Chengyang, C.; Eldeeb, H.; Yang, Z.; Guorui, X.; Mohammed, O. Optimal Design Of High-Speed Solid-Rotor Cage Induction Motors Considering Ferromagnetic Materials Behavior And Manufacturing Process. *IEEE Trans. Ind. Appl.* **2020**, *56*, 4345–4355. [CrossRef]
9. Gulbahce, M.; Kocabas, D. High-speed solid rotor induction motor design with improved efficiency and decreased harmonic effect. *IET Electr. Power Appl.* **2018**, *12*, 1126–1133. [CrossRef]
10. Chong, D. Modeling and Analysis of a High-Speed Solid-Rotor Induction Machine. Ph.D. Thesis, Lappeenranta–Lahti University of Technolog, Espoo, Finland, 2020.
11. Pyrhönen, J.; Nerg, J.; Kurronen, P.; Lauber, U. High-speed high-output solid-rotor induction-motor technology for gas compression. *IEEE Trans. Ind. Electron.* **2010**, *57*, 272–280. [CrossRef]
12. Shaofeng, C.; Yaofei, H.; Zhixun, M.; Guozhen, C.; Shuai, X.; Jikai, S. Influence Analysis of Structural Parameters on the Performance of 120° Phase Belts Toroidal Winding Solid Rotor Induction Motor. *Energeies* **2020**, *13*, 5387. [CrossRef]
13. Feng, H.; Si, J.; Wu, W.; Dong, L.; Cheng, Z. Equivalent Circuit Parameter Calculations and Characteristics Analysis of 2-DoF Direct Drive Induction Motor with a Slotted Solid Rotor. *Appl. Sci.* **2019**, *9*, 2191. [CrossRef]
14. Gieras, J.; Saari, J. Performance Calculation for a High-Speed Solid-Rotor Induction Motor. *IEEE Trans. Ind. Electron.* **2011**, *59*, 2689–2700. [CrossRef]
15. Siyuan, G.; Libing, Z.; Tong, Y. An analytical method for determining circuit parameter of a solid rotor induction motor. In Proceedings of the 15th International Conference on Electrical Machines and Systems (ICEMS), Sapporo, Japan, 21–24 October 2012.
16. Markovic, M.; Perriard, Y. An analytical solution for the torque and power of a solid-rotor induction motor. In Proceedings of the IEEE International Electric Machines & Drives Conference, Niagara Fails, ON, Canada, 15–18 May 2011. [CrossRef]
17. Räisänen, V.; Suuriniemi, S.; Kurz, S.; Kettunen, L. Rapid Computation of Harmonic Eddy-Current Losses in High-Speed Solid-Rotor Induction Machines. *IEEE Trans. Energy Conv.* **2013**, *28*, 782–790. [CrossRef]
18. Wu, W.; Si, J.; Feng, H.; Cheng, Z.; Hu, Y.; Gan, C. Rotor Eddy Current Loss Calculation of a 2DoF Direct-Drive Induction Motor. *Energies* **2019**, *12*, 1134. [CrossRef]
19. Di, C.; Petrov, I.; Pyrhönen, J.; Chen, J. Accelerating the Time-Stepping Finite-Element Analysis of Induction Machines in Transient-Magnetic Solutions. *IEEE Access* **2019**, *7*, 122251–122260. [CrossRef]
20. Di, C.; Petrov, I.; Pyrhönen, J. Extraction of Rotor Eddy-Current Harmonic Losses in High-Speed Solid-Rotor Induction Machines by an Improved Virtual Permanent Magnet Harmonic Machine Model. *IEEE Access* **2019**, *7*, 27746–27755. [CrossRef]

21. Garbiec, T. Fast Computation of Performance Characteristics for Solid-Rotor Induction Motors With Electrically Inhomogeneous Rotors. *IEEE Trans. Energy Conv.* **2016**, *31*, 1688–1696. [CrossRef]
22. Lin, J.D.; Zhou, P.; Chen, N.; Lu, C.; Christini, M. Fast Methods for Reaching AC Steady State in FE Transient Analysis. In Proceedings of the IEEE International Electric Machines and Drives Conference (IEMDC), Miami, FL, USA, 21–24 May 2017. [CrossRef]
23. Dyck, D.N.; Weicker, P.J. Periodic Steady-State Solution of Voltage-Driven Magnetic Devices. *IEEE Trans. Magn.* **2007**, *43*, 1533–1536. [CrossRef]
24. Rosu, M.; Zhou, P.; Lin, D.; Ionel, D.M.; Popescu, M.; Blaabjerg, F.; Rallabandi, V.; Staton, D. *Multiphysics Simulation by Design for Electrical Machines, Power Electronics and Drives*; John Wiley & Sons: Hoboken, NJ, USA, 2017.
25. De Gersem, H.; Hamayer, K. Air-Gap Flux Splitting for the Time-Harmonic Finite-Element Simulation of Single-Phase Induction Machines. *IEEE Trans. Magn.* **2002**, *38*, 1221–1224. [CrossRef]
26. De Gersem, H.; De Brabandere, K.; Belmans, R.; Hameyer, K. Motional time-harmonic simulation of slotted single-phase induction machines. *IEEE Trans. Energy Convers.* **2002**, *17*, 313–318. [CrossRef]
27. Räisänen, V.; Suuriniemi, S.; Kettunen, L. Generalized Slip Transformations and Air-Gap Harmonics in Field Models of Electrical Machines. *IEEE Trans. Magn.* **2016**, *52*, 1–8. [CrossRef]
28. Mezani, S.; Laporte, B.; Takorabet, N. Saturation and space harmonics in the complex finite element computation of induction motors. *IEEE Trans. Magn.* **2005**, *41*, 1460–1463. [CrossRef]
29. Ouazir, Y.; Takorabet, N.; Ibtiouen, R.; Touhami, O.; Mezani, S. Consideration of space harmonics in complex finite element analysis of induction motors with an air-gap interface coupling. *IEEE Trans. Magn.* **2006**, *42*, 1279–1282. [CrossRef]
30. Garbiec, T.; Jagiela, M.; Kulik, M. Application of Nonlinear Complex Polyharmonic Finite-Element Models of High-Speed Solid-Rotor Induction Motors. *IEEE Trans. Magn.* **2020**, *56*. article no. 7515304. [CrossRef]
31. Garbiec, T.; Jagiela, M. Accounting for Magnetic Saturation Effects in Complex Multi-harmonic Model of Induction Machine. *Energies* **2020**, *13*, 4670. [CrossRef]
32. Davat, B.; Ren, Z.; Lajoie-Mazenc, M. The movement in field modeling. *IEEE Trans. Magn.* **1985**, *21*, 2296–2298. [CrossRef]
33. Escarela-Perez, R.; Campero-Littlewood, E. Moving-band and sliding-surface combined technique for the simulation of rotor motion in transient FEM modelling of electrical machines. In Proceedings of the International Conference on Electrical Machines (ICEM), Espoo, Finland, 28–30 August 2000.
34. Singh, G.K. A research survey of induction motor operation with non-sinusoidal supply wave forms. *Electr. Power Sys. Res* **2005**, *75*, 200–213. [CrossRef]
35. Doggett, L.A.; Queer, E.R. Induction motor operation with non-sinusoidal impressed voltages. *AIEE Trans.* **1929**, *48*, 1217–1223.
36. Chalmer, B.J.; Sarkar, B.R. Induction motor losses due to non-sinusoidal supply waveforms. *Proc. Inst. Electr. Eng.* **1968**, *115*, 1777–1782. [CrossRef]
37. Klingshirn, E.A.; Jordan, H.E. Poly phase induction motor performance and losses on nonsinusoidal voltage sources. *IEEE Trans. Power Appar. Syst.* **1968**, *PAS-87*, 624–631. [CrossRef]
38. Available online: https://www.mathworks.com/products/matlab.html (accessed on 28 June 2021).
39. Available online: https://gmsh.info (accessed on 28 June 2021).

energies

MDPI

Article

Static and Dynamic Simulation of an Induction Motor Using Matlab/Simulink

P. F. Le Roux * and M. K. Ngwenyama

Department of Electrical Engineering, Tshwane University of Technology, Emalahleni 1034, South Africa; ngwenyamamk@yahoo.com
* Correspondence: lerouxpf@tut.ac.za

Abstract: Industries are adequately configured with the operational devices that are required to develop induction motors. Engineers should precisely comprehend the kind of equipment that is constructed, as with every other production system, and should start by having the goal in their perspective. An adaptable simulation of an induction motor with a protective scheme is presented. The adaptable simulation assists engineers in accurately designing motors that meet all protective standards for certain purposes. This work achieved simulations of induction motors in stable and unstable conditions. An extensive study was performed to determine the optimum design of an induction motor. This paper attempts to provide engineers with a thorough grasp of the adaptable modelling of an induction motor. In this work, a direct dq0-direct axis algorithm is presented to implement both static and dynamic modelling of a three-phase induction machine due to possible faults and high-performance requirements in induction machines. The proposed algorithm was tested against several conventional methods, and it was observed that under the stable condition of the machinery, the proposed algorithm could remove any developing faults. This conserves time and minimises the labour required of the operator, which makes the proposed algorithm more efficient. Furthermore, the machine is demonstrated in a steady-state performance with respect to the current, active power, efficiency, reactive power, power factor, and speed when the torque loads range from 0 to 125% of its nominal torque. The transient behaviour of the machine was shown through the current, electromagnetic torque, electromagnetic torque versus speed, and speed under no-load, half-load (50%), and full-load (100%) conditions. Finally, the results of the proposed technique were compared to the results of the measured parameters. It was observed that when the load changed from a half load (50%) to a full load (100%), then the supply voltage was suddenly halved with the load at full load (100%). It was observed that the proposed algorithm provides accurate estimates with a deviation of not more than +/−2% from the measured parameters.

Keywords: dynamic model; induction motor; Matlab/Simulink; rotor winding; stator winding

Citation: Le Roux, P.F.; Ngwenyama, M.K. Static and Dynamic Simulation of an Induction Motor Using Matlab/Simulink. *Energies* **2022**, *15*, 3564. https://doi.org/10.3390/en15103564

Academic Editor: Ryszard Palka

Received: 6 March 2022
Accepted: 18 April 2022
Published: 12 May 2022

1. Introduction

The most popular motors utilized in economic mobility automation applications and primary-supply residential electrical consumables are AC induction machines. The key benefits of AC induction machines are their simplicity and robust construction, competitive prices, minimal servicing, and straightforward integration into an AC power supply. There are many different kinds of AC induction machines accessible in the industry. Several machines are appropriate for various functions [1]. However, AC induction machines are convenient to construct compared to DC machines. Controlling the rpm and torque in different varieties of AC induction machines requires a deeper grasp of the configuration and features of such motors. However, DC motors are efficient at commencing and moderating speed. Such machines have a great torque concentration [2]. A DC machine works quietly and has a hugely variable speed. The electromagnetic disturbance is minimal, and the overcurrent or inrush tolerance is substantial. The construction or assembly of a

DC machine is one of its limitations. The commutator and the brush [3] have a rubbing connection, resulting in sparks and mechanical degradation. As a result, DC machines possess a comparatively limited operating lifespan, requiring a high service expense. This also casts uncertainties about the system's durability and safety. As a result, the usage of DC machines in some industrial applications is restricted nowadays [4].

Over the years, motors have revolutionised the mining and automation industry. Processes such as hoisting conveyor belt systems for moving minerals, e.g., gold, coal, diamonds, etc., from underground and opencast mines, depend largely on the utilisation of induction motors [5]. Thus, for the reliable operation of these machines, proper protection needs to be implemented for safe operation under load conditions. Any malfunction of an induction motor can be described as an electrical fault, environmental factor, or mechanical breakdown. Rotor bearings could result in overheating, wear, and tear due to mechanical stresses [6]. Drawing enormous magnitudes of currents ensure high temperatures. Modelling an induction motor is somewhat complex, stemming from its non-linear behaviour triggered by electromagnetic exhaustion and the significant temperature influence from the synchronous motor settings [7].

Furthermore, the shaft time constant of an induction motor can change due to rotor heat. Such characteristics render the mathematical modelling of induction motors somewhat insurmountable. Most researchers use simplified models that do not consider the factors mentioned above. Production of these machines is imperative and requires urgency in reproduction during their idle state [8].

Chitra and Prabhakar [9] presented a simulation of an induction machine by utilising the fuzzy logic approach. The authors applied the approach in their study in order to regulate the velocity of an induction machine to obtain the optimum torque with the least amount of loss. They used the field-oriented control approach to create a fuzzy logic controller that enables improved control of motor torque with greatly variable performance. Their simulated design was evaluated by utilising multiple Matlab toolboxes. They observed that the induction motor's efficiency increased in stable conditions. The results show that the suggested speed regulator was efficient and reliable.

Elnaghi et al. [10] proposed using a genetic algorithm (GA) to process experimental loads on an inductive machine. The principle of predicting motor parameters from testing data was demonstrated using a genetic algorithm-based technique. The specifications were determined using typical no-load and blocked rotor experiments. The cost equation—the graded sum of the stator currents and rotor velocity—was studied and improved for various motor parameter values. The impact of differential equations on the estimates was also shown. The estimated speed and torque parameters from the mathematical equation were compared to the experimental findings, and both exhibited a strong connection, proving the validity of the mathematical equation and the genetic algorithm method for improvement.

Sadasivan and Mammen [11] applied the same algorithm to obtain parameters that linked the proposed technique and the loading of the electric motor using the evaluation function. They used the genetic algorithm on three separate situations of simulated loading and found that the outcomes were superior in terms of the overall losses induced by the motor. The authors' technique proved to be effective in terms of parameter estimation.

Jirdehi and Rezaei [12] presented a simulation of an induction motor by utilising an artificial neural network (ANN) and an adaptive neuro-fuzzy inference system (ANFIS) to investigate variables that are often difficult to obtain. They used both methods to test 20 induction motors of varying power outputs. The experimental results consisted of the starting torque, current, maximum torque, full-load slip, efficiency, rated active power, and reactive power. The authors compared the findings produced by the proposed ANN and ANFIS models and the practical results. They discovered a good relationship between the projected values and the practical data. However, the proposed ANFIS model was more precise than the proposed ANN model.

Keerthipala et al. [13] explained the ANN algorithm and how it may be used to monitor an induction motor's torque and speed regulation using linear and non-linear models. The authors reported that the linear observer approach is simple to apply in real-time; however, it does not accurately estimate the rotor and vector angle since the induction machine generally works in the saturated region. The non-linear observer approach considers the impact of the magnetic saturation of the induction machine; however, it cannot be practically applied using conventional techniques because estimating the angle requires too much time. Their suggested technique compensates for the effect of saturation and estimates the angle in a few milliseconds, which is well within the real-time limit.

This study presents an adaptable simulation of an induction motor with a downstream protective scheme [9]. In this work, a direct dq0-direct axis algorithm is presented to implement both static and dynamic modelling of a three-phase induction machine due to possible faults and high-performance requirements in induction machines. The proposed algorithm was tested against several conventional methods, and it was observed that under a stable condition of the machinery, the proposed algorithm could remove any developing faults. This conserves time and minimises the labour required of the operator, which makes the proposed algorithm more efficient. Furthermore, the machine is demonstrated in a steady-state performance with respect to current, active power, efficiency, reactive power, power factor, and speed when the torque loads range from 0 to 125% of its nominal torque. The transient behaviour of the machine was shown through the current, electromagnetic torque, electromagnetic torque versus speed, and speed under no-load, half-load (50%), and full-load (100%) conditions. Finally, the proposed technique was compared to the results of the measured parameters. It was observed that when the load changed from half load (50%) to full load (100%), the supply voltage was suddenly halved with the load at full load (100%). It was observed that the proposed algorithm provides accurate estimates with a deviation of not more than +/−2% from the measured parameters.

2. Problem Statement

An induction motor is a complex machine to design and practically implement. Today, a vast majority of software is used by different manufacturing institutions to simulate the machine before the design is implemented. However, an induction motor's dynamic model is usually implemented in dq0-direct, quadrature, and zero-sequence axes. A static and dynamic motor model's implementation is a mathematical representation; this strategy eliminates human error, enables designs of the utmost performance, and provides highly efficient induction motors. This work implements both static and dynamic modelling of a three-phase induction machine due to possible faults and high-performance requirements for induction machines. Using the direct dq0-direct axis method, the dynamic model's differential equations were first derived and implemented in Matlab/Simulink; their performance was assessed in the steady-state.

3. Objectives of the Research

Induction motors help companies run their operations efficiently. It is imperative to extend studies that enhance the performance and accuracy of the designs for these organisations. The objectives of this work are to conduct comprehensive modelling of induction motors and [14]:

- To contribute to the development of induction motor models and protection systems.
- To simulate an adaptable simulation of a three-phase induction motor in Matlab/Simulink.
- To accurately size the motor parameters and increase performance thereof.
- To implement a dq0-axis reference frame modelling technique of a rotating machine.
- To implement a feeder protection system for the machine during its steady operation.

4. Methodology

To comprehend and engineer vector-controlled drives [15], an adaptable simulation of the induction machine that is responsive to operation and safety needs to be defined. Since every control must face possible changes and faults, the adaptable simulation of a motor is argued to be an imitation of an actual factory. Nonetheless, the simulation must include the significant factor impacts that emerge throughout steady-state and transient-state events. It should also be applicable for future modifications in supply voltages and currents. Due to the precision and convenience of the space vector and phasor hypothesis, such a simulation should be produced using this hypothesis and the two-axis principle [16] of synchronous generators. All of the techniques, as well as the simulations of induction motors, are discussed and criticised accordingly in this paper, along with the proposed dq0 axis reference frame strategy.

5. Simulations of Induction Motors

Algorithms for the regulation of motors with great velocity and performance are required in spinning machinery and electrical cars. Iron loss (ILS) influences flux measurement, variable detection, actual torque, and acceleration control. Motor losses are an essential measure of the simulation variables [17]. The models of induction motors considered in this state-of-the-art literature survey are discussed in line with the following criteria:

- Two-pole, three-phase windings are symmetrical.
- The slotting impacts are considered.
- Iron losses are not disregarded.
- The conductivity of the iron elements is limitless.
- The magnetisation in the airgap is circular.
- The stator and rotor windings are coordinated as a single and multi-turn full-pitch coil located on opposite sides of the air gap.

The dynamic models may be split into two major sections: the Γ Model and the Loss Model.

5.1. Γ Model

Several evaluations of three-phase induction motors have employed a traditional linear machinery design, which is generally in the internal and reciprocal inductances of the well-known Γ type of equivalent circuit [5,18]. Such a model is considerably more complicated than is required for a linear assessment. However, it is insufficient for usage when machine characteristics vary, such as when the rotor time changes with heat and reciprocal inductances are altered whenever the machine is overloaded. The inverse Γ model, as indicated in Figure 1, is very useful for learning and analysing vector control mechanisms [18], so it reduces complications when compared to the T-model. Once the machinery settings are altered, it then becomes a non-linear system. The extended Kalman filter (EKF) [19] is a randomised monitor for a non-linear system. The sound generators in the EKF take the measurement and simulation errors into account.

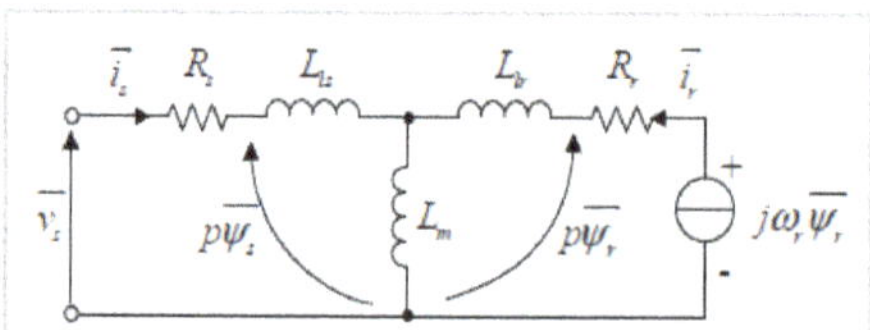

Figure 1. Γ Model of an induction machine [20].

5.2. Inverse Γ Model

The three-phase stator windings of an induction machine are intended to create a symmetrical waveform dispersed (MMF) in space across the airgap margin. The proportion of magnetic flux [21] must similarly be symmetrically provided when the airgap is homogeneous, and the impacts of slot distortions are ignored. It is further presumed that the drive's neutral connector is free, ensuring that phase voltages, currents, and flux connections are constantly symmetrical, and thus, the circuit contains no zero-phase sequence components. Considering variable stability, two readings of complex impedance or four variables may be monitored on the stator by running under no-loading and lockout rotor conditions. It is insufficient to provide five variables in these comparable systems. This is often corrected by randomly setting the magnetising inductance to be identical to the rotor inductance [22,23]. Figure 2a,b show design configurations that are well suited for learning and analysing vector control mechanisms. With this setup, the stator's current space vector $\vec{i_S}$ is managed in a manner in which the rotor flux connection via the magnetising current $\vec{i'_M}$ maintains a fixed value, thus supplying a rotor current vector $\vec{i'_R}$ in the space inversion, with $\vec{i_M}$ providing the appropriate power output.

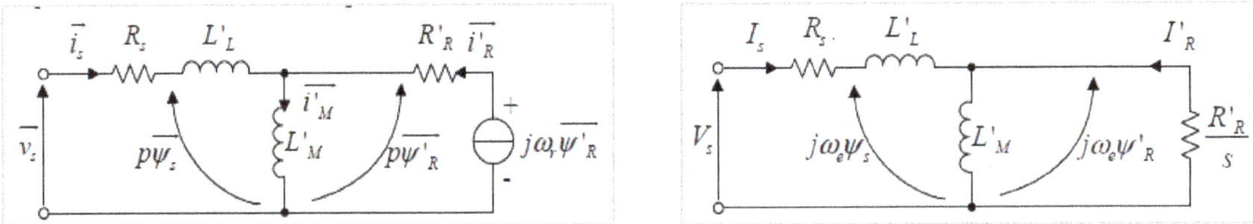

(a) (b)

Figure 2. (**a**) Transient circuit of the inverse Γ model; (**b**) steady-state inverse Γ model [20].

5.3. Loss Model

The induction motor simulation in [24] centres on an experimental three-phase stator and rotor circuit design using a magnetic coupling in the intermediate stage, thus neglecting core loss. As seen in Figure 3a, the concept serves as the backbone for major vector control derivations and a design based on the study of an electric drive. The stator-side resistance, leakage inductance, and mutual inductance [25] are denoted as R_S, L_{lS}, and L_{ms}. R'_r, L_{lr}', and L_{mr}' are the rotor-side resistance, leakage inductance, and mutual inductance, respectively, as with the stator side. The flux on each rotor or stator circuit is divided into the leakage and mutual components because solely the former reaches the magnetic connection field with which the stator and rotor interface. Figure 3b depicts a conventional steady-state per-phase circuit diagram, which takes core loss into account as the energy lost in r_{c}_ph. L_m is the steady-state magnetising inductance corresponding to $(1.5 \times L_{ms})$. It should be noted that the connection impacts of other phases are summed into L_m which is acquired from steady-state observations. As a result, the diagram is entirely irrelevant for transient applications [26]. Figure 4 depicts the suggested induction machine design, which is influenced by the preceding two conventional designs. The core loss is understood as a resistor R_c in parallel with L_{ms} in each stator phase. It should be noted that R_c is not equivalent to r_{c}_ph, although they were associated in [27].

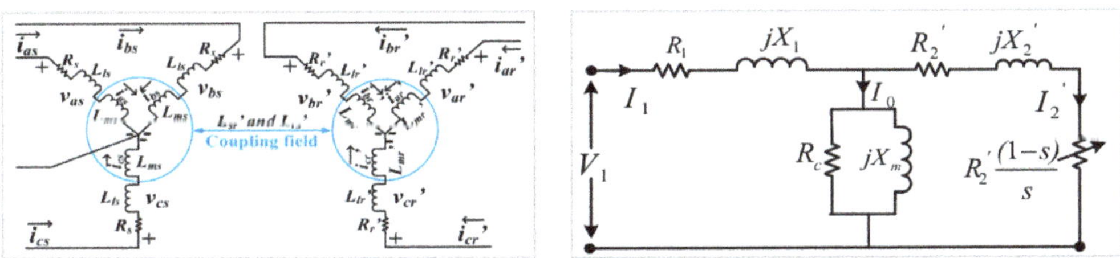

(a) (b)

Figure 3. (**a**) The classical induction machine model considers only copper loss; (**b**) the steady-state per-phase equivalent circuit [27].

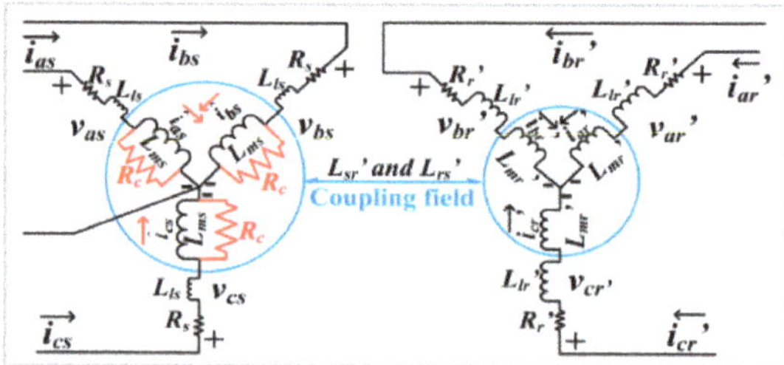

Figure 4. Loss model of an induction motor [27].

5.4. Simplified Model

Many stable algorithms contain an adaptable induction machine simulation that is derived from either or both of the equivalent topologies shown. A single-cage induction machine offers modest start-up torque multiplication. With that, the single-cage design is frequently inaccurate. Such issues can be addressed by employing a dual-cage or deeper bar rotor design. A shortage of data frequently limits the usage of dual-cage and deeper bar designs. Improving the torque-slip property of a single-cage design is a streamlined remedy for these instances. The rotor resistance is often changed with slip, which is simple to accomplish during each convergence phase of the model. An adaptable design simulation that relies on the equivalent circuits illustrated in [27] is often utilised for transient reliability research [27–29].

5.5. Simplified d–q Design

The steady-state equivalent diagram depicted in [29] serves as the foundation for the concept of the induction machine. The simplified model presented in [29] is utilised for the dynamic simulation diagram. The d–q model of this analysis was regarded as precisely acceptable for acting as a baseline design and was subsequently utilised to assess the adequacy of different designs when implemented for voltage balance analyses [30]. The variables of the designs are depicted in [29].

6. Proposed Design

Figure 5 presents the proposed modelling technique for an induction motor; the three-phase supply voltage is supplied to the machine, and then the voltages are transformed using the 0dq axis to produce the vectors V_q and V_d. Hence, the stator currents and voltages are produced; subsequently, the rotor parameters are also considered. The implementation is outlined in this paper.

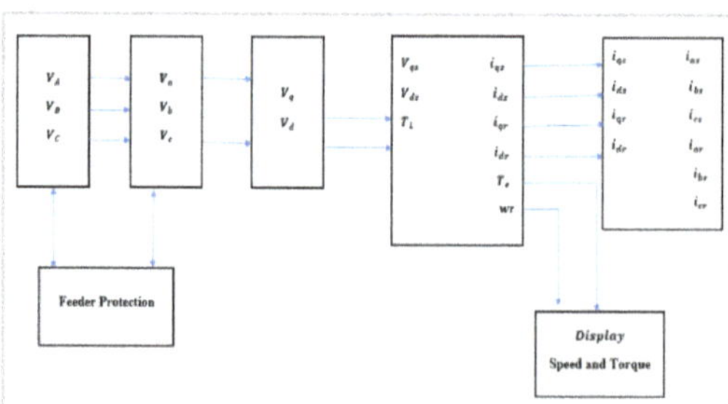

Figure 5. A proposed dynamic model of an induction motor.

Figure 6 presents the workflow of designing a dynamic and static model of an induction machine. The adaptable design is mathematically represented and subsequently simulated. The equations are implemented separately in terms of subsystems and then integrated to present the full model of the machine. Similarly, for a static model, the derivation is almost the same as that of the dynamic model, but it only changes when the speed is equal to zero (when the machine is stationary). The simulation is also given for the static model [5].

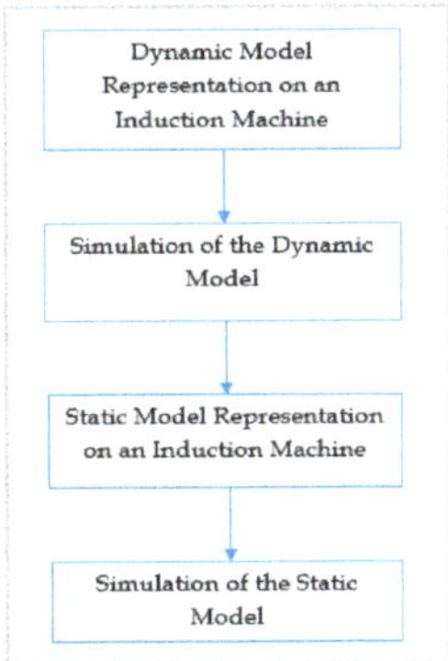

Figure 6. Design workflow.

7. Proposed Design

7.1. Dynamic Model

There is a very deep relationship between the stator and the rotor of an induction motor. If one is to extract parameters of the rotor, there is a need to know the relationship of the currents and voltage between these two elements of an induction machine. A d–q axis model of an induction machine is presented in Figure 7a. A q-axis equivalent circuit for an adaptable design of an induction machine is presented in Figure 7b. The stator can then be represented by (1) to (10), and the rotor's d–q transformation is represented by (11) to (18). The torque is then represented by the stator and rotor parameters (19) [31].

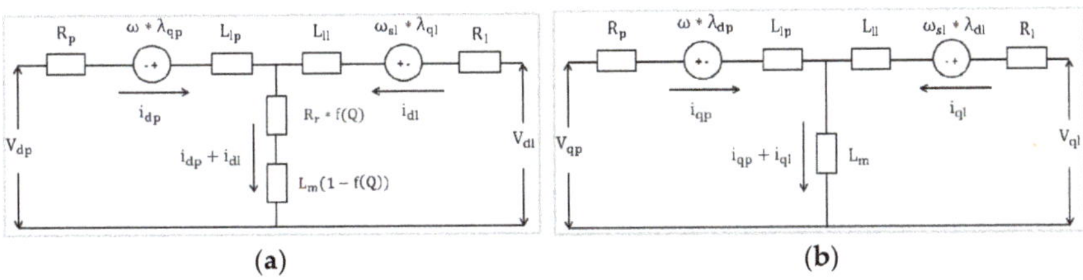

Figure 7. (**a**) The d-axis equivalent circuit of an induction motor. (**b**) The q-axis equivalent circuit of an induction motor.

Stator voltage modelling:

$$V_{sd} = \sqrt{\frac{2}{3}}\left[\cos(\theta_{ds} \times v_{ds}) + \left(\cos(\theta_{ds} \times \frac{2\pi}{3} \times v_b)\right) + \left(\cos(\theta_{ds} \times \frac{4\pi}{3} \times v_a)\right)\right] \quad (1)$$

$$V_{sq} = -\sqrt{\frac{2}{3}}\left[\sin(\theta_{ds} \times v_{ds}) + \left(\sin\theta_{ds} \times \frac{2\pi}{3} \times v_b\right) + \left(\sin\theta_{ds} \times \frac{4\pi}{3} \times v_a\right)\right] \quad (2)$$

V_{sd} and V_{sq} may now be simplified to:

$$V_{sd} = R_S \times i_{sd} + \frac{d}{dt}(\lambda_{sd}) - \omega_d \times \lambda_{sq} \quad (3)$$

$$V_{sq} = R_s \times i_{sq} + \frac{d}{dt}(\lambda_{sd}) - \omega_d \times \lambda_{sd} \tag{4}$$

Stator fluxes:

$$\lambda_{sd} = L_S \times i_{sd} + L_m \times i_{rd} \tag{5}$$

$$\lambda_{sq} = L_S \times i_{sq} + L_m \times i_{rd} \tag{6}$$

Stator currents:

$$i_{ds} = \frac{1}{Xl_S}(\lambda_{ds} - \lambda_{md}) \tag{7}$$

$$i_{qs} = \frac{1}{Xl_S}(\lambda_{qs} - \lambda_{mq}) \tag{8}$$

Stator voltages:

$$V_{sd} = R_S \times i_{sd} + \frac{d}{dt}(\lambda_{sd}) - \omega_d \times \lambda_{sq} \tag{9}$$

$$V_{sq} = R_S \times i_{sq} + \frac{d}{dt}(\lambda_{sq}) - \omega_d \times \lambda_{sd} \tag{10}$$

Mathematical model of the rotor:

$$V_{rd} = R_r \times i_{rd} + \frac{d}{dt}(\lambda_{rd}) - \omega_{dA} \times \lambda_{rq} \tag{11}$$

$$V_{rq} = R_r \times i_{rd} + \frac{d}{dt}(\lambda_{rd}) - \omega_{dA} \times \lambda_{rd} \tag{12}$$

Rotor flux equations:

$$\lambda_{rd} = L_r \times i_{rd} + L_m \times i_{sd} \tag{13}$$

$$\lambda_{rq} = L_r \times i_{rq} + L_m \times i_{sd} \tag{14}$$

Rotor currents:

$$i_{dr} = \frac{1}{Xl_S}(\lambda_{dr} - \lambda_{md}) \tag{15}$$

$$i_{qr} = \frac{1}{Xl_S}(\lambda_{qr} - \lambda_{mq}) \tag{16}$$

Rotor voltages:

$$V_{rd} = R_r \times i_{rd} + \frac{d}{dt}(\lambda_{rd}) - \omega_{dA} \times \lambda_{rq} \tag{17}$$

$$V_{rq} = R_r \times i_{rq} + \frac{d}{dt}(\lambda_{rq}) - \omega_{dA} \times \lambda_{rd} \tag{18}$$

Electromagnetic torque:

$$T_{em} = \frac{P}{2} \times L_m(i_{sq} \times i_{rd} - i_{sd} \times i_{rq}) \tag{19}$$

where

V_{qs}, V_{ds} are the q and d axes of the stator voltage(s).
V_{qr}, V_{dr} are the q and d axes of the rotor voltage(s).
$\lambda_{mq}, \lambda_{md}$ are the q and d axes magnetising the flux linkages.
R_S is the stator resistance.
R_r is the rotor resistance.
X_{ls} is the stator leakage reactance.
X_{lr} is the rotor leakage reactance.
p is the number of poles.
T_{em} is the magnetic torque.

Figure 8a,b present the implantation of the dynamic model of the induction motor. The stator supply voltages and currents are presented in Figure 8a,b. The rotor currents are presented in Figure 8b. The torque and speed are presented in Figure 9a,b [32], where Figure 10 illustrates the complete model of the dynamic induction motor.

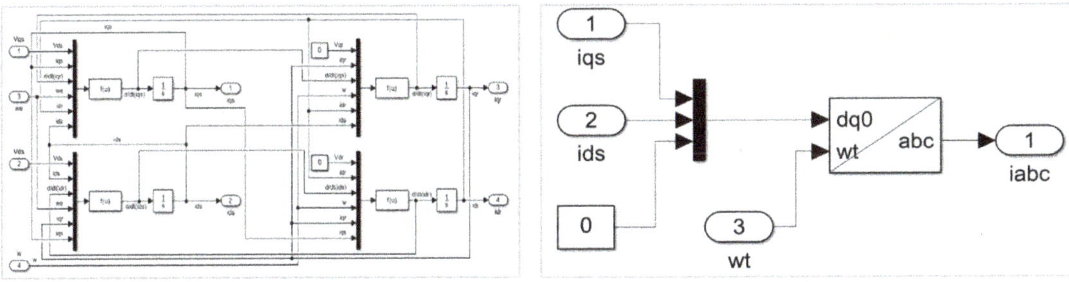

Figure 8. (**a**) Simulation of a dynamic model of an induction motor; (**b**) current equation of the stator.

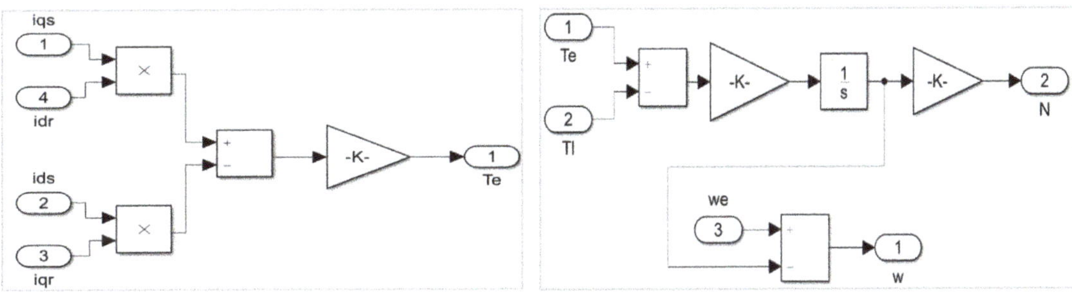

Figure 9. (**a**) Torque equation; (**b**) speed equation.

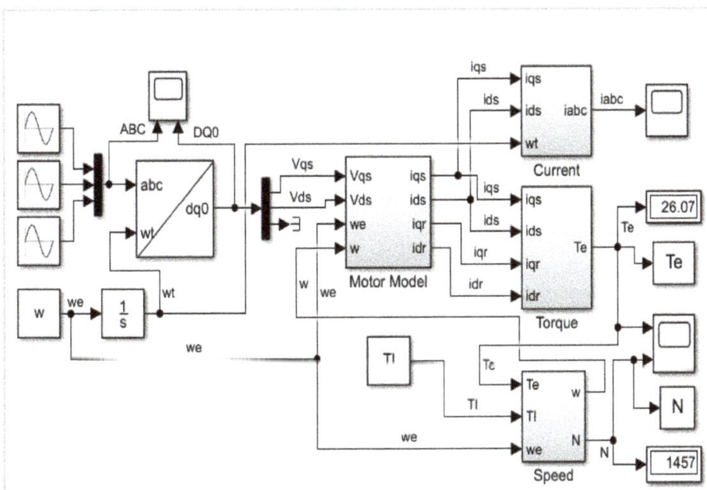

Figure 10. A complete model of the dynamic induction motor.

7.2. Static Model

The standard-frame velocity in a static-source framework is that of the stator, i.e., $\omega_s = \omega_r = 0$. Such a standard frame is chosen when potential asymmetrical differences in the stator are interrupted and the potential differences in the rotor are symmetrical. The potential difference equations of an induction machine in a static reference frame are determined by substituting $\omega_r = 0$ into (3). The resultant solution is the static design of an induction machine in which the rotor is removed. As a result, the machinery can be described as static.

It is noticeable that there is a major relationship between the stator and the rotor of an induction motor. As the subscripts, r and s indicate the stator (s) and rotor (r). The electromagnetic subscripts are i, v, and λ. The resistance is r, the leakage inductance is L_l, and the mutual inductance is L_m. The phase voltages are represented by a, b, and c.

The actual values of the induction motor's parameters can be derived from the model specified in Section 4 [26]. To determine the parameters, a no-load test and a load test must be performed, so the parameters can be determined as follows.

The no-load test is performed by supplying the voltage V_S at a rated frequency. The motor will rotate close to a synchronous speed, resulting in a close-to-zero slip [22].

Assuming that R_S, Ω, and $L_S H$ are much lower than the magnetising inductance $L_m H$, the following equation is derived [1].

$$L_m = \frac{V_S}{2\pi f_s I} \tag{20}$$

where V_S is the applied phase voltage in the stator, I is the current supplied to the stator, and f_s is the stator frequency.

Now that the magnetising inductance has been determined, the other parameters are extracted with the locked rotor test, and the resulting equivalent circuit is shown in Figure 11.

$$\begin{bmatrix} v_{ds} \\ v_{qs} \\ v_{dr} \\ v_{qr} \end{bmatrix} = \begin{bmatrix} R_S + sL_s & 0 & sL_m & 0 \\ 0 & R_s + sL_s & 0 & sL_m \\ L_m & R_s + sL_s & R_r + sL_r & \omega_r L_r \\ -\omega_r L_m & sL_m & -\omega_r L_r & R_s + sL_r \end{bmatrix} \begin{bmatrix} i_{ds} \\ i_{qs} \\ i_{dr} \\ i_{qr} \end{bmatrix} \tag{21}$$

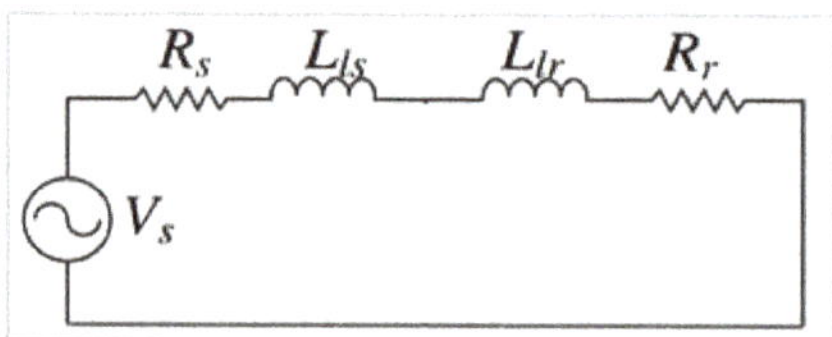

Figure 11. Equivalent circuit of the locked rotor.

Should $\omega_r = \omega_s = 0$, Equation (21) becomes:

$$\begin{bmatrix} v_{ds} \\ v_{qs} \\ v_{dr} \\ v_{qr} \end{bmatrix} = \begin{bmatrix} R_S + sL_s & 0 & sL_m & 0 \\ 0 & R_s + sL_s & 0 & sL_m \\ L_m & R_s + sL_s & R_r + sL_r & 0 \\ 0 & sL_m & 0 & R_s + sL_r \end{bmatrix} \begin{bmatrix} i_{ds} \\ i_{qs} \\ i_{dr} \\ i_{qr} \end{bmatrix} \tag{22}$$

where R_S and R_r are the resistances of the stator and rotor, L_S and L_r are the stator's and rotor's self-inductance and the stator's and rotor's speeds.

Figure 12 illustrates a static model of an induction motor.

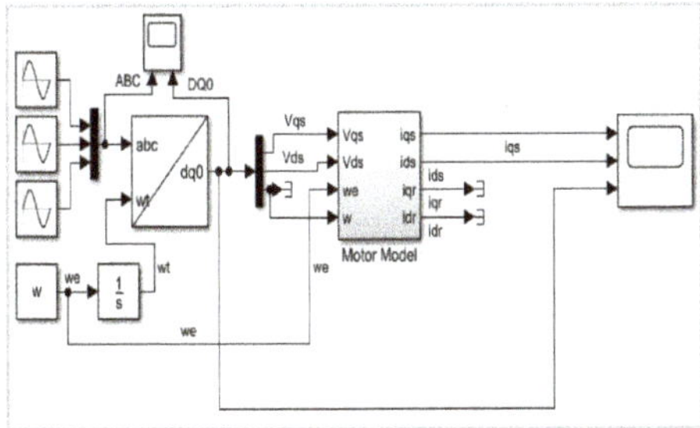

Figure 12. A static model of an induction motor.

8. Results

The following parameters were acquired from experimental work with a three-phase induction motor. These settings were then utilised to simulate and examine the behaviour of the induction motor using Matlab/Simulink.

8.1. The Machine's Steady-State Performance Behaviour When Loaded from 0 to 125% of the Rated Load, Shown in Both Tabular and Graphical Form with Current, Power Factor, Real Power, Reactive Power, Speed, Efficiency, and Power Factor versus the Percentage or Per-Unit Loading

The parameters indicated in Table 1 were used to simulate the induction machine using Matlab/Simulink were:

Table 1. Parameters used for induction motor simulation.

P_{rated}	7.5 kW	L_s	$42.5e-3$H
f	60 Hz	L_r	$41.8e-3$H
V_m	220 V	L_m	$41.2e-3$H
r_s	$288e-3\Omega$	L_{ls}	$L_s - L_m$
r_r	$158e-3\Omega$	L_{lr}	$L_r - L_m$
J	0.4 kg$\cdot$ m^2	P	4

The rated torque was not provided; therefore, before we can calculate the rated torque, the synchronous speed of the machine must first be calculated. The synchronous speed is calculated as follows:

$$N_s = \frac{120 \times f}{p} = 1800\ rp$$

Knowing the synchronous speed of a four-pole, 60 Hz machine, the rated torque may now be calculated.

$$T = \frac{P_{rated} \times 9.5493}{N_s} = 39.78875\ N.m$$

To simplify, 40 N.m was used as the rated torque value. Table 2 provides the results obtained from the induction machine.

As seen in Figure 13a, the higher the torque load is, the higher the current will be. In Figure 13b, we can see that the active power drawn by the induction machine is almost linear. Thus, the active power is proportional to the percentage of the loading.

As shown in Figure 14a, the efficiency of the induction motor is poor when the machine is lightly loaded. Theoretically, the optimal point must be at 100% loading; however, in this case, the optimal point is at 80–90%. This is mainly due to additional power losses (theoretical vs practical).

Figure 14b illustrates that the reactive power initially decreases. This is because the power factor is poor with no load and increases with the increase in the load; however, the reactive power drawn will also increase due to the increase in the load.

Table 2. The induction machine's results.

Torque Load (%)	Current (A)	P_{in} (W)	P_{out} (W)	Efficiency (%)	Qin (Var)	Power Factor (PF)	Speed (rpm)	Tem (N.m)
0	9.708	81.42	1.14×10^{-7}	1.40×10^{-9}	4530	0.01797	1800	6.05×10^{-10}
10	9.826	837.4	752.7	0.8988	4509	0.1826	1797	4
20	10.23	1598	1503	0.9401	4499	0.3348	1794	8
30	10.9	2365	2250	0.9411	4502	0.465	1790	12
40	11.78	3136	2994	0.9452	4517	0.5703	1787	16
50	12.85	3913	3736	0.9497	4544	0.6525	1784	20
60	14.06	4695	4475	0.9521	4584	0.7155	1780	24
70	15.39	5482	5210	0.9532	4638	0.7635	1777	28
80	16.81	6276	5943	0.9549	4705	0.8001	1773	32
90	18.3	7075	6672	0.9549	4786	0.8283	1770	36
100	19.86	7881	7399	0.9531	4881	0.8502	1766	40
110	21.48	8692	8122	0.9504	4990	0.8672	1763	44
120	23.14	9510	8841	0.9469	5115	0.8807	1759	48
125	23.99	9922	9199	0.9454	5184	0.8863	1757	50

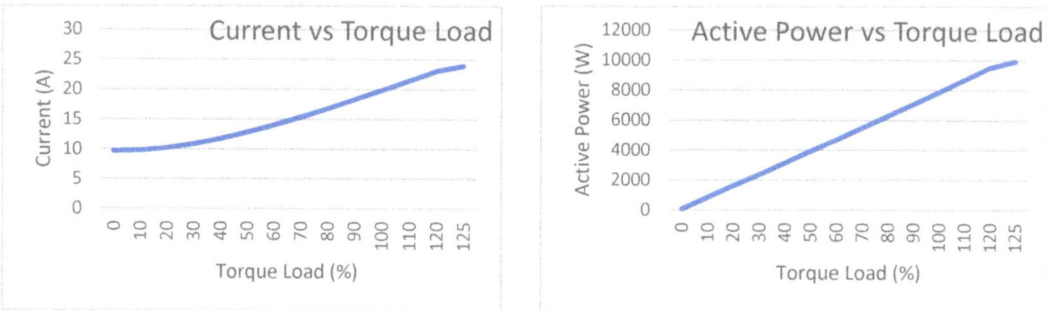

(a) (b)

Figure 13. (**a**) Current vs torque load; (**b**) active power vs torque load.

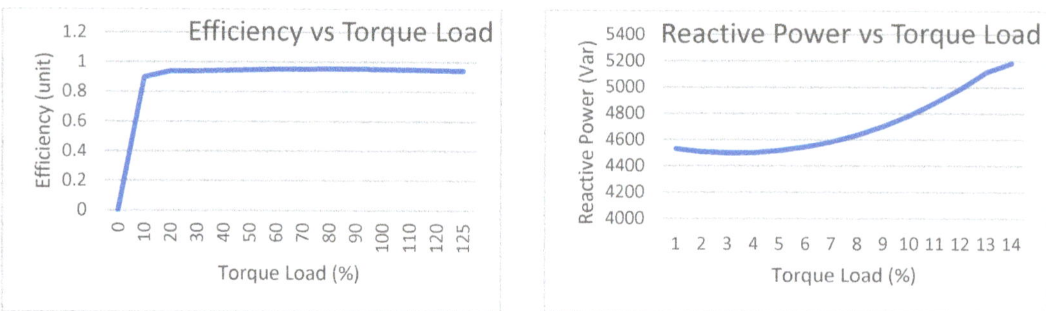

(a) (b)

Figure 14. (**a**) Efficiency vs torque load; (**b**) reactive power vs torque load.

As shown in Figure 15a, the power factor is similar to the efficiency. The lower the load is, the lower the power factor will be, and a very good power factor is reached at the rated torque.

As shown in Figure 15b, the induction machine's speed decreases as the torque load increases, starting from a value that is very close to the synchronous speed. Because the machine acts as an induction motor, it will always run below the synchronous speed.

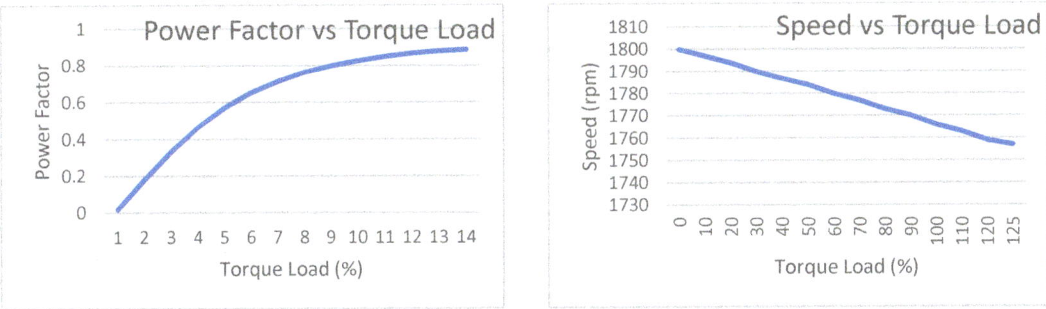

(**a**) (**b**)

Figure 15. Power factor vs torque load (**a**); speed vs torque load (**b**).

8.2. Transient Behaviour of the Current and Torque Versus Speed during Starting, Assuming That (i) the Machine Is Unloaded, (ii) the Machine Is 50% Loaded, and (iii) the Machine Is 100% Loaded

In Figure 16b, we see the three-phase current. One characteristic of an induction machine is that it has a very high starting current in the transient state.

As seen in Figure 17a, the electromagnetic torque oscillates during the transient state until the oscillation stops and moves towards the torque load.

In Figure 17b, we see the same oscillation behaviour in the transient state; however, the machine reaches an optimal point that is close to the synchronous speed due to lack of a load.

In Figure 18, we see a small disturbance in the speed. This is due to the oscillation of the electromagnetic torque. The motor reaches a steady-state close to the synchronous speed due to the lack of a load.

CRITERIA—0% Torque Load; thus, 0 N.m

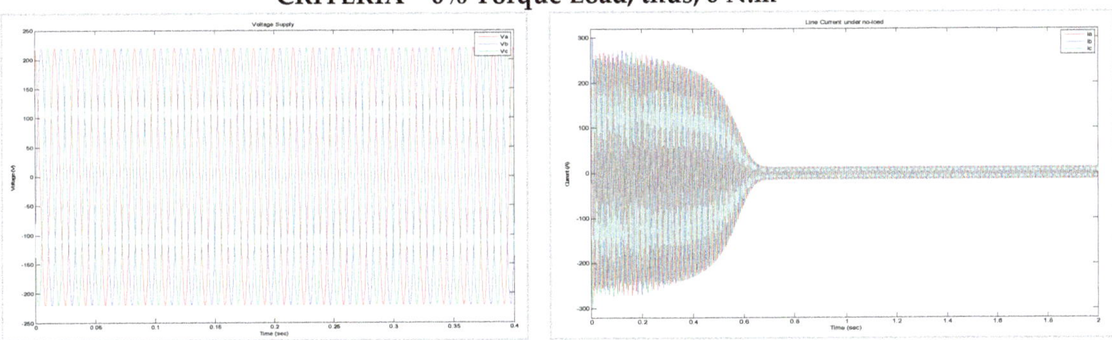

(**a**) (**b**)

Figure 16. (**a**) Voltage supply under no-load; (**b**) line currents under no-load.

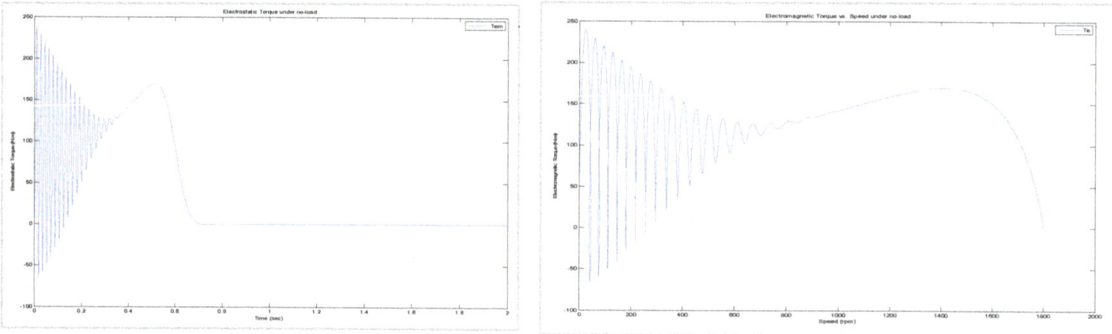

(a) **(b)**

Figure 17. (**a**) Electromagnetic torque under no-load; (**b**) electromagnetic torque vs speed under no-load.

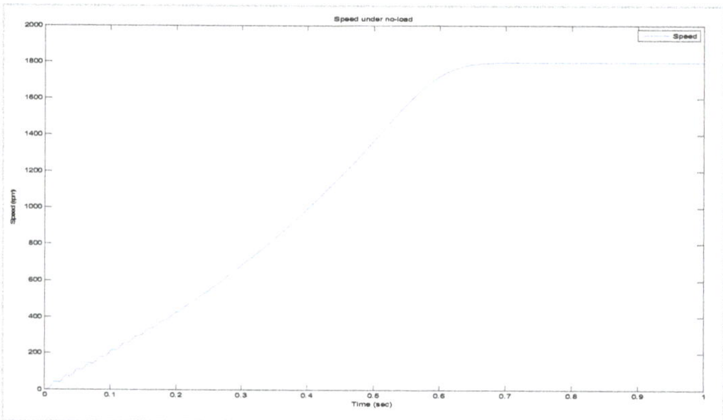

Figure 18. Speed under no load.

The supply stays the same because no impedances exist at the source in this simulation. Figure 19a–c illustrates that the steady-state reached approximately 0.15 s later than with no load. This was expected because of the increase in the load.

As seen in Figure 19a, the amplitude was not affected by the higher magnitude of the load. Therefore, the conclusion can be drawn that the machine current is only a function of the machine's parameters.

As seen in Figure 20, the optimal point is at a lower speed than with no load.

As seen previously, as the loading increases, it takes longer for the steady-state to be achieved. At a full load (100% torque load), reaching the steady-state took approximately 0.2 s longer than with the half load (50% torque load). This can be seen in Figure 21a–c.

CRITERIA—50% Torque Load; thus, 20 N.m

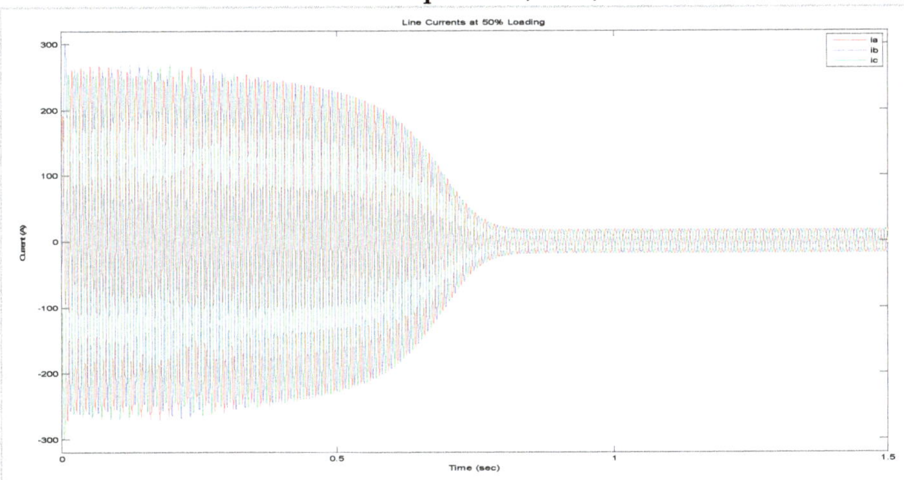

(**a**)

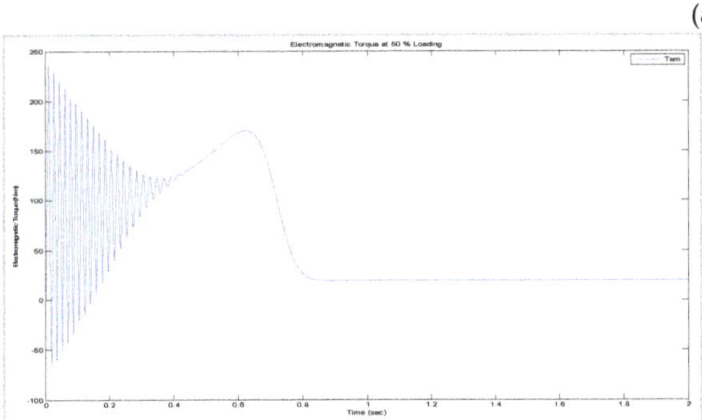

(**b**)

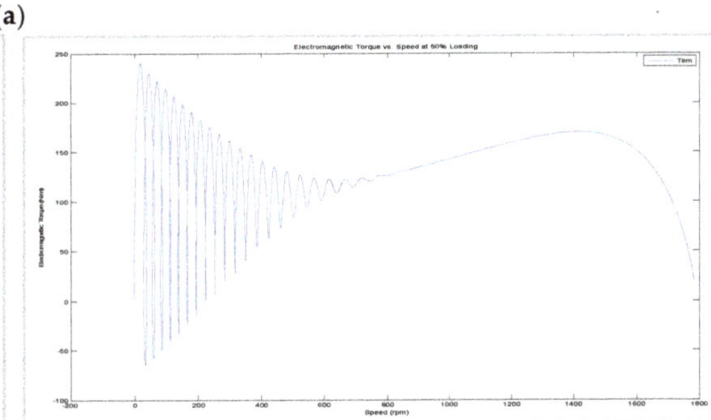

(**c**)

Figure 19. (**a**) Line currents at 50% loading. (**b**) Electromagnetic torque at 50% loading. (**c**) Electromagnetic torque vs. speed at 50% loading.

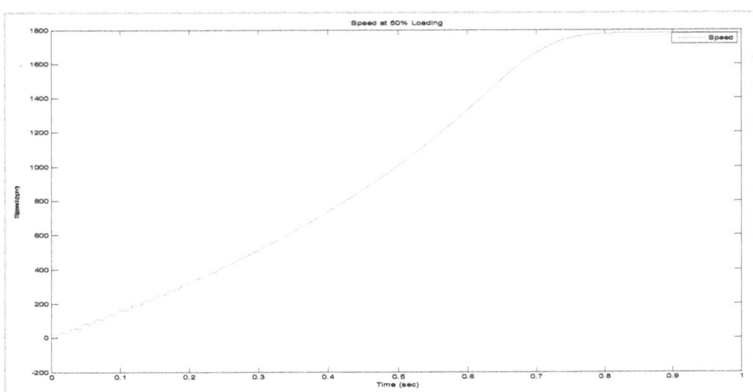

Figure 20. The speed at 50% loading.

CRITERIA—100% Torque Load; thus, 40 N.m

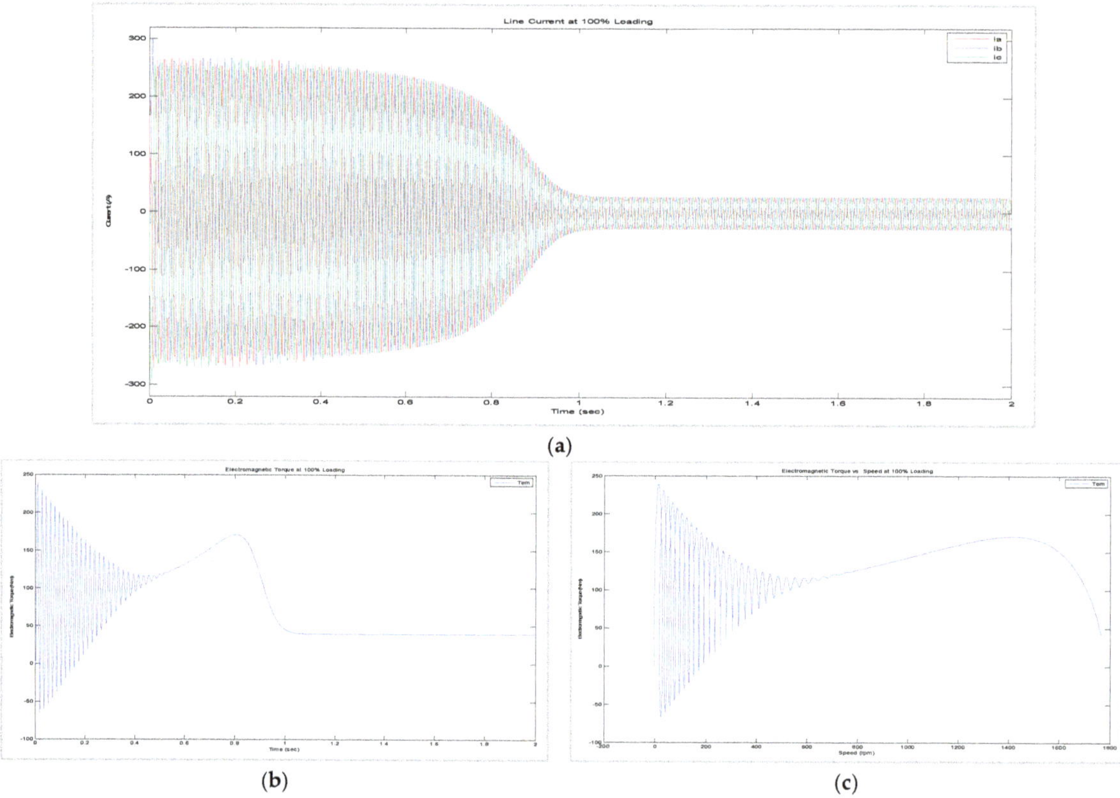

Figure 21. (**a**) Current at 100% loading. (**b**) Electromagnetic torque at 100% loading. (**c**) Electromagnetic torque vs speed at 100% loading. As seen in Figure 22b, the optimal point is at a lower speed than with no-load or half-load.

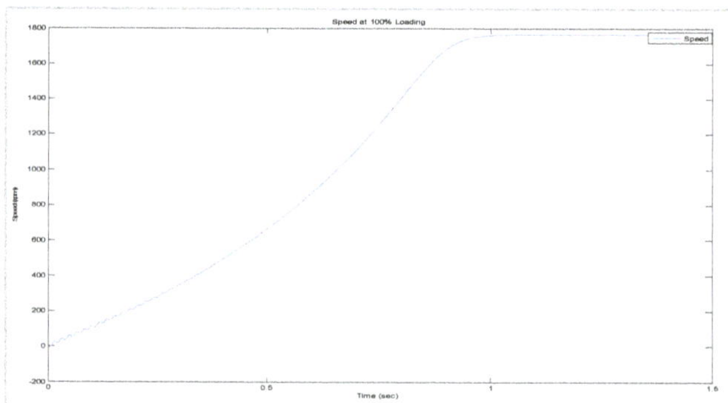

Figure 22. The speed at 100% loading.

8.3. Dynamic Behaviour of the Machine When (i) the Load Was Suddenly Changed from 50% to 100% and (ii) the Supply Voltage (for the Motors) or the Torque Input (for the Generators) Was Suddenly Halved While the Load Was Maintained at 100%

As seen in Figure 23a, we have a normal transient state; however, an increase in currents may be seen as the load changes from a half load to a full load (50% to 100%).

As seen in Figure 23b, we have a normal transient state; however, with an increase in the electromagnetic torque, the load changes from a half load to a full load (50% to 100%). It can be seen in Figure 23c that the optimal point of the full load (100%) is at a lower speed; however, there is a higher torque than with the half load.

In Figure 24, we see the decrease in the speed of the machine after 1 s when the load changes from 50% to 100%.

As seen in Figure 25a, the voltage supply is halved at 1.5 s. We can see in Figure 25b that the current at 1.5 s approximately doubles. At 1.5 s, the protection of the induction machine will operate with a disconnect from the supply voltage.

In Figure 26a, at 1.5 s, the machine cannot produce the electromagnetic torque required for the load torque; thus, the speed of the machine decreases. In Figure 26b, we see that no stable point has been reached.

Dynamic simulations: Load changes from 50% to 100% at 1 s

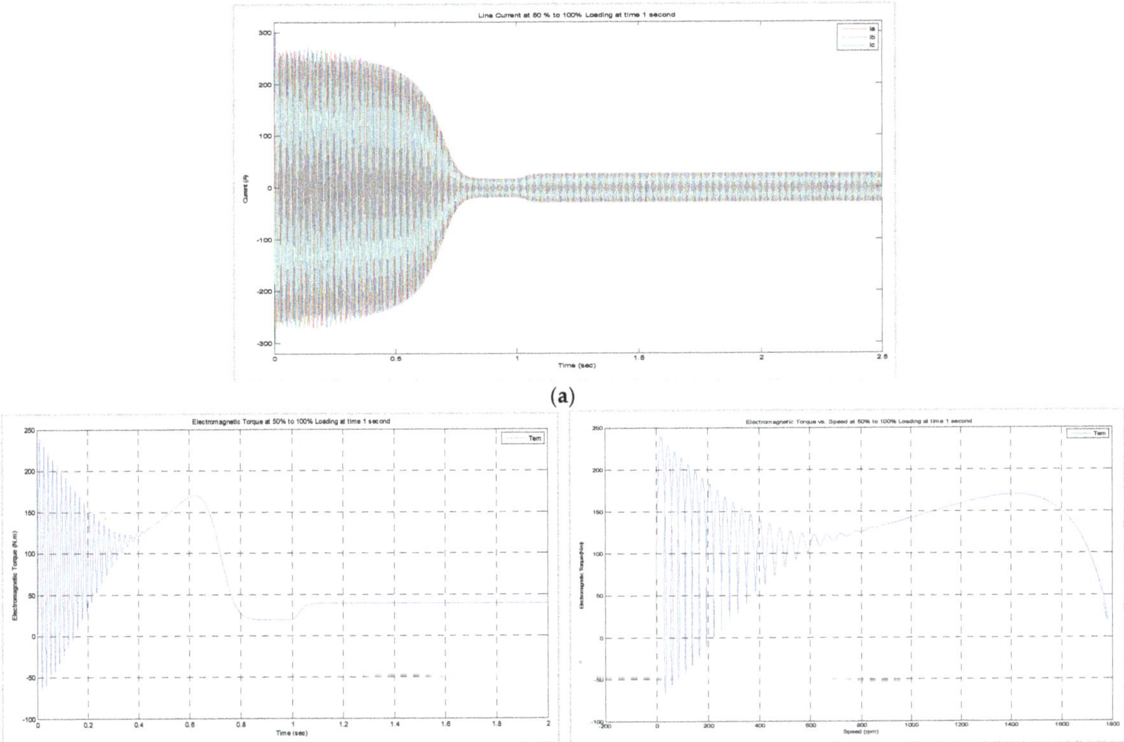

(a)

(b) (c)

Figure 23. (**a**) Line current at 50% to 100% loading at 1 s. (**b**) Electromagnetic torque at 50% to 100% loading at 1 s. (**c**) Electromagnetic torque vs. speed at 50% to 100% loading at 1 s.

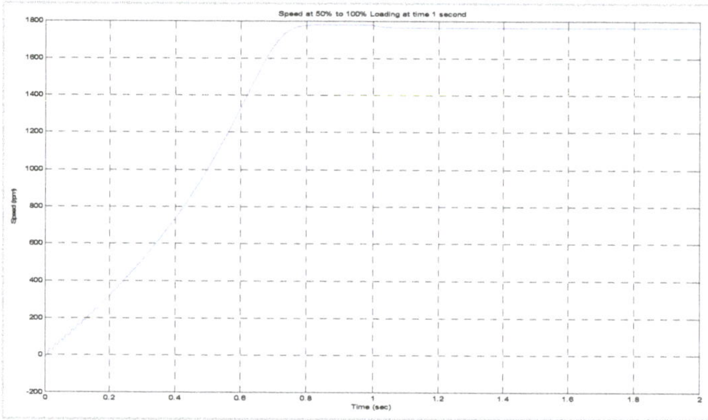

Figure 24. Speed at 50% to 100% loading at 1 s.

Supply voltage changes from 100% to 50% at 1.5 s

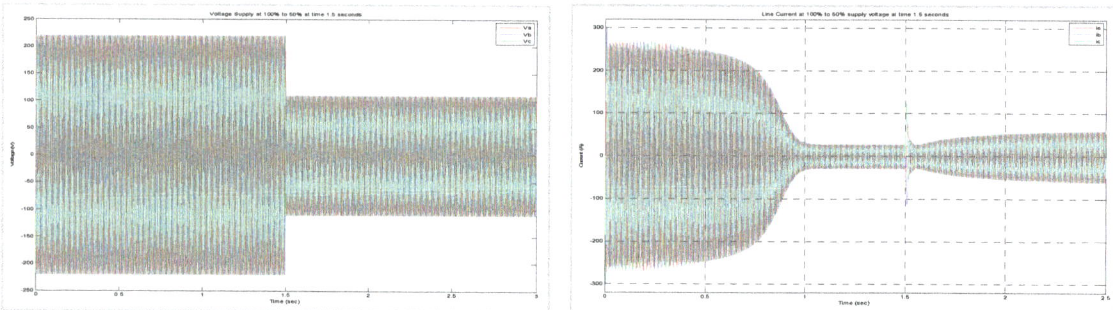

(a) (b)

Figure 25. (**a**) The change in the voltage supply from 100% to 50% at 1.5 s. (**b**) Line current when the voltage changes from 100% to 50% at 1.5 s.

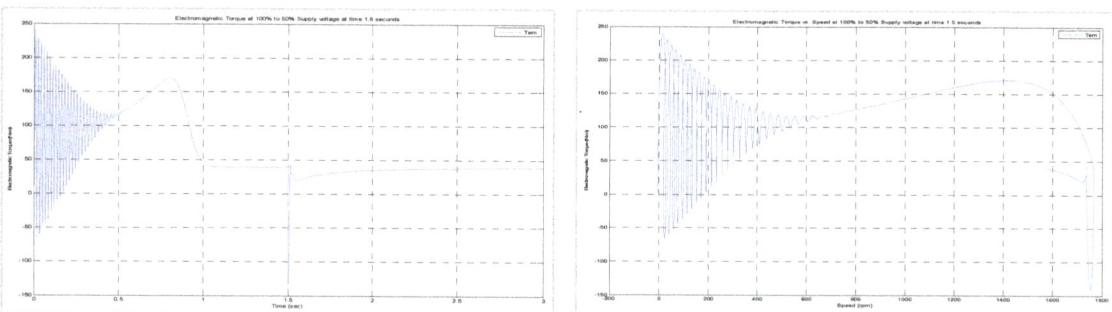

(a) (b)

Figure 26. (**a**) Electromagnetic torque when the supply voltage changes from 100% to 50% at 1.5 s. (**b**) Electromagnetic torque vs speed when the supply voltage changes from 100% to 50% at 1.5 s.

As seen in Figure 27, the speed decreases due to the supply voltage being halved at 1.5 s; thus, the electromagnetic torque is less than the loading torque. This means that the machine is not able to operate under these specific conditions.

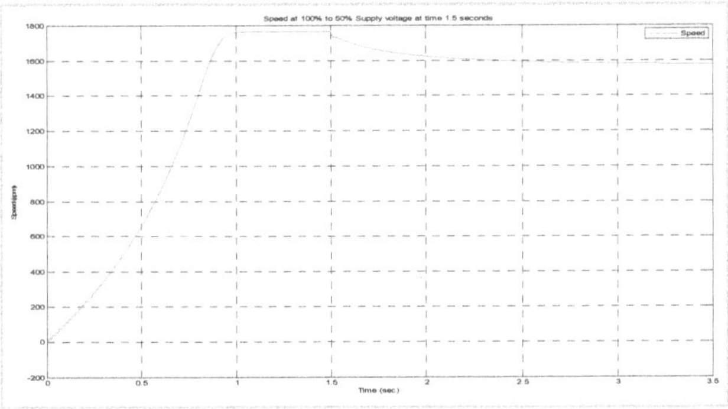

Figure 27. The speed when the supply voltage changes from 100% to 50% at 1.5 s.

9. Conclusions

The performance of the stator and rotor variables was effectively studied by utilising dynamic and static Simulink designs for the modelling of an induction machine. In contrast to several existing induction machine design applications, the operator has access to all internal parameters in this design in order to gain knowledge about the machine's operations. By utilising such designs, any machine control method can be modelled in the Matlab/Simulink software without estimation techniques. For every module, individual variable calculations were performed. Each designer's function was modelled, and critical variables were observed.

In this work, a dq0-direct axis algorithm was presented in order to implement both static and dynamic modelling of a three-phase induction machine due to possible faults and high-performance requirements for induction machines. The proposed algorithm was compared with several conventional methods. It was observed that under stable conditions of the machinery, the proposed algorithm could remove any developing faults. This conserves time and minimises the labour required of an operator, which makes the proposed algorithm more efficient. Furthermore, the machine demonstrated a steady-state performance with respect to the current, active power, efficiency, reactive power, power factor, and speed when the torque loads ranged from 0% to 125% of the nominal torque. The transient behaviour of the machine was shown through the current, electromagnetic torque, electromagnetic torque versus speed, and speed under no-load, half-load (50%), and full-load (100%) conditions. Finally, the results of the proposed technique were compared to the results of the measured parameters. It was found that when the load changed from a half load (50%) to a full load (100%), the supply voltage was suddenly halved with the load at full load (100%). It was observed that the proposed algorithm provides accurate estimates with a deviation of not more than +/−2% from the measured parameters.

Author Contributions: Conceptualisation and methodology, P.F.L.R. and M.K.N.; Software, P.F.L.R.; Validation, P.F.L.R.; Investigation, P.F.L.R.; Writing—original draft preparation, P.F.L.R. and M.K.N.; Writing—review and editing, P.F.L.R.; Supervision, P.F.L.R. All authors have read and agreed to the published version of the manuscript.

Funding: This research received no external funding.

Acknowledgments: The authors would like to acknowledge the reviewers for their academic and specialist assistance and beneficial remarks.

Conflicts of Interest: The authors declare no conflict of interest.

References

1. Rajput, S.; Bender, E.; Averbukh, M. Simplified algorithm for assessment equivalent circuit parameters of induction motors. *IET Electr. Power Appl.* **2020**, *14*, 426–432. [CrossRef]
2. Ahmed, M.; Vahidnia, A.; Datta, M.; Meegahapola, L. An adaptive power oscillation damping controller for a hybrid AC/DC microgrid. *IEEE Access* **2020**, *8*, 69482–69495. [CrossRef]
3. Iegorov, O.; Iegorova, O.; Miroshnyk, O.; Savchenko, O. Improving the accuracy of determining the parameters of induction motors in transient starting modes. *Energetika* **2020**, *66*, 15–23. [CrossRef]
4. Wang, H.; Chau, K.; Lee, C.H.; Cao, L.; Lam, W.H. Design, Analysis, and Implementation of Wireless Shaded-Pole Induction Motors. *IEEE Trans. Ind. Electron.* **2020**, *68*, 6493–6503. [CrossRef]
5. Rahaman, N.; Govindraju, H. Modeling and Simulation of a Three-Phase Electric Traction Induction Motor Using MATLAB/Simulink. *Int. J. Electr. Electron. Comput. Syst.* **2014**, *2*, 5.
6. Kozjaruk, A.E.; Vasilev, B.U.; Shtop, S.A.; Serdukov, N.A. Currents in bearings of induction motors of electric drives with semiconductor converter. In Proceedings of the 17th International Ural Conference on AC Electric Drives (ACED), Ekaterinburg, Russia, 26–30 March 2018; pp. 1–5.
7. Maleki, M.G.; Chabanloo, R.M.; Farrokhifar, M. Accurate coordination method based on the dynamic model of overcurrent relay for industrial power networks taking contribution of induction motors into account. *IET Gener. Transm. Distrib.* **2019**, *14*, 645–655. [CrossRef]
8. Abdelwanis, M.I.; Selim, F.; El-Sehiemy, R. An efficient sensorless slip dependent thermal motor protection schemes applied to submersible pumps. *Int. J. Eng. Res. Afr.* **2015**, *14*, 75–86. [CrossRef]
9. Chitra, V.; Prabhakar, R. Induction motor speed control using fuzzy logic controller. *World Acad. Sci. Eng. Technol.* **2006**, *23*, 17–22.
10. Elnaghi, B.E.; Mohammed, R.H.; Dessouky, S.S.; Shehata, M.K. Load test of induction motors based on PWM technique using genetic algorithm. *Int. J. Eng. Manuf.* **2019**, *9*, 1. [CrossRef]
11. Sadasivan, J.; Mammen, O. Genetic algorithm based parameter identification of three phase induction motors. *Int. J. Comput. Appl.* **2011**, *31*, 51–56.
12. Jirdehi, M.A.; Rezaei, A. Parameters estimation of squirrel-cage induction motors using ANN and ANFIS. *Alex. Eng. J.* **2016**, *55*, 357–368. [CrossRef]
13. Keerthipala, W.; Duggal, B.; Chun, M.H. Torque and speed control of induction motors using ANN observers. *Int. Conf. Power Electron. Drives Energy Syst. Ind. Growth Proc.* **1998**, *1*, 282–288.
14. Chen, J.; Severson, E.L. Design and modeling of the bearingless induction motor. In Proceedings of the IEEE International Electric Machines & Drives Conference (IEMDC), San Diego, CA, USA, 12–15 May 2019; pp. 343–350.
15. Swami, H.; Jain, A.K. An Improved Scalar Controlled Drive Based on Steady State Model of Vector Controlled Drive for Squirrel Cage Induction Motor. In Proceedings of the IEEE 30th International Symposium on Industrial Electronics (ISIE), Kyoto, Japan, 20–23 June 2021; pp. 1–6.
16. Levi, E.; Vukosavic, S.; Jones, M. Vector control schemes for series-connected six-phase two-motor drive systems. *IEEE Proc.-Electr. Power Appl.* **2005**, *15*, 226–238. [CrossRef]
17. Goolak, S.; Gubarevych, O.; Yermolenko, E.; Slobodyanyuk, M.; Gorobchenko, O. Mathematical modeling of an induction motor for vehicles. *East.-Eur. J. Enterp. Technol.* **2020**, *2*, 104. [CrossRef]
18. Prabakaran, S.; Venkatesan, S. Analysis of 3 phase Induction Motor Protection Using Numerical Relay. *Int. J. Eng. Tech.* **2018**, *4*, 513–519.
19. Yildiz, R.; Barut, M.; Demir, R. Extended Kalman filter based estimations for improving speed-sensored control performance of induction motors. *IET Electr. Power Appl.* **2020**, *14*, 2471–2479. [CrossRef]
20. Zhefeng, M.L.; Zhang, G.; Diao, L.; Liu, Z. Extended Kalman Filter based on inverse Γ model of induction motor. In Proceedings of the IEEE Vehicle Power and Propulsion Conference, Harbin, China, 3–5 September 2008; pp. 1–5.
21. Khamehchi, S.; Mölsä, E.; Hinkkanen, M. Comparison of standstill parameter identification methods for induction motors. In Proceedings of the IEEE 9th International Symposium on Sensorless Control for Electrical Drives (SLED), Helsinki, Finland, 13–14 September 2018; pp. 156–161.
22. Bimal, K. *Modern Power Electronics and AC Drives*; Prentice-Hall: Hoboken, NJ, USA, 2001.
23. Tang, Y.; He, X.; Cao, J. Fast calculation method of transient temperature rise of motor for electro-mechanical braking. *Sci. Prog.* **2021**, *104*, 368504211024553. [CrossRef] [PubMed]
24. Krause, P.C.; Wasynczuk, O.; Sudhoff, S.D.; Pekarek, S.D. *Analysis of Electric Machinery and Drive Systems*; John Wiley & Sons: Hoboken, NJ, USA, 2013.
25. Wang, L.; Aleksandrov, S.; Tang, Y.; Paulides, J.J.; Lomonova, E.A. Fault-tolerant electric drive and space-phasor modulation of flux-switching permanent magnet machine for aerospace application. *IET Electr. Power Appl.* **2017**, *11*, 1416–1423. [CrossRef]
26. Duan, N.; Ma, X.; Lu, S.; Wang, S.; Wang, S. Simulation Calculation of Loss of Induction Traction Motor. In *The Proceedings of the 9th Frontier Academic Forum of Electrical Engineering*; Springer: Berlin/Heidelberg, Germany, 2021; pp. 599–606.
27. Liu, Y.; Bazzi, A.M. A detailed induction machine core loss model in the arbitrary reference frame. In Proceedings of the IEEE Applied Power Electronics Conference and Exposition (APEC), Charlotte, NC, USA, 15–19 March 2015; pp. 2617–2622.
28. Shepard, R.A. A Computational Investigation of Cathode Materials for Next-Generation Secondary Batteries. Ph.D. Thesis, State University of New York at Binghamton, Binghamton, NY, USA, 2021.

29. Claassens, A.J. Transient Modelling of Induction Motors in a Petrochemical Plant Using Matlab. Master's Thesis, Stellenbosch University, Stellenbosch, South Africa, 2008.
30. Smith, J.R.; Chen, M.J. *Three-Phase Electrical Machine Systems: Computer Simulation*; Research Studies Press Taunton: Taunton, UK, 1993.
31. Pereira, L.; Kosterev, D.; Mackin, P.; Davies, D.; Undrill, J.; Zhu, W. An interim dynamic induction motor model for stability tudies in the WSCC. *IEEE Trans. Power Syst.* **2002**, *17*, 1108–1115. [CrossRef]
32. Diaz, A.; Saltares, R.; Rodriguez, C.; Nunez, R.F.; Ortiz-Rivera, E.I.; Gonzalez-Llorente, J. Induction motor equivalent circuit for dynamic simulation. In Proceedings of the IEEE International Electric Machines and Drives Conference, Miami, FL, USA, 3–6 May 2009; pp. 858–863.

energies

MDPI

Article

Efficiency Improvement by Deriving the Optimal Operating Slip Frequency of a Linear-Induction-Style Maglev Train

Sang-Uk Park [1], Hyung-Soo Mok [1], Jae-Won Lim [2], Hyun-Uk Seo [3] and Sang-Hun Oh [1,*]

1 Department of Electrical Engineering, Konkuk University, Seoul 05029, Korea; shyman01@konkuk.ac.kr (S.-U.P.); hsmok@konkuk.ac.kr (H.-S.M.)
2 Department of AI Machinery, System Engineering Research Division, Korea Institute of Machinery and Materials, Daejeon 34103, Korea; einses@kimm.re.kr
3 Department of Robotics & Mechatronics, Korea Institute of Machinery and Materials, Daejeon 34103, Korea; shu@kimm.re.kr
* Correspondence: osh@muha.kr

Received: 5 November 2020; Accepted: 9 December 2020; Published: 11 December 2020

Abstract: While urban maglev trains have the advantage of being optimized for urban environments where noise is low and dust is less generated, their driving efficiency is low when compared to rotary induction motors owing to the structural limitations of linear devices. To compensate for these disadvantages, various studies on train control schemes have been conducted. Representative control methods include improving the efficiency of using slip frequency by directly controlling the propulsion force using vector control. However, this method has limitations in its use as it relates to the normal force that affects the train's levitation system. Therefore, in this study, mathematical analysis was conducted for each factor that mutually affects the control of the train. On this basis, the magnitude of the normal force related to the safety of the train is limited. Operating efficiency was improved by varying the slip frequency according to the operating conditions of the train. In addition, for verification, the effect was proved through a comparative experiment using an 18 ton class maglev train running at Incheon International Airport.

Keywords: LIM; slip frequency; linear induction motor; automatic train operation

1. Introduction

The linear induction motor (LIM) has been widely studied as a transportation system running in urban areas owing to its low noise, environmentally friendly factors that do not generate dust, and its excellent performance on slopes and around sharp curves. An LIM is a system that levitates and is propelled through the interaction of the rails and vehicles using the power of an electromagnet. It comprises, primarily, a levitation system and a propulsion system. In the levitation system, as shown in Figure 1b, the train guide generates an attraction force through the lower part of the rail to levitate the train. As shown in Figure 1a, the propulsion system generates magnetic flux using an electromagnet mounted on the train, linking it to the rail. Subsequently, the linkage magnetic flux generates a counteracting flux in the direction of the train on the rail. Consequently, the train and rail are attracted and repelled by the correlation between the magnetic flux generated by the electromagnet mounted on the train and the counteracting magnetic flux of the rail. The LIM generate thrust for propulsion through attraction and repulsion—a normal force being generated in the rail direction. Therefore, to generate the thrust required to propel the train, a normal force that does not contribute to the propulsion of the train is generated. In addition, because the normal force is generated in the opposite direction of the levitation force of the train, the levitation system must overcome gravity and normal forces, and float the train. In other words, the unnecessarily generated normal force is a factor that destabilizes the levitation system of the train, it being a potential safety problem due to train levitation failure. It also induces additional energy consumption in both the propulsion system and the levitation system, thereby reducing efficiency [1]. Therefore, for the efficient operation of trains, a train control technique that reflects the characteristics of linear devices is required.

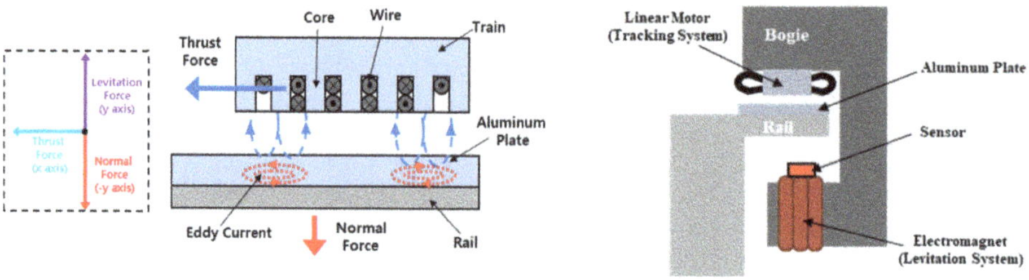

(**a**) (**b**)

Figure 1. Structure of maglev train: (**a**) Structure of trains and rails and (**b**) Structure of levitation system.

For LIM control, a control method using slip frequency and a method using indirect vector control were widely studied. The slip-frequency control method was used because of its independence from parameter fluctuations and ease of implementation. Because the linear motor is based on an induction motor, the size of final load R_L fluctuated according to slip, as shown in Figure 2a. Accordingly, the ratio of the current for magnetization and the current for propulsion fluctuated. As shown in Figure 2b, the magnitude of the input current required for operation based on the slip increased when the slip was large, and decreased when the slip was small [2–5].

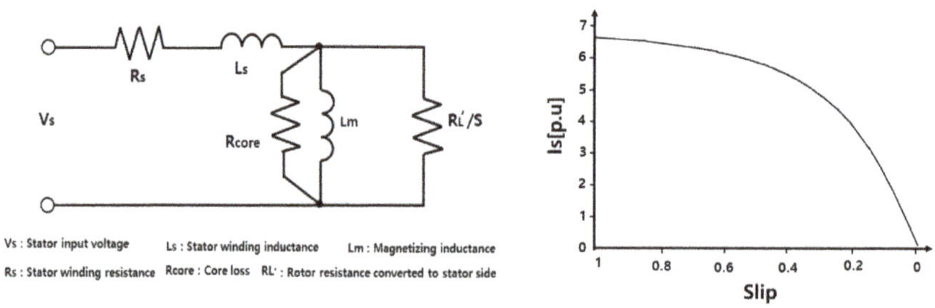

(**a**) (**b**)

Figure 2. Efficiency fluctuation due to slip: (**a**) Induced motor equivalent circuit and (**b**) Input current and slip.

Because the LIM is a system based on an induction motor that cannot directly control the slip, the slip frequency (having a proportional relationship to the slip) was used. Accordingly, a study was conducted on a method of improving efficiency through the size of the slip frequency [6–8]. However, slip is a factor related to the normal force that affects the potential failure of the train. There is also a problem in that normal force increases when the size of slip frequency decreases [9].

In [8], a fixed slip-frequency control method was proposed that used fixed high slip frequency that did not fail to levitate the train. However, by using the same slip frequency in the operating bands of all trains, a problem occurred in that operating efficiency was lowered by using the same slip frequency, even in sections where high slip frequency was not required (on the basis of train operating conditions). Subsequently, a study of a variable slip-frequency control method was conducted to change the slip frequency on the basis of the operating conditions of the train to lower the slip frequency while limiting the normal force of the LIM [10].

Second, as a method, an indirect vector control method was proposed that is widely used in rotary induction motors with fast response and excellent performance [11]. However, for indirect vector control, the air-gap magnetic flux must be kept constant, but is difficult to apply in the LIM because the air-gap magnetic flux fluctuates during train operation owing to the characteristics of linear devices. In [12–15], a method was presented using the current of the d axis, which is the axis where the magnetic flux of the motor is generated during the vector control of an induction motor. The attenuated magnetic flux was compensated by controlling d-axis current i_d. However, this method also had a problem, in that i_q associated with the thrust force fluctuated to maintain the slip frequency constant when i_d was changed to compensate for the attenuated magnetic flux, as shown in slip angular velocity Equation (1) of indirect vector control (here, i_q means the current in the q axis generating torque in the d-q axis for vector control):

$$\omega_{sl} = \frac{1}{T_r} + \frac{i_q}{i_d}, \tag{1}$$

where ω_{sl} is the slip angular velocity, $T_r = \frac{L_r}{R_r}$, L_r is rotor winding impedance, R_r is rotor winding resistance, i_d is the d-axis current (air-gap magnetic flux), and i_q is the q-axis current (thrust).

Accordingly, in [16], a control method using both indirect vector control and variable slip-frequency control was proposed. When the i_d value was changed to compensate for the air-gap magnetic flux, ω_{sl} changed using i_q such that the slip value was within the allowable range. However, because this method was not a result derived through mutual mathematical analysis of train operating conditions, it was difficult to guarantee safety because the exact normal force was unknown. In addition, because all input values for each condition must be derived through direct experiments, the process costs much time and money. For this reason, maglev trains currently in operation utilize a fixed slip-frequency control method that can guarantee train safety. Therefore, in order to improve train efficiency while ensuring safety, it is necessary to analyze the mutual influence through mathematical analysis of the slip, normal force, and propulsion force. On the basis of the analyzed data, if the calculated slip frequency is instantaneously changed on the basis of the operating conditions of the train, it is possible to safely and efficiently operate the train (the proposed method increases efficiency by using the ratio of slip frequency, normal force, and traction force, which are the characteristics of electromagnetic-suspension-type LIM. Therefore, it is difficult to apply this method to types of maglev trains with different structures and driving methods).

The remainder of this paper is organized as follows. In Section 2, the mathematical relationship between normal/propulsion force and slip frequency is analyzed through an investigation of the relationship among normal force, propulsion force, slip, and slip frequency. After that, through the derived equation, the change in efficiency within the limited normal force is presented. Consequently, a control algorithm for controlling the proposed method is presented. In Section 3, the effect is shown through simulation. In Section 4, experimental evaluation conducted using actual vehicles running on the island of Yeongjong, Korea is summarized. Lastly, Section 5 presents our conclusions.

2. Control Method

2.1. Conventional Method

As described above, in a maglev train, both propulsion and normal force fluctuate according to slip. In particular, levitation in the normal force affecting the levitation system of the train becomes a factor that can hinder the safety of the train. Therefore, to control the train, normal force must be controlled by slip. However, because the LIM is a system based on an induction motor and cannot directly control the slip, it was controlled using slip frequency that has a proportional relationship with the slip, as shown in Equation (2):

$$f_{sl} = f_e S \tag{2}$$

$$f_e = f_m + f_{sl} \tag{3}$$

where f_e is the synchronization frequency, which means the supply frequency of the AC voltage supplied to the stator winding; f_m is the rotor frequency, which is the physical rotation frequency; f_{sl} is the slip frequency, which is the difference between the synchronization frequency and the rotor frequency; and S is the slip, which is the ratio of the slip frequency to the synchronization frequency, and it can be expressed as $s = f_{sl}/f_e$.

Table 1 shows an example of fluctuations in slip and normal force when slip frequency is fixed and the synchronous speed of the train is changed using Equation (2) and Figure 3. Figure 3 shows the normal force value derived through finite-element analysis.

Table 1. Correlation between elements in fixed slip-frequency control.

f_{sl} (Hz)	f_e (Hz)	s	Normal Force (N)
	100	0.1	−4500
10	50	0.2	−3000
	33	0.3	−2000

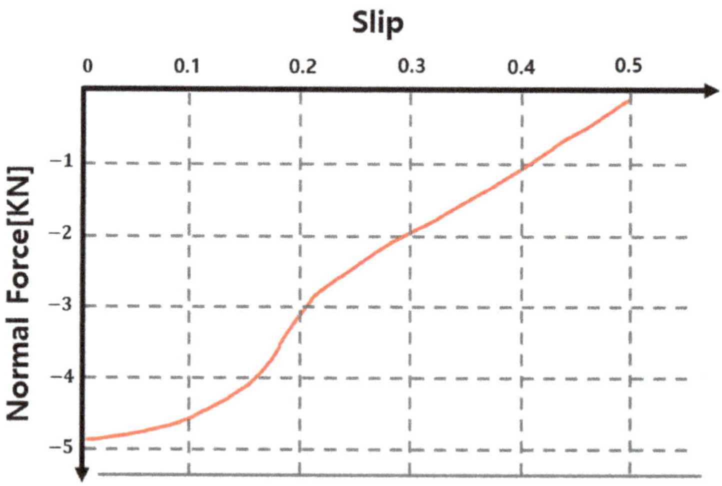

Figure 3. Normal force by slip frequency through finite-element analysis.

Table 1 shows that, as the synchronous frequency associated with the train's synchronous speed decreased, slip increased, maintaining a fixed slip frequency. As shown in Figure 3, as slip increased, vertical force changed in the direction of decreasing, increasing the safety of the train [5,13]. If the slip frequency were fixed, Equation (3) shows that the synchronous speed also had its maximal value at the point where the speed of the train was maximal, and the greatest vertical force occurred at this moment (here, the maximal synchronous speed was the maximal change speed of the magnetic field generated by the stator when the train was running at maximal speed. The greatest normal force was at the moment when the magnitude of the levitation force that made the levitation system unstable was the greatest), that is, if slip magnitude and the normal force at which the levitation did not fail at the maximal speed of the train, maximal slip frequency could be derived from the synchronous speed of the train. If the derived slip frequency were fixed, the train would be in an area where there was less risk of accidents caused by vertical force when the train is running at full speed. This means that the risk of accidents caused by vertical force would also be low in the section where the speed of the train was decreasing. So, the train could safely run in all speed zones. For train safety, this method is applied to currently running maglev trains. However, this method ignores the train's operating conditions and uses a fixed slip frequency, which leads to low-efficiency train operation due to large slip at low speeds. Consequently, if the ratio of the fluctuations of the propulsion force and normal force of the train based on the slip frequency is known, the efficiency of the train can be improved by adjusting slip frequency within the normal force that operates safety zones according to the required propulsion force.

2.2. Proposed Method

The equations of normal force and levitation force used in the control of a maglev train can be obtained through FEM analysis [17] and the structural properties of the target vehicle. Equations (4) and (5) show the relationship between normal force and slip frequency, and driving force and slip frequency, respectively:

$$F_N = \frac{l\tau\mu_0}{2}\frac{1-(R_mS)^2}{(sinh\beta g)^2+(R_mScosh\beta g)^2}(Z_mI_m)^2 \tag{4}$$

$$F_T = l\tau\mu_0\frac{R_mS}{(sinh\beta g)^2+(R_mScosh\beta g)^2}(Z_mI_m)^2 \tag{5}$$

where $R_m = \sigma_t\mu_0\lambda f$ is the magnetic Reynolds number, $\sigma_t = \sigma t_{eff}$ is the effective conductivity, t_{eff} is the effective thickness of the secondary conductor, $\beta = \frac{\pi}{\tau}$ is the air:gap:wavelength ratio, $S = 1 - \frac{V_m}{V_{sy}} = \frac{f_{sl}}{f}$ is the slip, V_m is a variable representing the speed of the train, V_{sy} is a variable representing synchronous speed, Z_m is maximal winding density (the maximal winding density per unit length of the train core), I_m is the maximal current (the maximal current input for the required current to operate the system), g is the effective void, τ is the pole spacing, f is the power frequency, σ is the conductivity of the secondary conductor, f_{sl} is the slip frequency, and l is the primary core width.

By deriving the relational expression between driving force and normal force from the relational expressions in Equations (4) and (5), the ratio of normal force and driving force, as shown in Equation (6), can be derived using the equation related to slip frequency:

$$\frac{F_N}{F_T} = -\frac{1}{2}(R_mS - \frac{1}{R_mS}) \tag{6}$$

where $R_m = \sigma_t\mu_0\lambda f$ and $S = \frac{f_{sl}}{f}$. From this result, R_mS is slip frequency.

If this is summarized in terms of slip frequency, it can be expressed as Equations (7) and (8). Through Equation (8), the maximal normal force can be obtained from the currently running fixed slip-frequency value. If this maximal normal force is then substituted for the required thrust, the maximal usable slip frequency range can be calculated for each operating condition. That is, if the normal force and the propulsion force at the maximal propulsion force of a currently running train are substituted, slip frequency can be determined and controlled so as to be limited to a range that does not fail in levitation.

$$0 = (\sigma_t\mu_0\lambda)^2 f_{sl}{}^2 + 2\frac{F_N}{F_T}(\sigma_t\mu_0\lambda)f_{sl} - 1 \tag{7}$$

$$f_{sl} = \frac{-2\frac{F_N}{F_T}\sigma_t\mu_0\lambda \pm \sqrt{(2\frac{F_N}{F_T}\sigma_t\mu_0\lambda)^2 + 4(\sigma_t\mu_0\lambda)^2}}{2(\sigma_t\mu_0\lambda)^2} \tag{8}$$

Figure 4 shows the ratio of normal force/thrust force according to slip frequency using Equation (6). Table 2 uses Figure 4 and Equation (8) to calculate the slip frequency in which the maximal normal force is generated within the range of levitation not failing when propulsion force is changed (if slip frequency is negative, it operates as a braking mode; if it is positive, it operates as a powering mode for thrust).

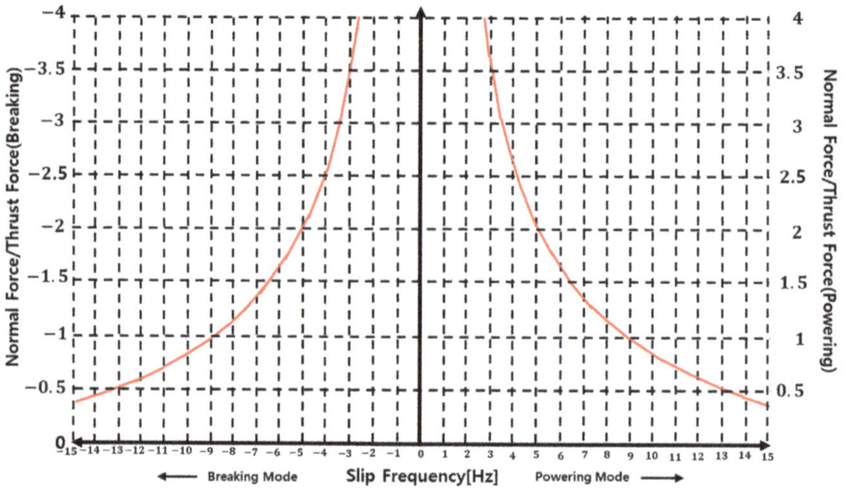

Figure 4. Ratio of normal force and thrust force according to slip frequency.

Table 2. Normal force margin ratio by slip frequency at each thrust.

Thrust	Margin	Slip Frequency (Hz)
100%	1	13.5
75%	1.33	12.5
50%	2	9
30%	3.33	6.5

Table 2 shows that, when the maximal driving force was 100%, when the driving force decreased, the normal force also decreased. Therefore, when using the same slip frequency of 13.5 Hz at approximately 75%, 50%, and 30% thrust, the amount of normal force generated on the basis of maximal thrust is reduced. Assuming that the margin ratio of the normal force is 1 at a slip frequency of 13.5 Hz, margin rates at each operating condition are 1.33, 2, and 3.33, respectively. If slip frequency was lowered by this margin factor, efficient operation would be possible within the range of normal force that did not affect safety. If the optimal slip frequency suitable for train operation conditions were derived in this way, a train-operation pattern capable of varying the appropriate slip frequency during train operation could be obtained.

2.3. Algorithm

Figure 5 shows a block diagram of the proposed control algorithm. Looking at the left side of Figure 5, given thrust command $F_T{}^*$ and train speed v_m, the recommendation calculator determines the slip-frequency command that can be used within the range not exceeding the normal force limit through the previously proposed method.

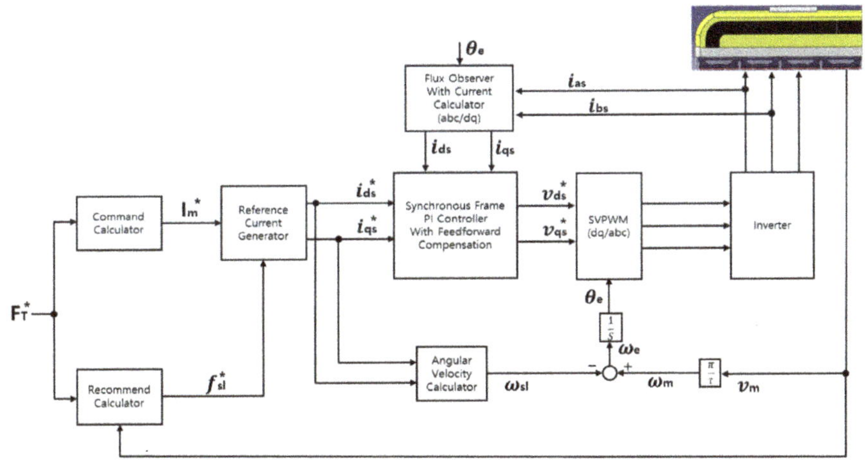

Figure 5. Control algorithm of proposed method.

At this time, the normal force-limit value was derived from finite-element analysis and the initial train test. If derived slip-frequency command $f_{sl}{}^{*}$ and command $\mathrm{I_m}{}^{*}$ of the total current required to drive the system are given as "reference current generator", i_d and i_q for vector control are derived. At this time, total current command $\mathrm{I_m}{}^{*}$ is determined through Equation (5), and the method of deriving i_d and i_q using "reference current generator" is as follows.

$$\omega_{sl} = \frac{R_r}{L_r}\frac{i_q}{i_d} \tag{9}$$

$$I_{rms}{}^2 = i_d{}^2 + i_q{}^2 \tag{10}$$

Equation (9) shows the relationship between slip frequency and d-q axis reference currents i_d and i_q in the induction motor, and Equation (10) shows the relationship between I_m and i_d, i_q derived through the propulsion force. These two equations are used to determine the i_d, i_q current command in the "reference current generator".

$$i_d = \frac{R_r}{L_r}\frac{i_q}{\omega_{sl}} \tag{11}$$

$$I_{rms}{}^2 = (\frac{R_r}{L_r})^2(\frac{i_q}{\omega_{sl}})^2 + i_q{}^2 \tag{12}$$

If Equation (9) is summarized for i_d as in Equation (11) and substituted into Equation (10), Equation (12) representing the relationship between i_q and I_m can be obtained. Reorganizing this for i_q can be expressed as Equation (13):

$$i_q = \frac{L_r}{R_r}\omega_{sl}\sqrt{\frac{1}{(1+(\frac{L_r}{R_r}\omega_{sl}))^2}}I_{rms} \tag{13}$$

$$i_d = \sqrt{\frac{1}{(1+(\frac{L_r}{R_r}\omega_{sl}))^2}}I_{rms} \tag{14}$$

By arranging i_d in the same way, we can obtain Equation (14). The current command of the d-q axis can be derived from constants R_r and L_r of the train, total current I_m required by the driving force, and slip angular velocity ω_{sl} that can be obtained from the slip frequency. Therefore, when the thrust and normal force-limit values, which are the driving conditions of the train, are proposed, it is possible to directly control the i_q current, involved in the propulsive force of the train, and i_d, involved in the lift of the train.

3. Simulation

In order to perform the simulation under the same conditions as those of the actual train, the data of a train that was actually operated were used as input values, as shown in Figure 6a, through which the current command and slip required for control of the slip frequency was derived and utilized. The black waveform represents the speed pattern of the train, and the blue waveform represents the driving-force waveform of the train. For efficiency comparison, total power consumption and generated normal force were compared between the conventional method using a slip frequency of 13.5 Hz, and the proposed method in which the slip frequency fluctuated during train operation based on the operating conditions. In the proposed method, the margin rate of the limited normal force was set to vary between 9.5 and 13.5 Hz according to the applied operating conditions.

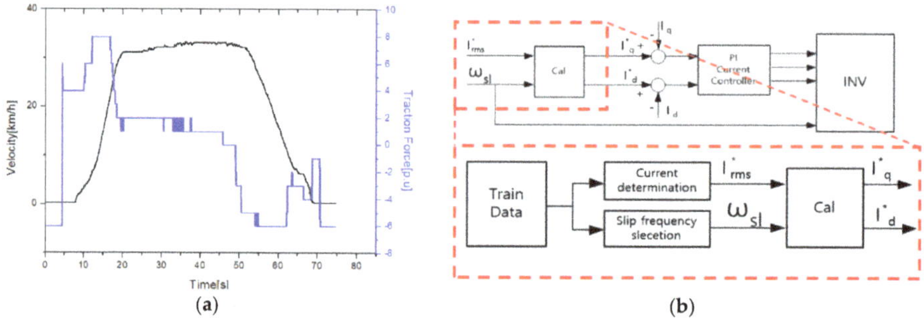

Figure 6. Simulation conditions and calculation blocks: (**a**) Train operation pattern (train data) and (**b**) Block diagram using train pattern.

Figure 7 compares the results of the existing control method and the proposed method. Figure 7a shows the accumulated power consumption while the train was running. The existing control method of the red curve consumed approximately 140 Wh of power while operating under the same conditions and section, whereas the proposed method of the black curve consumed approximately 116 Wh of power, an improvement of approximately 24 Wh, which is a reduction in power consumption and an efficiency improvement of approximately 19.6%. Figure 7b shows the normal force change during train operation. When a margin ratio of approximately 30% was compensated for safety from the actual train's limited normal force, the limit value was about −2.5 kN, called the critical normal force. If the normal force falls below this value, levitation fails. Looking at the waveform in Figure 7b, the maximal generated normal force of the conventional method was −2.04 kN, and the maximal generated normal force of the proposed method was −2.45 kN. Both systems were in the safe area. Therefore, when using the proposed method, efficiency increased by approximately 17.14%, but it was confirmed that the efficiency improvement of the proposed method was effective because the train was running within a safe range of normal force.

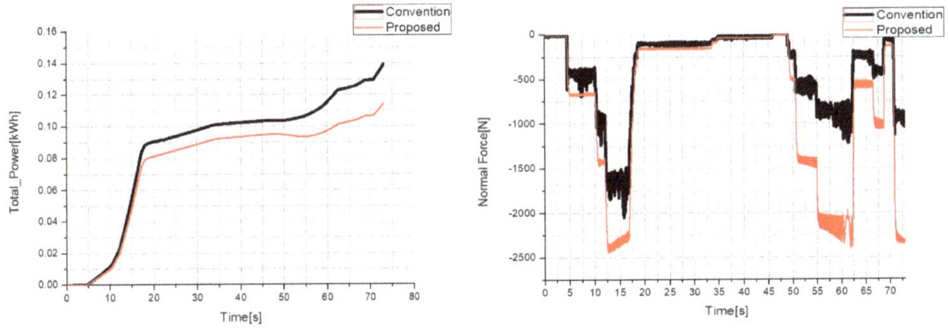

(a) (b)

Figure 7. Comparison of existing and proposed condition data: (**a**) Accumulated power consumption and (**b**) Normal train force.

4. Experiment

Figure 8a shows the LIM train of Incheon International Airport in Korea used for the experiment. In order to compare train efficiency, the power system installed on the train and inverter power were directly measured. Figure 8b shows the train's operating route used in the experiment. Five sections were operated over one round trip from Station 101 (the start station) to Station 106 (the end station).

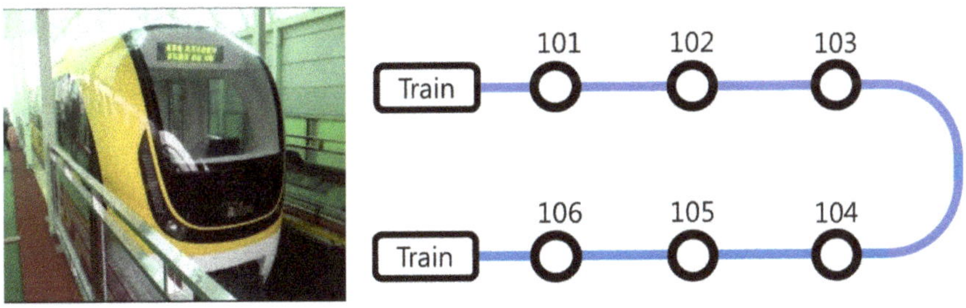

(a) (b)

Figure 8. Actual vehicle test conditions and used vehicle: (**a**) Experimental linear induction motor (LIM) and (**b**) Train test-run section.

Table 3 shows the specifications of the trains used in the experiment; the train was composed of 1 car and 2 trains. For comparison, the widely used 13.5 Hz slip-frequency fixed control method and the proposed slip-frequency variable-vector control method were compared. To increase the reliability of the experiment, it was conducted in triplicate, and results were calculated using the average. Lastly, the train was operated using the automatic-train-operation (ATO) method [18], an automatic train control system that propels, rides, and brakes trains according to given commands. The ATO was used for train operations because it reduces the deviation of experiment results using train drivers and quickly responds to slip-frequency fluctuations during operation.

Table 3. Experimental vehicle specifications.

Traction System		Levitation System	
Parameter	**Value**	**Parameter**	**Value**
Length	1785 mm	Length	2600 mm
Number of slots	53	Number of yokes/poles	4
Number of poles	8	Number of coils/poles	2
Air gap	11 mm	Turns/coil	193
Turns/coil	5	Air gap	8 mm
Al plate thickness	5 mm	Pole width	32 mm
Required thrust	60.4 kN	Rated life force	33 kN
Maximal speed	110 km/h		

Figures 9 and 10 show the accumulated power consumption of each part according to actual train operation. Figure 9 shows the operating results from Station 101 to Station 106, and Figure 10 shows the train operating results from Station 106 to Station 101. Figures 9 and 10a show the results of the conventional train control method, and Figures 9 and 10b show those of the proposed control method. In each curve, the green line represents the total amount of consumed power to float the train, the blue line represents the total power consumption used to propel the train, and the red line represents the sum of the power consumption of propulsion. Lastly, the black line indicates the speed of the train. The moment when the curve changed in value was when the train was running between the stations, and the moments when the curve had no value indicate the waiting time after arriving at the station. To exclude the effect of energy consumption caused by differences in waiting times at each station on the results, the consumed energy during the waiting time at each station was removed from the actual comparison. In addition, the test train was used when comparing the actual consumed energy twice as often as the measured value in the two trains.

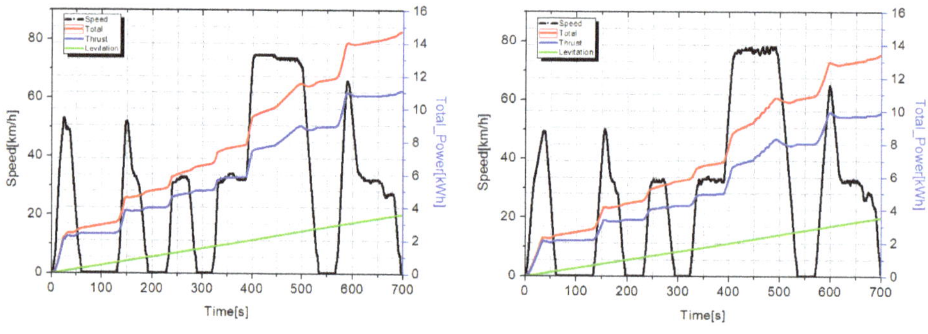

Figure 9. Comparison of accumulated power consumption between existing and proposed methods (101 → 106): (**a**) Conventional method and (**b**) Proposed method.

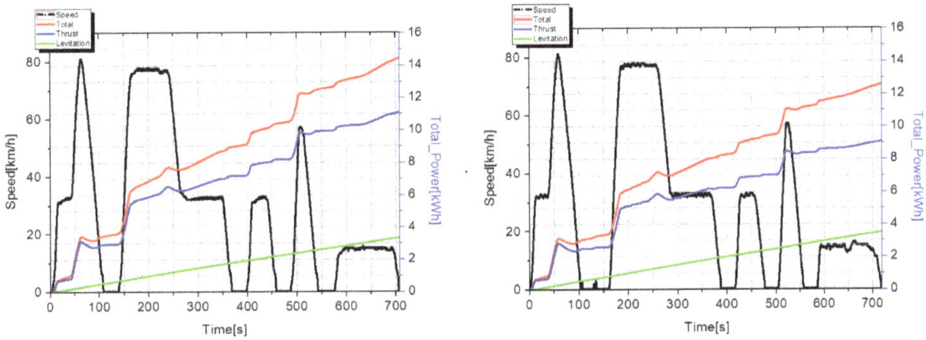

(**a**) Conventional method. (**b**) Proposed method.

Figure 10. Comparison of accumulated power consumption of existing and proposed methods (106 → 101): (**a**) Conventional method and (**b**) Proposed method.

Table 4 shows the results of comparing the two methods in consideration of the elimination of reverse waiting time, and (1) quantity and (2) schedule. When moving from Station 101 to Station 106, the total power consumption of the existing method was 27.74 kWh, and the total power consumption of the proposed method was 25.12 kWh. When using the proposed method, there was approximately 2.62 kWh (9.45%) increased efficiency. When moving from Station 106 to Station 101, the total power consumption of the existing method was 27.8 kWh, the total power consumption of the proposed method was 23.64 kWh; when using the proposed method, there was approximately 4.16 kWh (14.96%) increased efficiency. As a result, efficiency increased by approximately 6.78 kWh (12.2%) when using the proposed method, to 55.54 and 48.76 kWh, respectively, from Station 101 to Station 106.

Table 4. Comparison of power-consumption results.

	Operation from 101 to 106		Operation from 106 to 101	
	Conventional	Proposed	Conventional	Proposed
Total (kWh)	27.74	25.12	27.8	23.64
Thrust (kWh)	22.1	19.6	22.1	18.54
Levitation (kWh)	5.64	5.52	5.7	5.1
Reduction rate (%)		9.45%		14.96%

5. Conclusions

In this study, as part of our research on improving the operating efficiency of a maglev train using an LIM, the relationship between train slip frequency, normal force, and propulsion force was analyzed through a mathematical study. Using the analytical results, the slip frequency having the optimal efficiency was derived on the basis of the train's operating conditions while limiting the normal force to the extent to which the levitation system of the train did not fail. Subsequently, slip frequency was changed according to the operating conditions of the train in real time. Through the ATO driving system, a simulation test in which slip frequency was varied on the basis of the driving conditions of the train while it was running, and an experiment using an actual train were conducted. As a result of the simulation test for one operating section in which the actual train was running, when the proposed method was used rather than the existing fixed system slip frequency of 13.5 Hz, a cumulative power-consumption decrease of approximately 24 Wh and an efficiency gain of approximately 17.14% were achieved. These results confirmed that the efficiency improvement using the proposed method was significant. In the case of the experiment, when the proposed method was compared with the existing fixed system slip frequency of 13.5 Hz, the cumulative power consumption decreased by approximately 6.78 kWh and efficiency increased by approximately 12.2%. Through this, we verified that the proposed method is more efficient than the existing method is (the proposed method uses LIM characteristics, which are suitable for low- and medium-speed types. Therefore, it is difficult to apply to maglev trains with different structures and principles, such as superconducting-repulsion or permanent-magnet types).

Author Contributions: Conceptualization, S.-U.P.; methodology, S.-U.P.; software, H.-U.S.; validation, S.-U.P.; formal analysis, S.-U.P.; investigation, S.-U.P.; resources, J.-W.L.; data curation, S.-U.P.; writing—original-draft preparation, S.-U.P.; writing—review and editing, H.-S.M. and S.-H.O.; visualization, S.-U.P.; supervision, H.-S.M.; project administration, S.-H.O.; funding acquisition, J.-W.L. All authors have read and agreed to the published version of the manuscript.

Funding: This research was funded by the Korea Institute of Machinery and Materials (KIMM).

Conflicts of Interest: The authors declare no conflict of interest.

References

1. Boldea, I.; Nasar, S.A. *Linear Induction Motor. Linear Motion Electromagnetic Devices*; Taylor & Francis: New York, NY, USA, 1997; pp. 35–102.
2. Xu, W.; Zhu, J.G.; Zhang, Y.; Li, Z.; Li, Y.; Wang, Y. Equivalent Circuits for Single-Sided Linear Induction Motors. *IEEE Trans. Ind. Appl.* **2010**, *46*, 2410–2423. [CrossRef]
3. Nozaki, Y.; Yamaguchi, T.; Koseki, T. Equivalent circuit model with parameters depending on secondary speed for high-speed transport—A smart combination of classical modeling and numerical electromagnetic calculation. In Proceedings of the 19th International Conference on Magnetically Levitated Systems Linear Drives, Dresden, Germany, 13–15 September 2006; pp. 537–543.
4. Duncan, J. Linear induction motor-equivalent-circuit model. *IEE Proc. B Electr. Power Appl.* **1983**, *130*, 51–57. [CrossRef]
5. Wang, K.; Li, Y.; Ge, Q.; Shi, L. Indirect field-oriented control of linear induction motor based on optimized slip frequency for traction application. In Proceedings of the 2016 18th European Conference on Power Electronics and Applications, Karlsruhe, Germany, 5–9 September 2016.
6. Lu, C.; Dawson, G.E.; Eastham, T.R. Dynamic performance of a linear induction motor with slip frequency control. In Proceedings of the Canadian Conference on Electrical and Computer Engineering, Vancouver, BC, Canada, 14–17 September 1993.
7. Wallace, A.; Parker, J.; Dawson, G. Slip control for LIM propelled transit vehicles. *IEEE Trans. Magn.* **1980**, *16*, 710–712. [CrossRef]
8. Lu, Q.; Li, Y.; Ye, Y.; Zhu, Z.Q. Investigation of Forces in Linear Induction Motor under Different Slip Frequency for Low-Speed Maglev Application. *IEEE Trans. Energy Convers.* **2013**, *28*, 145–153. [CrossRef]
9. Yang, Z.; Zhao, J.; Zheng, T.Q. A novel traction and normal forces study for the linear induction motor. In Proceedings of the International Conference on Electrical Machines and Systems, Wuhan, China, 17–20 October 2008.
10. Park, S.U.; Zun, C.Y.; Park, D.-Y.; Lim, J.; Mok, H.S. A Study on Improvement of Operation Efficiency of Magnetic Levitation Train Using Linear Induction Motor. *IEEE Trans. Energy Convers.* **2013**, *28*, 145–153. [CrossRef]
11. Ogino, Y.; Murakami, Y.; Nakaoka, M. High performance ultra-low speed control of inverter-fed linear induction motor using vector control scheme. In Proceedings of the IEEE Industry Applications Conference Twenty-Eighth IAS Annual Meeting, Toronto, ON, Canada, 2–8 October 1993.
12. Sung, J.-H.; Nam, K. A new approach to vector control for a linear induction motor considering end effects. In Proceedings of the IEEE Industry Applications Conference. Thirty-Fourth IAS Annual Meeting, Phoenix, AZ, USA, 3–7 October 1999.
13. Kang, G.; Nam, K. Field-oriented control scheme for linear induction motor with the end effect. *IEEE Proc. Electr. Power Appl.* **2005**, *152*, 1565–1572. [CrossRef]
14. Silva, E.F.; Santos, E.B.; Machado, P.C.M.; Oliveria, M.A.A. Vector control for linear induction motor. In Proceedings of the IEEE International Conference on Industrial Technology, Maribor, Slovenia, 10–12 December 2003.

15. Yaohua, L.; Jinqi, R. Drive system of single side linear induction motors. In Proceedings of the International Conference on Electrical Machines and Systems, Wuhan, China, 17–20 October 2008.
16. Wang, K.; Li, Y.; Ge, Q.; Shi, L. An Improved Indirect Field-Oriented Control Scheme for Linear Induction Motor Traction Drives. *IEEE Trans. Ind. Electr.* **2018**, *65*, 9928–9937. [CrossRef]
17. Ooi, B.-T.; White, D. Traction and Normal Forces in the Linear Induction Motor. *IEEE Trans. Power Appl. Syst.* **1970**, *89*, 638–645. [CrossRef]
18. Xiaomin, Z.; Xiang, L. The modeling of test systems of Automatic Train Operation (ATO) in Urban rail transit based on LABVIEW. In Proceedings of the International Conference on Computer Application and System Modeling, Taiyuan, China, 22–24 October 2010.

Publisher's Note: MDPI stays neutral with regard to jurisdictional claims in published maps and institutional affiliations.

energies

Review

Linear Induction Motors in Transportation Systems

Ryszard Palka * and Konrad Woronowicz

Faculty of Electrical Engineering, West Pomeranian University of Technology, Sikorskiego 37, 70-313 Szczecin, Poland; Konrad.Woronowicz@zut.edu.pl
* Correspondence: Ryszard.Palka@zut.edu.pl; Tel.: +48-91-449-4870

Abstract: This paper provides an overview of the Linear Transportation System (LTS) and focuses on the application of a Linear Induction Motor (LIM) as a major constituent of LTS propulsion. Due to their physical characteristics, linear induction motors introduce many physical phenomena and design constraints that do not occur in the application of the rotary motor equivalent. The efficiency of the LIM is lower than that of the equivalent rotary machine, but, when the motors are compared as integrated constituents of the broader transportation system, the rotary motor's efficiency advantage diminishes entirely. Against this background, several solutions to the problems still existing in the application of traction linear induction motors are presented based on the scientific research of the authors. Thus, solutions to the following problems are presented here: (a) development of new analytical solutions and finite element methods for LIM evaluation; (b) comparison between the analytical and numerical results, performed with commercial and self-developed software, showing an exceptionally good agreement; (c) self-developed LIM adaptive control methods; (d) LIM performance under voltage supply (non-symmetrical phase current values); (e) method for the power loss evaluation in the LIM reaction rail and the temperature rise prediction method of a traction LIM; and (f) discussion of the performance of the superconducting LIM. The addressed research topics have been chosen for their practical impact on the advancement of a LIM as the preferred urban transport propulsion motor.

Keywords: linear induction motors; finite element analysis; end effect

Citation: Palka, R.; Woronowicz, K. Linear Induction Motors in Transportation Systems. *Energies* **2021**, *14*, 2549. https://doi.org/10.3390/en14092549

Academic Editors: Frede Blaabjerg and Andrea Mariscotti

Received: 10 March 2021
Accepted: 26 April 2021
Published: 29 April 2021

1. Introduction

A traction Linear Induction Motor (LIM) has been deployed worldwide in numerous transit systems and in driverless, elevated guideway systems requiring all weather operations under very short headways. LIM-based urban transport has proven to be, by far, the least expensive in terms of operations and maintenance (including the energy costs). LIMs are also found in other various applications, ranging from small-power industrial material handling and amusement park roller-coaster propulsion to very high-output military aircraft launchers; advanced research is underway to investigate LIMs as potential power conversion devices for ocean wave energy recovery [1].

Because of low operating costs and extremely high reliability, LIM-propelled systems have become an ever more frequent part of the public transport offering. LIM-based public transit systems have already been in operation for a few decades, and they are serving such cities as Yokohama, Vancouver, Toronto, Tokyo, Osaka, Seoul (Yongin), New York, Moscow (Moscow Monorail), Kuala Lumpur, Guangzhou, Fukuoka, and Beijing. References for the applications are readily available by any browser search using such keywords as "Linear Metro". For economic reasons, the operation of these systems as well as most other LIM-based systems has been based on a single-sided LIM [1,2]. Although, in comparison to its rotary counterpart, a traction LIM has necessarily a large air gap between the stator and its rotor equivalent and thus is less efficient on the motor-to-motor comparison bases, the LIM-based systems show better system-level performance resulting from several characteristic features specific to the LIMs. First, in comparison with rotary induction motors, a mechanical gear box that introduces high energy losses and accounts for a significant life cycle cost and is a potential source of serious reliability issues is eliminated. The LIM has no moving parts, and the propelling force is directly applied to the vehicle in the direction of motion, thus avoiding the losses introduced by the mechanical gear box. In most instances, rotary-motor-based transportation systems rely on adhesion between the wheels and the running rail. Relying on adhesion limits their acceleration and deceleration performance under wet or otherwise contaminated rail conditions. LIM systems do not suffer such disadvantages because their tractive effort is developed as a direct electrodynamic force between the LIM primary and the reaction rail, and this allows for an adhesion independent, reliable operation under all environmental conditions. This also means that LIM-propelled trains (objects) can accelerate at any rate and achieve their nominal speeds sooner, which limits the high current/tractive effort demand period and decreases the overall energy consumption. Because of their flat form, LIMs occupy significantly smaller vertical space, which enables a lower profile steerable bogie construction and, consequently, a lower vehicle cross-sectional area, thus decreasing the potential tunnel construction costs and energy consumption resulting from the motion air resistance [3,4]. In addition, LIM-based vehicles can run on steeper grades (due to direct forces) and negotiate sharper curves (due to steerable bogies), providing more flexibility in the elevated guideway structure design, which helps to reduce civil and land release costs. A large urban center allows for more targeted interconnection of the various multi-modal transportation systems, which is not always achievable with rotary-motor-based rail vehicles. The uniqueness of Linear Motor (LM) systems led to a heightened interest in LIM technology and resulted in a number of research projects aimed at improving LIM performance and further decreasing the operating cost.

In this paper, an overview and categorization of linear transportation systems is completed, in which LIMs are found to mainly be used for traction and braking. Next, the major characteristics of a LIM are described and associated with the system performance. Finally, a series of important practical research works carried out by the authors and aimed at advancing LIM system performance are reported, highlighting challenges associated with improving LIM performance in such areas as performance prediction, LIM adaptive control, LIM thermal protection, and the application of superconductivity. The conclusions summarize the authors' experience in the subject matter and highlight the areas of advancement for future research.

2. Overview of Linear Transportation Systems

Linear transportation systems (LTSs) can be divided into some fundamental groups according to their support, guidance, and drive solutions. The most important groups are listed below.

2.1. Levitated LTS

Levitated LTSs mainly use electromagnetic suspension. These systems, Maglevs, can be divided into Permanent Magnet (PM)- and superconductor-based levitating systems. According to the postulate on the stability of bodies in various static force fields given in 1842 by Earnshaw [5], no object placed in an inverse square law force field (e.g., magnetic field) could be in a stable equilibrium; thus to achieve stability, PM flux should be modulated by the respective control coil currents that depend on the size of the air gap. Such controlled electromagnets are utilized in many different levitation systems, for both side and vertical stabilization [6]. Efficient operation of these systems depends on the advanced optimization of the magnetic field distribution within the air gap by means of proper geometry design and proper selection of material and power supply characteristics [7,8].

The first Maglev line to open to public traffic was the Birmingham Maglev in 1984 (propelled by a LIM), the second was the M-Bahn in Berlin (also 1984), and the third was the Shanghai Transrapid Maglev (the latter two using a long-stator synchronous linear-motor-based propulsion).

Braunbeck [9] extended stability investigations to the systems containing diamagnetic materials. The characteristic feature of a diamagnetic material is that it opposes the external field variations, the feature exhibited by the superconductors. Since their inception, superconducting levitation systems have been used in many different industrial applications. The principle of these systems is based on the interaction between the magnetic field and high-temperature superconductors [10,11]. Two examples of superconducting levitation systems are the Miyazaki and Yamanashi Maglevs. In the early stages of Maglev development, at the Miyazaki test track (1977), a purely repulsive electrodynamic suspension system was used [12]. A major advantage of the repulsive electrodynamic suspension system is its inherent stability—a decreasing distance between the track and the vehicle results in strong reactive forces bringing the system to its original position. The magnetic field can be produced by either superconducting magnets (as in JR-Maglev) or an array of permanent magnets (as in Inductrack). The disadvantage of the electrodynamic suspension is that the repulsive forces are speed dependent and are low at low vehicle speeds. For this reason, the vehicle must use support wheels until it reaches take-off speed.

In [13], the hyperloop all-in-one advanced LIM system (propulsion, levitation, and guidance) was proposed. The Superconducting Transverse Flux Linear Motor with integrated levitation, guidance, and propulsion system was described in [14]. Another superconducting levitation system for linear drives was proposed in [15].

In 2015, an SC-Maglev train operated by the Central Japan Railway Company (JR Central) broke the train speed world record by clocking in at 603 km/h (374 mph); a new Chinese Maglev system intended for speeds up to 620 km/h was unveiled in January 2021 by CRRC in Chengdu.

2.2. Non-Levitated LTS

This type of LTS uses conventional wheels and is the most typical solution of linear transportation systems [1,16–18]. According to the electrical drive system, these linear transportation systems can be driven by DC motors and synchronous or induction motors—conventional or superconducting.

2.3. LTS with Synchronous Motors

Various levitated LTSs using conventional synchronous motors based propulsion were described in [19], e.g., they are permanent-magnet-excited machines. A high-temperature superconducting linear synchronous motor was described in [20]. The application of superconducting linear flux-switching permanent magnet motors was discussed in [21].

2.4. LTS with Induction Motors

Linear induction machines were discussed as the most promising solution for LTSs in [1,15,17,18]. In the late 1940s, the British electrical engineer Eric Laithwaite, Professor at Imperial College London, developed the first full-size working prototype of the linear induction motor. The first commercial Maglev transport system in the world, the Birmingham Maglev, which opened in 1984, was also propelled with a linear induction motor. The latest problems related to the application of LIMs are discussed in Section 3.

2.5. LTS with Superconducting Induction Motors

When it comes to LTSs, superconductivity can not only be used for levitation but also for generating a tractive effort [10,21–26]. Superconducting motors have their windings made of low-temperature, conventional, or high-temperature superconductors. A typical linear induction motor with an iron core and copper winding can produce only a limited thrust because of the flux saturation of the iron core. High-temperature superconducting windings can generate a strong magnetic field and, consequently, large thrust. High flux density and high thrust can be produced over a wide gap range because of extremely high ampere turns (see Section 4.5).

Some general problems connected with the electromagnetic fields related to high-speed LTSs driven by different linear motors (synchronous, induction, superconducting) were described in [27].

3. Linear Transportation Systems Using Induction Motors

A LIM can be obtained from its rotary counterpart, the induction motor, by an imaginary process of cutting the rotary's stator and rotor in a radial plane and unrolling it, at the same time as replacing a cage or a winding with a conducting sheet, as in Figure 1.

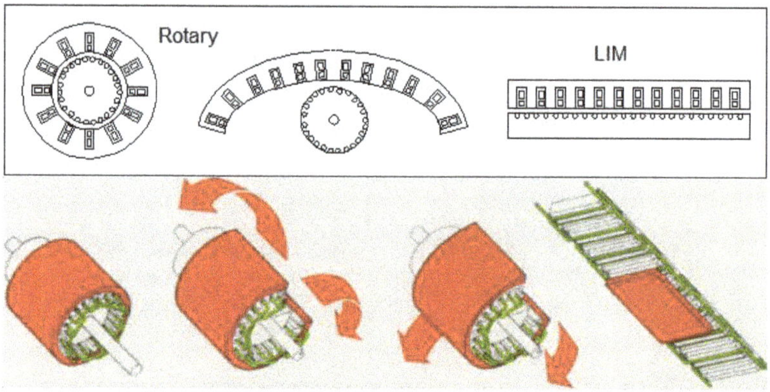

Figure 1. Conversion of rotary machine into LIM.

Should the second primary be added to the single-sided LIM in Figure 1, a double-sided LIM would be formed. Depending on the relative length of the secondary and primary, the LIM can be categorized as a short-secondary (Figure 2) or a short-primary LIM.

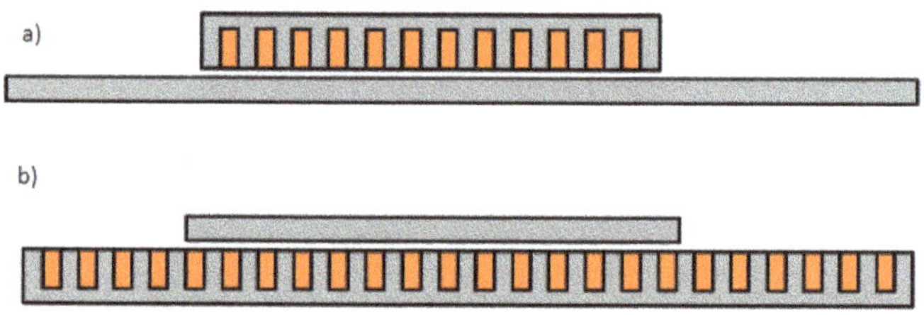

Figure 2. Short-primary LIM (**a**); short-secondary LIM (**b**).

The rotary motor can be thought of as "infinite" in that its primary winding generated magnetic field is continuous and has no beginning or end around its circumference. Unlike the rotary, the short-primary LIM has a finite length. Thus, only the part of the secondary side that is immediately below the primary is subjected to a primary generated magnetic field. During motion, the new unexcited parts of the secondary side equivalent "rotor" continually enter under the LIM primary magnetic field generated by a distributed Magnetomotive Force (MMF). This process generates a continuous electromagnetic response in the new incoming segments of the secondary, the Reaction Rail (RR), in a form of induced MMF, thereby resisting the immediate establishment of the magnetic flux under the front end of the LIM primary. Subsequently, the reaction rail MMF decays but at a lower rate dictated by the "rotor time constant" of the motor. Figure 3a shows a short-primary LIM, and Figure 3b shows the reaction rail. The RR consists of a series of aluminum top cap segments, connected for electric continuity, and underlying iron bars, the Back Iron (BI), corresponding to the conventional rotor winding and rotor laminations, respectively.

(a) (b)

Figure 3. LIM primary (supplied part) (**a**); LIM secondary (reaction rail) (**b**) [28].

LIMs can be further classified into a number of other topologies, but so far only the single-sided, short-primary linear induction machine has been successfully used in urban transportation systems [1,2,15]. It seems to be a natural choice since the cost of building an active multiphase primary along a multikilometer guideway would render such systems uneconomical. In most existing urban applications, the primary is suspended under the bogie, over a track-installed reaction rail consisting of either solid or laminated mild steel BI covered with an aluminum extrusion supported on an assembly that permits the transfer of forces to the guideway. In most current applications, mainly in South-East Asia, Canada, and the USA, the guideway is of an elevated, right-of-way type requiring a minimal footprint and does not affect other modes of ground transportation.

As already mentioned, the relative motion between the finite-length primary and the infinite secondary induces a dynamic end effect by creating end-effect currents in the aluminum top cap that demagnetize the oncoming end of the motor. The currents produce additional forces and losses that exist even at synchronous speed and increase with vehicle velocity. The static end effect, another LIM-related phenomenon, occurs because of the phase impedance imbalance caused by the finite length of the phase winding. The effect is amplified by the dynamic end effect, which distorts the air-gap magnetic flux density, having a direct effect on the flux linked to the phase windings. The transverse edge effect is yet another phenomenon characteristic of LIMs. Its major source is the longitudinal component of the top cap induced current. The two major impacts of the transverse edge effect are an increase of the equivalent secondary resistance and an uneven flux distribution across the LIM's primary. Because of the dynamics of the vehicle as well as the RR's limited construction accuracy, the reaction rail is usually offset from the longitudinal symmetry line of the primary side of the LIM, leading to decentralized transverse forces and potential lateral instability. The asymmetrical construction of the reaction rail necessitated by the vicinity of switches aggravates this effect.

Many constraints exist in the high-speed urban electric traction LIM application, which requires a large distance between the LIM primary and the secondary side RR. Running rail and truck deflection, rail canting, and wheel wear are the major reasons for using a large air gap with the LIM. For a gap length of ten to fifteen millimeters, the ratio between the air-gap width and the pole pitch is significant and leakage flux is considerable. The values of up to 100 Hz are not uncommon in today's applications of urban traction LIMs. At the operational slip of around 10–15%, the skin effect in the aluminum cap is not completely negligible. Finally, there are unbalanced normal forces, attractive and repulsive, that add additional complexity in the analysis of the optimal gap and the construction of the motor, as they affect the distance between the lowest point of the primary and the top of the reaction rail top cap.

Because of the differences between the LIM and the rotary machine, unconventional analysis techniques and modeling methods have been developed in an attempt to account for the number and magnitude of LIM-characteristic phenomena.

Many methods of LIM calculation, optimization, and control are identical (or very similar) with the methods applicable to rotating induction machines. The electromagnetic calculations of the rotary motor are reasonably simple because of the motor's "infinite" character and the possibility of applying symmetry boundary conditions, thus limiting the solution region and speeding up the calculations even further. Two dimensional calculations assure sufficient accuracy for the performance prediction of a standard rotary motor. However, the LIM is not symmetrical. The phenomena occurring in the front end of the motor are different than in the receding end and therefore the symmetry boundaries cannot be used, which leads to longer calculation times.

3.1. *Analytical Solutions Applied for the LIM Evaluation*

LIMs are made of several major components such as the magnetic steel primary core, distributed three-phase copper winding with a three-phase excitation terminal, aluminum top cap, and magnetic BI. The primary is finely laminated and during LIM electromagnetic calculations, for simplicity, the conductivity of the primary can be set to zero. However, the BI conductivity cannot be assumed as zero because the BI is usually only coarsely laminated—mainly for economic reasons. Instead, for the case of the slab-shaped BI, the BI conductivity must be set to its true material value and a respectively lower value for the case of coarse laminations. This value of conductivity is usually determined based on matching the test LIM's longitudinal performance with the simulation model results. Theoretical methods of assessing the BI conductivity have been proposed but, so far, they only serve as a guidance for the empirical derivation. Figure 4 shows the three-dimensional model of the LIM [29].

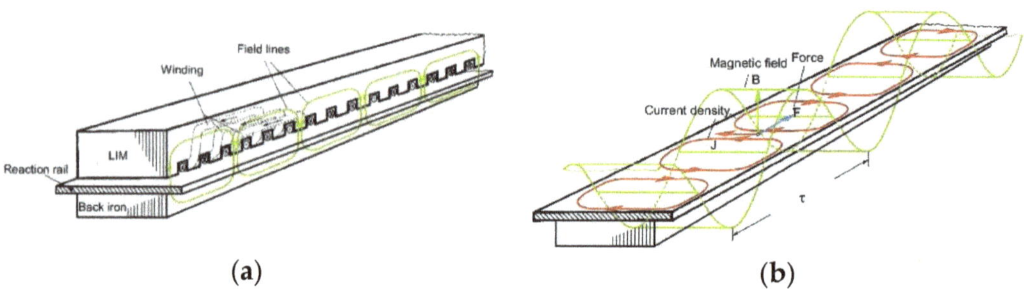

Figure 4. Three-dimensional model of the LIM [29].

The development of analytical methods aimed at solving the LIM problems (separation of variables method, integral equations method) took place in the 1970s. Currently, the main method of analysis is Finite Element Analysis (FEA). Many recently published papers contain both analytical and numerical solutions, where the analytical calculations have mostly been made for validating the results of the numerical models.

Despite significant progress in numerical algorithm efficiency and modern computer speeds, the 3D simulation models of large LIMs are still prohibitively time consuming to solve. The following simplifying assumptions are usually made to the LIM to simplify the calculation process:

1. A 2D analysis can be used;
2. The iron magnetization curve is linear;
3. The conductivity of the reaction rail is constant;
4. Motion in the x-direction only is allowed.

The equation that describes the electromagnetic field distribution within the 2D LIM model has the following form:

$$\frac{\partial^2 A}{\partial x^2} + \frac{\partial^2 A}{\partial y^2} = \mu(-J + j\omega\sigma A + \sigma v_x \frac{\partial A}{\partial x}) \tag{1}$$

where A is the z-component of the magnetic vector potential, J is the z-component of current density, ω is the angular frequency, v_x is the velocity in the x-direction, and σ and μ are the conductivity and permeability, respectively.

Analytical solutions for the evaluation of the LIM properties have been discussed in many papers. In most of them, the standard mathematical approach to LIM modeling is to define the currents of the primary windings as sinusoidally distributed current sheets [17,30–35].

Mathematical and experimental research of coils or filaments moving above a conducting plate (limited to DC excitation) were performed in the majority of the Maglev application studies [36–40]. The system of a stationary filament or coil with AC excitation above a conducting plate has also been studied [41]. In [42], a method of LIM winding optimization has been discussed.

The finite length of a LIM results in a number of well-recognized effects such as the previously mentioned highly speed dependent end effect responsible for the demagnetization of the front end of the machine [17,35]. The end effect has been described by many authors. One description in particular [26], based on an average model of the non-saturated LIM and later expanded to include saturation [2], has been adopted for practical, real-life LIM control algorithms [29,43].

The end effect causes a drop in the effective magnetizing inductance [44,45] and a total motor impedance, which results in a more complex LIM control scheme [43]. Analytical equations defining the end-effect induced magnetizing inductance correction factor are derived in [46]. Extensive studies on the compensation of the magnetizing inductance, reported in [29], propose the adaptation of traction LIM control methods to the depth of demagnetization of the machine in real time. Correction of the end effect for vector controls was also proposed by others, as in [43] and [46], with the equivalent circuit of the LIM based on the formula proposed by Duncan [46]. The equivalent circuit concept developed by Duncan inspired many subsequent investigations [47–52]. In [43] and [53], the control circuit is based on d-axis and q-axis equivalent circuits with parameter correction based on Duncan's method. A fuzzy logic controller is suggested in [54] based on the flux linkages of the primary and secondary. LIM performance calculations of both the motoring and braking forces, impacted by the end effect, based on the space-vector equivalent circuit of the LIM, are calculated in [55–57]. Applications of the field-oriented controls and a model reference adaptive speed observer for LIMs are also reported in [58,59], respectively. The transverse edge effect and the saturation effect are added to Duncan's model in [60]. The last examples prove that any improvement in Duncan's approach have a high potential to enhance subsequent control schemes.

Experimental measurements carried out on a traction LIM demonstrate that the leakage inductance can be, depending on the design of the reaction rail, as high as ten percent of the secondary inductance. The effects of the secondary leakage become significant as the LIM speed increases. The non-immediate current response of the RR due to its non-zero leakage inductance causes the demagnetization to be less pronounced. The assumption, in [26], that only the constant magnetizing current of the primary elicits the RR current response has been adapted by many researchers. In [28], a more accurate assumption is made. It is observed that the induced current in the RR is excited by the sum of the averaged primary magnetizing current and the RR induced current. The resultant steady-state current response is obtained by an application of a recursive algorithm converging to a finite sum of the infinite series. The final form of the sum of the infinite series represents the magnetizing current correction for the equivalent LIM circuit. As has been demonstrated by comparing experimental results with measurement, a derived magnetizing inductance correction factor predicts the LIM performance much better.

The Fourier series method was applied in [61,62] for the LIM evaluations. Instead of simplifying the primary excitation to a form of current sheets, the authors modeled the primary excitations as discrete coils. The discrete coils approach leads to a more intuitive and realistic model of the LIM and allows for the representation of spatial harmonics arising from a discrete current distribution. Frequency domain solver FEA simulations are also used here to validate and cross check the analytical model results. This is important as it establishes the practicality of the FEA frequency domain computation as a preferred replacement for the time-consuming transient computations. First, the evaluation is done by analytically solving a simple pair of filaments moving relative to a RR constructed of aluminum and iron plates and carrying a harmonic current. This solution becomes a building block for modeling a complete LIM. Because of speed and convenience, the analytical model is a practical and efficient way of rapidly assessing the impact of design changes on the performance of the LIM and helps to qualify the adapted FEA solution method.

Figure 5a shows a steady-state 2D calculation model of a discretely distributed coil with AC current excitation at an arbitrary frequency, moving over and in parallel to a conducting plate with an arbitrary speed [61], and Figure 5b shows the extension of this model.

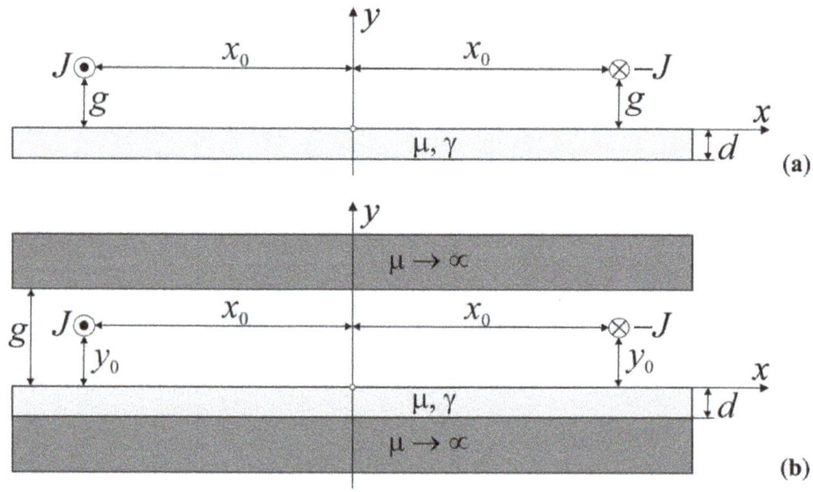

Figure 5. Two filaments $2x_0$ apart, placed at a distance g above the conducting plate of thickness d [61] (**a**). The back iron and primary iron are additionally modeled as two infinite layers with zero conductivity and infinite permeability (**b**).

The approach presented in [61] is similar to the algorithm given in [63,64], where Fourier transform and a mixture of magnetic scalar and vector potential formalism were used. However, the solution presented in [61] makes use of the Fourier series instead of transform and of vector potential formalism without the need of scalar potential to achieve the desired results. The analytical treatment of a 2D LIM model with a primary source moving relative to a conducting plate has many applications, including Maglev and traditional linear propulsion machines.

The analytical approach presented in [62] is similar to the work first presented in [32] and [61], where models of the LIM with current sheet finite primary excitations were presented using Fourier transforms and series methods, respectively. Two papers, [61,62], satisfy the analytical solution for the entire LIM by referring to a vector potential formalism only. The validation of results obtained in [61] and [62] was performed by means of FEA, described in the next chapter.

3.2. FE Methods Applied for Linear Induction Motors

In recent decades, LIM modeling and analysis started relying more on FEA simulations instead of analytical solutions [55,65–69]. Electromagnetic FEA calculations are crucial to optimizing LIM system performance as they can provide results necessary for predicting the end-effect shaped mechanical characteristic—force versus speed. This characteristic is crucial in designing efficient LIM controls as well as traction vehicle functionality. To simplify the FEA model and to minimize the time to numerical solution, the symmetrical three-phase current is typically used. In addition, the non-linear magnetizing characteristics of the LIM primary and back iron are simplified by linearization.

The most typical LIM analysis is the static analysis. One of the challenges that must be solved in the numerical FEA calculation is the proper evaluation of the penetration of the electromagnetic field into the moving and conducting reaction rail. Such modeling and analysis can be extremely difficult and time consuming as it requires a proper choice of the FE mesh, which depends on the velocity of the LIM and slip [70–72].

Because of non-linearity of the magnetizing characteristics and the continuous quasi steady state of LIM operation, time domain (transient) analysis must be performed to achieve a steady state. A transit LIM is a large machine, more than 2 m in length and over 60 cm wide, and even for 2D calculation it demands an extremely high number of mesh nodes. At high speeds, to achieve satisfactory computation accuracy, the time step of a transient analysis must necessarily be small and with the addition of a large distance the LIM must traverse before the steady state has been achieved, which significantly increases the solution space, the transient solution often becomes impractical. To overcome this problem, a recently developed feature of the Maxwell2D software, the translational motion periodic Master-Slave boundary, has been used to make the necessary calculations to render the LIM performance characteristics [73]. A time decomposition method, patented by ANSYS, is yet another attempt to improve on a solution time in electromagnetic transient analysis but even with these advances the time to solution for large a LIM is prohibitively long.

The analytically calculated forces acting on a coil in the horizontal and vertical directions as a function of frequency for different velocities are shown in Figure 6. The comparison between the analytical results and the FEA simulations, performed with COMSOL and ANSYS Maxwell2D software, shows an exceptionally good agreement.

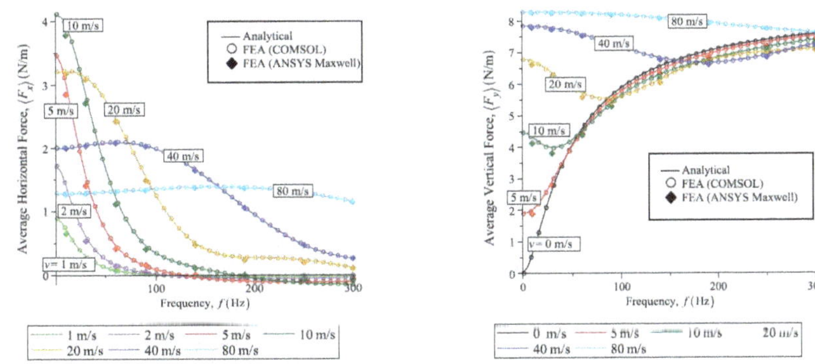

(a) **(b)**

Figure 6. Average horizontal force density (**a**) and average vertical force density (**b**) acting on the coil as a function of frequency for different LIM velocities [61].

LIM 2D models are shown in Figure 7. The analytical approach presented in [62] applies to a simplified 2D LIM model, as shown in Figure 7a. The source vector potential is obtained by summing vector potentials of all current-carrying wires of the LIM winding. The real coils of the motor are modeled to retain their position and currents and to form a complete three-phase, six-pole winding. The modeled winding is a two-layer type, but the analytical model treats the respective top and bottom layer currents as positioned at the same distance from the conducting plate. This was done to make sure that the magnetic reluctance for currents corresponding to two different layers but located in the same slot are identical, which closely approximates the conditions of the real motor.

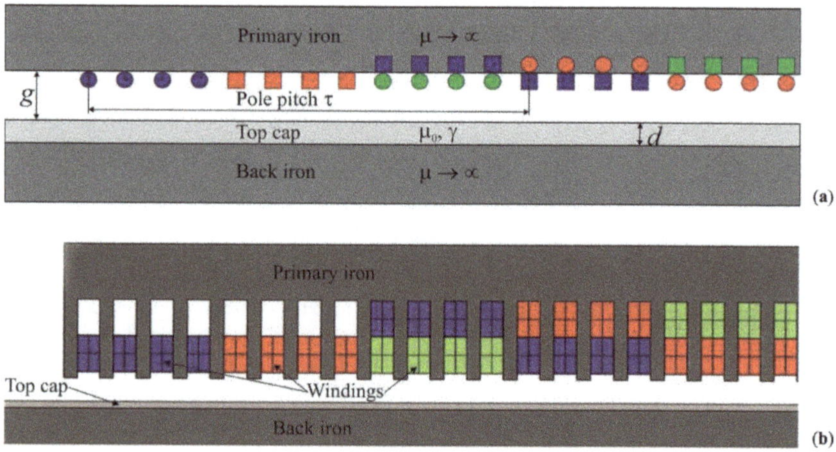

Figure 7. Evolution of 2D models of the LIM used for the numerical field evaluation. Idealized coils of a three-phase, six-pole machine [62] (**a**). Full model of the LIM [70] (**b**).

To further investigate and evaluate the applicability of the analytical solution of a simplified model as a LIM performance prediction tool, a 2D LIM model with teeth and a finite primary, as shown in Figure 7b, was developed and calculated using the FEA simulations.

For comparison, as shown in Figure 8, the performance characteristics obtained by the analytical approach were overlaid with the results generated by FEA simulations (COMSOL and ANSYS Maxwell2D). The agreement between these different calculation methods confirms the accuracy of the applied analytical and numerical methods and models.

The electric vector potential formalism was chosen for the calculations of the back iron power loss [29,74,75] (see Section 4.3). The same formalism was used to determine the magnetic field in the end regions of the induction motors as well as the motors' impedances [74,75]. This approach was also used for the formulation of the 3D equation for the scalar potential describing distribution of the electromagnetic field. The equations were solved analytically (separation of variables method) and numerically (FEA), which made it possible to determine the impedance of the windings for different boundary conditions defined on the surface of the region of analysis.

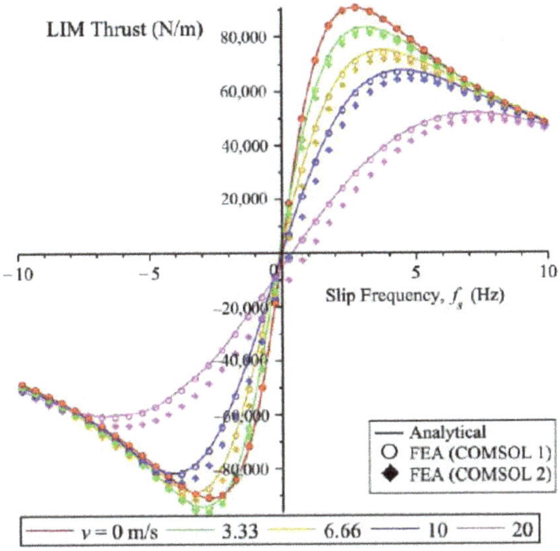

Figure 8. Average thrust per unit depth versus slip frequency for a LIM from Figure 7 [62].

4. Selected Problems of LIM Applications

The following section presents some of the LIM problems that were addressed and solved by the authors. The subject transit LIM is a six-pole, double-layer back wound motor with a pole pitch of 45 cm, a 9 mm mechanical gap, and a 13.5 mm magnetic gap. The reaction rail is made of a 4.5 mm thick aluminum screen over an inch-thick back iron.

4.1. LIM Performance Control; Adaptive Control

The flux vector-oriented control is one of the most advanced and widely accepted methods used for the rotary machine torque control. It was first conceptualized by Blaschke [76] in 1972 and has been a subject of interest of many researchers ever since, e.g., [77–85]. With progress in microprocessor techniques and power electronics, the flux vector-oriented control has become a method of choice for most industrial applications, especially in the development of electric traction propulsion systems, historically based mostly on DC motors, in an effort to replace them with the much less expensive and more robust induction machines. Vector control signifies the independent, or decoupled, control of flux and torque of the motor through coordinated change in the supply voltage and frequency [83]. Flux level control is essential to avoid saturation and minimize core losses under various steady-state operating conditions. As the flux variation tends to be slow, especially with the current control, maintaining constant flux may provide precise torque response and, consequently, a desired speed response.

It is possible to distinguish three flux linkages in the induction machine complex form equations. These flux linkages are the stator flux linkage, the main, or air gap, flux linkage, and the rotor flux linkage. The current decoupling network simplifies only for the rotor flux orientation, whereas the voltage-decoupling network simplifies for the stator flux orientation. Only for a constant rotor flux orientation, the mechanical characteristic does not have a peak value and is a straight line. This linear characteristic is ideal for control application. For a given stator flux in the flux-weakening region and under steady-state operation, however, the stator flux is superior in terms of torque per unit stator current.

Direct Torque Control (DTC) is yet another vector control technique. It was introduced by Depenbrock [84,85] and Takahashi [86] and has been developed by others [87,88]. The fundamental premise of DTC is that a specific DC-link voltage and a specific stator flux establish a unique frequency of inverter operation. This is because the time required by the time integral of the DC-link voltage to integrate up to the reference flux level is unique and represents the half-period time of the frequency of operation. Despite its simplicity, DTC can produce a fast torque response and is robust with respect to transient perturbations and motor parameter detuning [14]. It must also be noted that beside the already mentioned advantages, DTC does not use a modulator and does not employ current control loops, inherent to the vector-oriented flux control. However, during steady-state operation, a pulsation of torque, flux, and current may occur, affecting speed estimation and increasing the acoustic noise. This method is not established so well as the flux-oriented control and has not been applied in LIM controls; however, based on the up-to-date progress in its development, it shows exceedingly high application potential, particularly in the area where parameter sensitivity can be an issue.

Industrial applications of LIM motors require a relatively simple control algorithm because the parameters of industrial process LIMs are well known or can be measured in an off-line experiment. This is not so in urban transit applications since the motors are usually required to operate at peak thrust and the main parameters responsible for the precise peak tracking—the rotor resistance, R_2, and the mutual inductance, M—vary in a very wide range. Thus, the controller of a transit traction LIM should be capable of tracking the maximum available thrust, independent of the air-gap length or the reaction rail construction properties and temperature. Several LIM control methods have been reported thus far, most of them based on the concept of vector control [35]; however, none of them attempt to resolve the parameter adaptation issue. To solve this problem a modified flux vector control technique has been applied [29,43].

The thrust calculated in the rotor flux reference frame compares to measurement only if the rotor parameters, R_2 and L_2, are correctly estimated and their values do not change due to physical or environmental conditions. When these conditions are met, the secondary flux aligns with a d-axis and the back Electro-Motoric Force (EMF) naturally aligns with a q-axis. Should the rotor resistance, R_2, change its value from the set point, the secondary flux would become misaligned and so a non-zero, q-axis component would develop; this means that more voltage is demanded from the supply inverter. This increased voltage generates a negative EMF d-component by advancing the rotor flux. Although the magnitude of the primary current vector remains constant and the secondary flux has increased, the angle between the two vectors has changed and is no longer optimal. A change of the machine secondary resistance from the reference value detunes the controls and a non-zero d-component of the EMF is generated. The optimal operation can be achieved again with the adapted rotor resistance reference value but at a different synchronous frequency. For the magnetizing inductance change, regardless of the cause, e.g., change in air gap, change in the reaction rail geometry, or change in the lamination coarseness or magnetization characteristics of the RR, a q-component of the secondary flux is generated. The induced voltage develops a negative d-component, such as in the case of the secondary resistance detuning. If the value of the reference magnetizing inductance is corrected to equal that of the motor, the controls would become tuned in again and a developed thrust would be optimal, although it would be lower. Since the secondary resistance compensation loop that corrects a d-component of the induced voltage is active, the secondary flux will be regulated to align with the d-axis.

To verify the above parameter compensation control concept in the simulation software, a d–q model of the LIM is first derived, see Figure 9. This LIM signal network clearly shows the impact of parameter detuning on the rotor flux and slip frequency estimation.

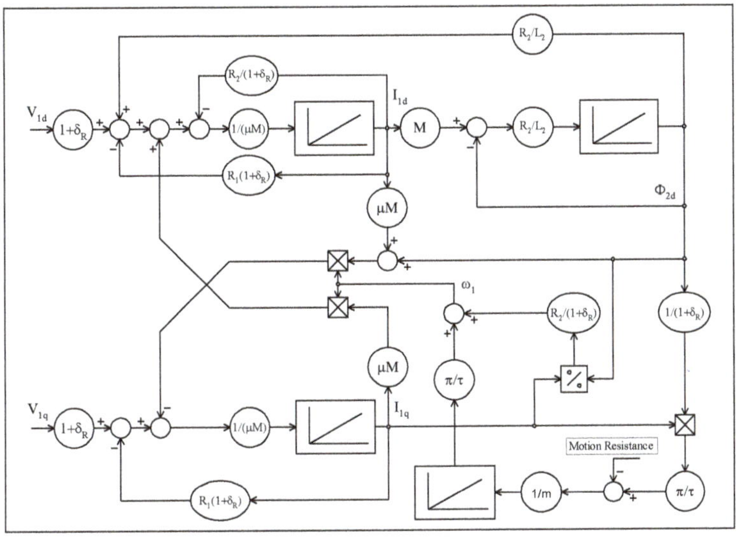

Figure 9. Rotor flux-oriented control LIM model [29,30].

To correlate the model and control variables of the *d–q* system with the real-time three-phase values, standard Clarke and Park transformations are applied [17]. Figure 10a,b shows the simulated response of the system to a step change of R_2 and M both with and without the adaptive compensation control loops.

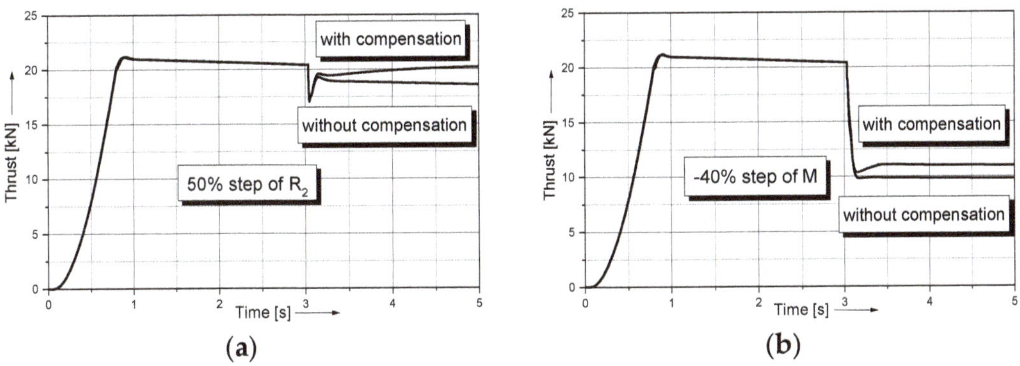

Figure 10. Response to a step of the secondary resistance R_2 (**a**); response to a step of the mutual inductance M (**b**) [29].

The adaptive algorithm improves the performance of the system by a significant margin by improving the mechanical output. The proposed algorithm successfully addresses the problem of LIM parameter detuning while preserving all the positive features of the rotor flux referenced vector control. The verification of the simulation results was performed by comparing the calculated output with measurements from a transit test vehicle fitted with the subject test LIM. This method exhibits the robustness necessary in severe transient conditions associated with the application of the LIM in transportation systems.

4.2. LIM Driven from the Voltage Inverter

In all typical LIM simulation models, the symmetrical three-phase current is fed into the three-phase winding to simulate a constant current mode; however, this does not reflect reality when the LIM is driven from the voltage inverter. The differences in slip versus thrust characteristics between the simplified approach and the approach where an asymmetry of phase currents arises naturally from the real supply conditions are presented in [70].

Typically, the LIM is powered from a PWM voltage inverter, converting thrust command into current at a desired frequency. However, as the phase impedances are unequal and the three-phase currents differ in their phase and magnitude, the negative sequence currents are produced leading to a decreased motor performance. In theory, if the LIM phase impedances were known, the phase currents could be equalized, although not entirely, by a proper phase voltage control, but at a price of increased voltage harmonics. The electromagnetic fields shown below (Figure 11) are calculated considering the natural asymmetry of phase currents under symmetrical voltage excitation.

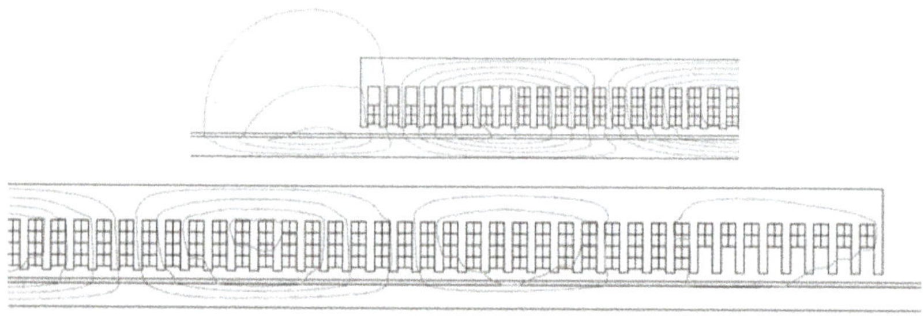

Figure 11. Magnetic field distribution within the LIM [70].

As can be seen from Figure 11, the magnetic field shows significant asymmetry on both ends of the machine. This results in asymmetric coupling and an asymmetric back electromotive force that leads to unequally coupled impedances and the asymmetry of phase currents. Because the phase currents are magnetically coupled with one another and additionally coupled with the induced currents of the reaction rail, these impedances are frequency and speed dependent; thus, their determination can be very involving.

The exemplary performance characteristics of the subject LIM in current and voltage modes for different speeds are shown in Figure 12.

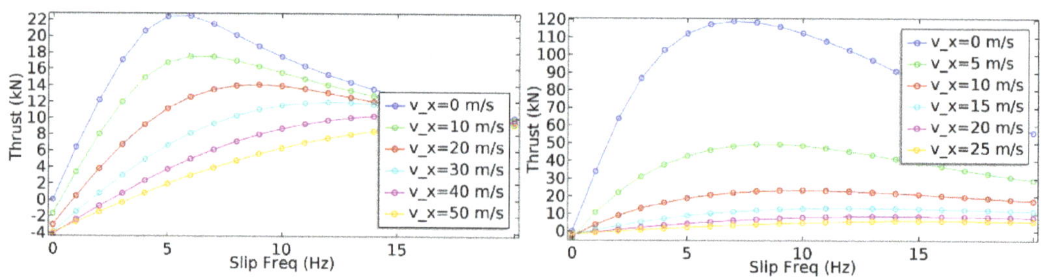

(a) (b)

Figure 12. LIM characteristics obtained for the current supply (I = 550 A) (**a**) and the voltage supply (V = 460 V) (**b**) [70].

The characteristic increase of peak-thrust slip frequency that can be seen in the figure above results from the end-effect induced magnetizing impedance change. As can be seen from above figure, it is important to account for the asymmetry of phase currents when determining the LIM performance. To do so, the electromagnetic transient FEA simulation with the symmetrical three-phase voltage source would have to be used. However, that would become prohibitively time consuming due to a need of remeshing a large solution space at every time step. Alternatively, the quasi-steady-state transient solution can be achieved by simulating in the frequency domain but only if the software allows for the modification of Ampere's law.

4.3. Losses in the Reaction Rail

In order to determine the transverse effects of the current flowing in the reaction rail, another numerical FE model should be applied. This can be done using the electric vector potential T, defined by the formula $rotT = J$, where J denotes the current density vector [74,75]. The differential equation for the electrical vector potential can be written as follows (movement only in x-direction is allowed):

$$\frac{\partial^2 T}{\partial x^2} + \frac{\partial^2 T}{\partial y^2} = \sigma(\frac{\partial B}{\partial t} + v_x \frac{\partial B}{\partial x}) \quad (2)$$

Solving Equation (2) with different values of σ, v_x, and f, it is possible to calculate the current density distribution (and other important parameters of the LIM, such as power losses and forces) for any combination of these parameters. Figure 13 shows an example of the current density distribution in the aluminum rail of the LIM for different slip values of constant frequency.

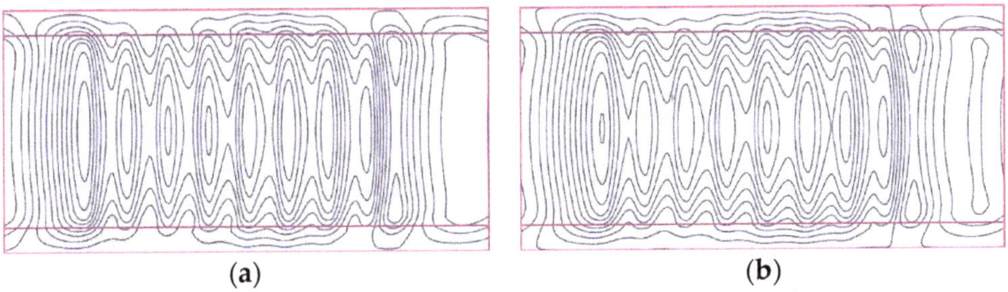

Figure 13. Current density distribution in the rail of the LIM with constant frequency and different slip values [29].

The electric vector potential method can also be applied for the calculation of eddy current distribution in the copper sheet of the rotor due to skewed armature slots (Figure 14a) and for the analysis of influence of the rotor slits on the eddy current distribution in the rotor of rotating induction machines (Figure 14b).

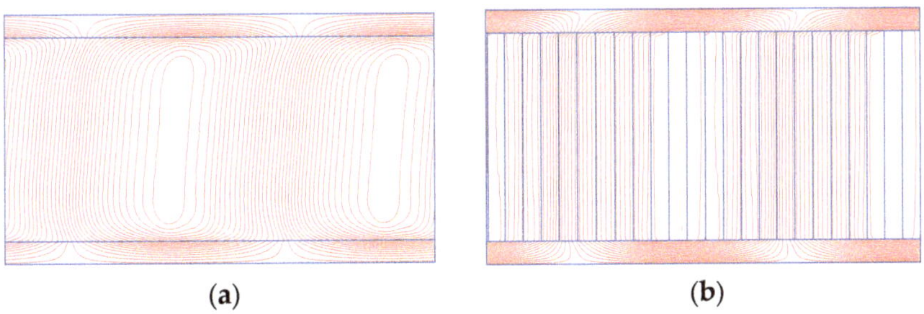

Figure 14. Eddy current distribution in the copper sheet of the rotor due to skewed armature slots (**a**). Influence of the rotor slits on eddy current distribution in the rotor (**b**) [89,90].

The method presented here has a broader meaning. It is a practical tool that enables the analysis of the power losses in the LIM reaction rail and their minimization. Similar results have also been presented in [55].

4.4. Real-Time Temperature Rise Prediction of a Traction LIM

The enclosed structure of the LIM primary, its continuous movement over the reaction rail, and a large gap between the primary and the reaction rail effectively isolate the primary from the heat flux generated in the RR. For all electric and electronic components and devices for which the heat losses can be determined solely from the electric current measurement, such as cables, inductors, relay coils, linear motors, actuators, etc., the maximum nominal temperature rise above ambient can be directly associated with the value of nominal current. However, the device's heat-producing current often varies, due to load changes, ranging from zero to values exceeding the value of the nominal current by a large factor. Such complex load cycles can produce highly variable thermal cycling and result in uncontrolled over-temperature, which negatively impacts the life cycle of a device. During the acceleration and deceleration of a LIM-powered electric traction vehicle, the instantaneous current of the motor exceeds, by a high margin, the nominal thermal value and then decreases below that value during coasting and drops to zero at station stops, therefore undergoing severe thermal cycling. The basic heat equation describes the balance between the dissipated, stored, and radiated heat. The dissipated heat will be partially transferred into the surrounding ambient and partially stored in the heated device by increasing its temperature. So long as the heat power is constant, the temperature rise will achieve a maximum value of ΔT_{max}, and the thermal transient state will be described by the following equation [91]:

$$\Delta T = \Delta T_{max}\left(1 - e^{-\frac{t}{\tau}}\right) + \Delta T_0 e^{-\frac{t}{\tau}} \tag{3}$$

where ΔT_0 is the temperature rise over ambient at $t = 0$, ΔT_{max} is the maximum temperature rise over ambient for $t \to \infty$, and τ is a characteristic constant (thermal time constant).

Equation (3) is frequently used to determine an initial temperature rise estimation of a device undergoing thermal cycling by solving it for the thermal cycle average rms current. However, this approach leads to an error in temperature prediction as it does not account for the temperature dependence of a heat power source. This problem was solved in [91] by correcting and then solving the conventional differential equation/algorithm and providing an exact solution that utilized a single heat-run-test data point. The temperature prediction performance of the basic and improved equations was analyzed by calculating the temperature rise of a linear motor subjected to a typical, nominal service duty cycle. The simulation took into consideration the result of a thermal test performed during a thermal qualification of the subject LIM. The simulation results were further compared against the results of temperature measurement taken from a thermal sensor imbedded in a production LIM's winding while operating the LIM-powered vehicle with the nominal load on the existing Vancouver Expo line system. The experimental data in the form of motor temperature and phase current were collected and overlayed with the simulation results. The results (see Figure 15) confirm that the improved algorithm (blue) predicts the LIM temperature rise (green) with much better accuracy than the basic algorithm (red). The measured temperature is still lower than the improved prediction, mainly due to an additional cooling effect resulting from the increased convection of a moving train. Because of a high degree of prediction accuracy and minimal application cost, the improved software algorithm found practical application and had been installed fleet wide.

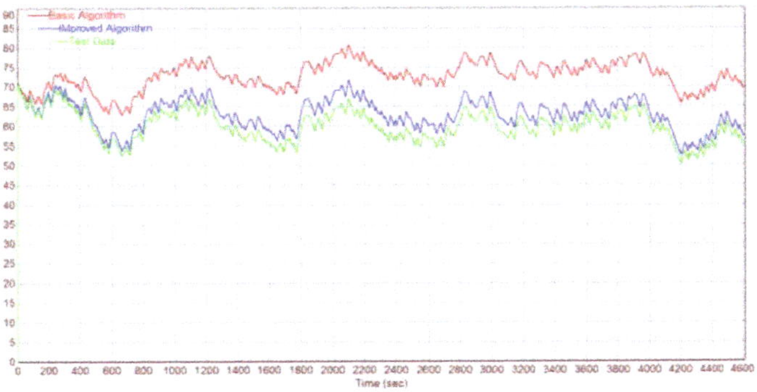

Figure 15. Comparison of different algorithms of temperature rise versus measured temperature results [91].

The efficiency of a traction LIM motor, as has already been mentioned, is rather low and in the case of a loss of forced cooling high losses can drive the temperature of the center of the winding quickly to above the maximum allowed level. Precise detection of this process is important as it allows the train control system to maintain the LIM in operation for as long as possible without compromising the insulation life cycle. Figure 16 shows the typical dependencies of power factor and efficiency for a traction LIM applied on most urban transportation systems in the world. These results were confirmed by experimental measurements taken by one of the authors for the motor, which now operates on a Rapid KL Rail System in Kuala Lumpur, Malaysia. It characterizes similarly built traction LIMs of comparable size and cooling efficiency.

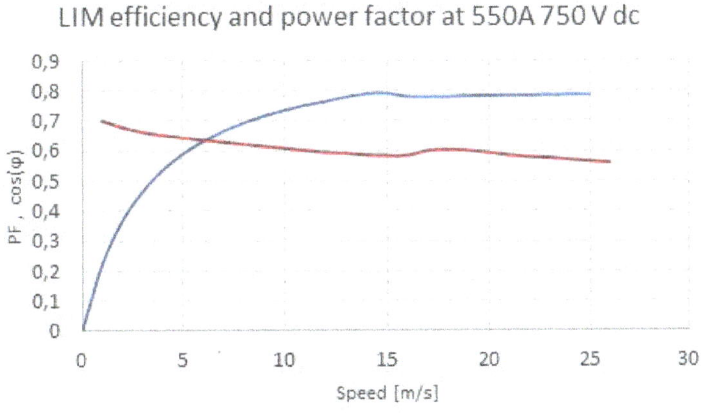

Figure 16. Power factor and efficiency curves for one example of a traction LIM.

4.5. Superconductig LIM

A typical linear induction motor with an iron core and copper windings can only produce limited thrust because of the saturation of the iron core and, in particular, its teeth. Superconducting Linear Induction Motors (SLIMs) are a promising alternative to conventional traction solutions. A novel high-temperature superconducting (HTSC) SLIM was proposed in [26]. This SLIM uses stacks of second generation (2G) superconducting tapes. Such a SLIM, capable of high ampere turns, can generate a strong magnetic field and, consequently, exceptionally large thrust as it can achieve high flux densities over wide air gaps. Commercial 2G HTS tapes utilizing yttrium- and gadolinium-based ceramics

(YBCO) can operate up to critical temperatures of 77 K, which can be provided by liquid nitrogen refrigerants. They can carry a critical current of 600 A at 77 K and self-field. These properties make 2G tapes a promising material for use in power electric equipment, including rotating and linear induction machines [23–25].

The conventional variant of such a SLIM has already been analyzed in [29,61,70]. The only additional component of the SLIM is the cryostat, as in [24].

To determine all the crucial parameters of the SLIM, the FEA has been applied. One of the challenges that must be solved is a proper evaluation of penetration of the strong electromagnetic field into the moving and conducting reaction rail. Such modeling and analysis can be extremely difficult and time consuming as it requires a proper choice of the FE mesh (depending on the velocity and slip) [70]. Because of the strong saturation of the laminated magnetic core, classical LIM construction methods may be put into question.

The current density within the HTSC coil was modeled according to the power law for superconducting windings [14]. Figure 17a shows an example of the magnetic field distribution within the subject SLIM, and Figure 17b shows the thrust characteristics for different speeds.

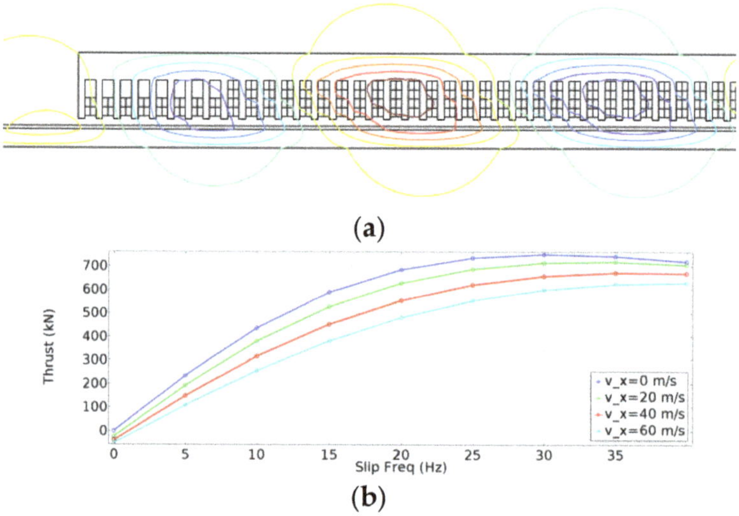

Figure 17. Magnetic field distribution within the end part of the SLIM (**a**). SLIM characteristics for different speeds (**b**) [26].

As can be seen from the figures above, the superconducting LIM has significantly increased thrust values compared to a conventional solution. An important computational problem here is the correct modeling of the superconducting windings in the LIM, as well as the correct consideration of the strong saturation of the magnetic circuit.

5. Conclusions

The paper provides an overview of different linear transportation systems and focuses on the applications that use linear induction motors. Against this background, the authors have presented and discussed new practical methodologies capable of solving some important transportation LIM problems. Despite the LIM's lower efficiency, when compared with the rotary motor, for many applications the LIM system is a superior transport solution that successfully competes against the conventional, rotary-motor-based alternative. This is because the efficiency is a broader concept and the efficiency of a transportation system must be analyzed in the context of the application. For elevated, driverless automated systems, the LIM is indeed a superior solution because it does not rely on adhesion, has no moving parts, and provides the lowest life cycle and operation and maintenance costs.

Based upon the experience gathered in the subject area of LIMs, the authors believe that future research work should concentrate on increasing the motor efficiency by improving the construction materials and production technology and researching the application of high-temperature superconductivity. This progress must be accompanied by the improvement in efficiency of predictive algorithms and more efficient FEA methods.

LIMs have been known and widely researched by the scientific community but mostly as standalone electric motors. However, LIMs often work as parts of an overly complex transportation system. Thus, future research should take into account the complex interaction of the LIM with its specific system environment.

Author Contributions: Conceptualization, R.P. and K.W.; investigation, R.P. and K.W.; resources, R.P. and K.W.; writing—original draft preparation, R.P. and K.W.; writing—review and editing, R.P. and K.W. All authors have read and agreed to the published version of the manuscript.

Funding: This research received no external funding.

Institutional Review Board Statement: Not applicable.

Informed Consent Statement: Not applicable.

Data Availability Statement: Not applicable.

Conflicts of Interest: The authors declare no conflict of interest.

References

1. Hellinger, R.; Mnich, P. Linear motor-powered transportation: History, present status, and future outlook. *Proc. IEEE* **2009**, *97*, 1892–1900. [CrossRef]
2. Woronowicz, K.; Safaee, A. A novel linear induction motor equivalent-circuit with optimized end effect model. *Can. J. Electr. Comput. Eng.* **2014**, *37*, 34–41. [CrossRef]
3. Woronowicz, K. Urban Mass Transit Systems: Current Status and Future Trends. In Proceedings of the ITEC 2014, Cologne, Germany, 13 August 2014.
4. Byers, D.C. The Linear Induction Motor in Transit, a System View. In Proceedings of the 5th International Conference on Automated People Movers, Paris, France, 10–14 June 1996.
5. Earnshaw, S. On the Nature of the Molecular Forces which regulate the Constitution of the Luminiferous Ether. *Trans. Camb. Philos. Soc.* **1842**, *7*, 97–112.
6. Weh, H.; Steingröver, A.; Hupe, H. Design and Simulation of a Controlled Permanent Magnet. *Int. J. Appl. Electromagn. Mater.* **1992**, *3*, 199–203.
7. Palka, R. Synthesis of magnetic fields by optimization of the shape of areas and source distributions. *Electr. Eng.* **1991**, *75*, 1–7. [CrossRef]
8. Sikora, R.; Palka, R. Synthesis of Magnetic-Fields. *IEEE Trans. Magn.* **1982**, *18*, 385–390. [CrossRef]
9. Braunbeck, W. Freely Suspended Bodies in Electric and Magnetic Fields. *Z. Phys.* **1938**, *112*, 753–763.
10. May, H.; Palka, R.; Portabella, E.; Canders, W.-R. Evaluation of the magnetic field – high temperature superconductor interactions. *Int. J. Comput. Math. Electr. Electron. Eng.* **2004**, *23*, 286–304. [CrossRef]
11. Patel, A.; Hopkins, S.; Giunchi, G.; Figini Albisetti, A.; Shi, Y.; Palka, R.; Cardwell, D.; Glowacki, B. The Use of an MgB_2 Hollow Cylinder and Pulse Magnetized (RE)BCO Bulk for Magnetic Levitation Applications. *IEEE Trans. Appl. Supercond.* **2013**, *23*, 6800604. [CrossRef]
12. Masada, E. Development of Maglev and Linear Drive Technology for Transportation in Japan. In Proceedings of the MAGLEV'95, Bremen, Germany, 26–29 November 1995; ISBN 3-8007-2155-4.
13. Ji, W.; Jeong, G.; Park, C.; Jo, I.; Lee, H. A Study of Non-Symmetric Double-Sided Linear Induction Motor for Hyperloop All-In-One System (Propulsion, Levitation, and Guidance). *IEEE Trans. Magn.* **2018**, *54*, 1–4. [CrossRef]
14. Ma, G.; Wang, Z.; Liu, K.; Qian, H.; Wang, C. Potentials of an Integrated Levitation, Guidance, and Propulsion System by a Superconducting Transverse Flux Linear Motor. *IEEE Trans. Ind. Electron.* **2018**, *65*, 7548–7557. [CrossRef]
15. Lee, C.Y.; Jo, J.M.; Han, Y.J.; Bae, D.K.; Yoon, Y.S.; Choi, S.; Park, D.K.; Ko, T.K. Estimation of Current Decay Performance of HTS Electromagnet for Maglev. *IEEE Trans. Appl. Supercond.* **2010**, *20*, 907–910.
16. Woronowicz, K. Linear motor drives and applications in rapid transit systems. In Proceedings of the Transportation Electrification Conference and Expo (ITEC), Detroit, MI, USA, 15–18 June 2014.
17. Nasar, S.A.; Boldea, I. *Linear Motion Electric Machines*; John Wiley & Sons: New York, NY, USA, 1976.

18. Cao, R.; Lu, M.; Jiang, N.; Cheng, M. Comparison between Linear Induction Motor and Linear Flux-Switching Permanent-Magnet Motor for Railway Transportation. *IEEE Trans. Ind. Electron.* **2019**, *66*, 9394–9405. [CrossRef]
19. Weh, H. Linear Electromagnetic Drives in Traffic Systems and Industry. In Proceedings of the LDIA'95, Nagasaki, Japan, 31 May–2 June 1995.
20. Cao, R.; Lu, M. Investigation of High Temperature Superconducting Linear Flux-Switching Motors With Different Secondary Structures. *IEEE Trans. Appl. Supercond.* **2020**, *30*, 1–5. [CrossRef]
21. Ma, G.; Gong, T.; Zhang, H.; Wang, Z.; Li, X.; Yang, C.; Liu, K.; Zhang, W. Experiment and Simulation of REBCO Conductor Coils for an HTS Linear Synchronous Motor. *IEEE Trans. Appl. Supercond.* **2017**, *27*, 1–5. [CrossRef]
22. Liu, B.; Badcock, R.; Shu, H.; Fang, J. A Superconducting Induction Motor with a High Temperature Superconducting Armature: Electromagnetic Theory, Design and Analysis. *Energies* **2018**, *11*, 792. [CrossRef]
23. Liu, J.; Ma, H.; Cheng, J. Research on Superconducting Induction Maglev Linear Machine for Electromagnetic Launch Application. *IEEE Trans. Plasma Sci.* **2017**, *45*, 1644–1650. [CrossRef]
24. Sotelo, G.G.; Sass, F.; Carrera, M.; Lopez-Lopez, J.; Granados, X. Proposal of a Novel Design for Linear Superconducting Motor Using 2G Tape Stacks. *IEEE Trans. Ind. Electron.* **2018**, *65*, 7477–7484. [CrossRef]
25. Chen, X.; Zheng, S.; Li, J.; Ma, G.T.; Yen, F. A linear induction motor with a coated conductor superconducting secondary. *Phys. C Supercond. Appl.* **2018**, *550*, 82–84. [CrossRef]
26. Palka, R.; Woronowicz, K.; Kotwas, J. Analysis of the Performance of Superconducting Linear Induction Motors. In Proceedings of the CEFC 2020, Pisa, Italy, 16–19 November 2020.
27. Kircher, R.; Klühspies, J.; Palka, R.; Fritz, E.; Eiler, K.; Witt, M. Electromagnetic Fields Related to High Speed Transportation Systems. *Transp. Syst. Technol.* **2018**, *4*, 152–166. [CrossRef]
28. Woronowicz, K.; Safaee, A. A novel linear induction motor equivalent-circuit with optimized end-effect model including partially-filled end slots. In Proceedings of the Transportation Electrification Conference and Expo (ITEC), Detroit, MI, USA, 15–18 June 2014.
29. Woronowicz, K.; Palka, R. Optimised Thrust Control of Linear Induction Motors by a Compensation Approach. *Int. J. Appl. Electromagn. Mech.* **2004**, *19*, 533–536. [CrossRef]
30. Woronowicz, K.; Palka, R. An advanced linear induction motor control approach using the compensation of its parameters. In *Electromagnetic Fields in Electrical Engineering*; IOS Press: Amsterdam, The Netherlands, 2002; pp. 335–338.
31. Gieras, J.; Dawson, G.; Eastham, A. Performance calculation for single-sided linear induction motors with a double-layer reaction rail under constant current excitation. *IEEE Trans. Magn.* **1986**, *22*, 54–62. [CrossRef]
32. Mendrela, E.A.; Gierczak, E. Two-dimensional analysis of linear induction motor using Fourier's series method. *Electr. Eng.* **1982**, *65*, 97–106. [CrossRef]
33. Nasar, S.A.; del Cid, L. Certain approaches to the analysis of single-sided linear induction motors. *Proc. Inst. Electr. Eng.* **1973**, *120*, 477–483. [CrossRef]
34. Freeman, E.M.; Lowther, D.A. Normal force in single-sided linear induction motors. *Proc. Inst. Electr. Eng.* **1973**, *120*, 1499–1506. [CrossRef]
35. Yamamura, S. *Theory of Linear Induction Motors*, 2nd ed.; Halster Press: New York, NY, USA, 1979.
36. Borcherts, R.H.; Reitz, J.R. High-speed transportation via magnetically supported vehicles: A study of the magnetic forces. *Transp. Res.* **1971**, *5*, 197–209. [CrossRef]
37. Fink, H.J. Instability of vehicles levitated by eddy current repulsion-case of an infinitely long current loop. *J. Appl. Phys.* **1971**, *42*, 3446. [CrossRef]
38. Reitz, J.R. Force on a rectangular coil moving above a conducting slab. *J. Appl. Phys.* **1972**, *43*, 1547. [CrossRef]
39. Langerholc, J. Torques and forces on a moving coil due to eddy currents. *J. Appl. Phys.* **1973**, *44*, 1587. [CrossRef]
40. Lee, S.-W.; Menendez, R. Force on current coils moving over a conducting sheet with application to magnetic levitation. *Proc. IEEE* **1974**, *62*, 567–577.
41. Tegopoulos, J.A.; Kriezis, E.E. *Eddy Currents in Linear Conducting Media*; Elsevier Science: New York, NY, USA, 1985.
42. Adamiak, K. A method of optimization of winding in linear induction motor. *Electr. Eng.* **1986**, *69*, 83–91. [CrossRef]
43. Kim, D.-K.; Kwon, B.-I. A novel equivalent circuit model of the linear induction motor based on finite element analysis and its coupling with external circuits. *IEEE Trans. Magn.* **2006**, *42*, 3407–3409. [CrossRef]
44. Gieras, J.F.; Dawson, G.E.; Easthan, A.R. A new longitudinal end effect factor for linear induction motor. *IEEE Trans. Energy Convers.* **1987**, *2*, 152–159. [CrossRef]
45. Satvati, M.R.; Vaez-Zadeh, S. End-effect compensation in linear induction motor drives. *J. Power Electron.* **2011**, *11*, 697–703. [CrossRef]
46. Duncan, J. Linear induction motor—Equivalent-circuit model. *IEEE Proc. B Electr. Power Appl.* **1983**, *130*, 51–57. [CrossRef]
47. Xu, W.; Zhu, J.G.; Zhang, Y.; Li, Y.; Wang, Y.; Guo, Y. An Improved Equivalent Circuit Model of a Single-Sided Linear Induction Motor. *IEEE Trans. Veh. Technol.* **2010**, *59*, 2277–2289. [CrossRef]
48. Xu, W.; Zhu, J.G.; Zhang, Y.; Li, Z.; Li, Y.; Wang, Y.; Guo, Y.; Li, Y. Equivalent Circuits for Single-Sided Linear Induction Motors. *IEEE Trans. Ind. Appl.* **2010**, *46*, 2410–2423. [CrossRef]
49. Lv, G.; Zeng, D.; Zhou, T. An Advanced Equivalent Circuit Model for Linear Induction Motors. *IEEE Trans. Ind. Electron.* **2018**, *65*, 7495–7503. [CrossRef]

50. Zeng, D.; Lv, G.; Zhou, T. Equivalent circuits for single-sided linear induction motors with asymmetric cap secondary for linear transit. *IEEE Trans. Energy Convers.* **2018**, *33*, 1729–1738. [CrossRef]
51. Dmitrievskii, V.; Goman, V.; Sarapulov, F.; Prakht, V.; Sarapulov, S. Choice of a numerical differentiation formula in detailed equivalent circuits models of linear induction motors. In Proceedings of the 2016 International Symposium on Power Electronics, Electrical Drives, Automation and Motion (SPEEDAM), Capri, Italy, 22–24 June 2016; pp. 458–463.
52. Zare-Bazghaleh, A.; Naghashan, M.; Khodadoost, A. Derivation of Equivalent Circuit Parameters for Single-Sided Linear Induction Motors. *IEEE Trans. Plasma Sci.* **2015**, *43*, 3637–3644. [CrossRef]
53. Sung, J.-H.; Nam, K. A new approach to vector control for a linear induction motor considering end effects. In Proceedings of the IEEE Industry Applications Society, Thirty-Fourth IAS Annual Meeting, Phoenix, AZ, USA, 3–7 October 1999; Volume 4, pp. 2284–2289.
54. Huang, C.-I.; Hsu, K.-C.; Chiang, H.-H.; Kou, K.-Y.; Lee, T.-T. Adaptive fuzzy sliding mode control of linear induction motors with unknown end effect consideration. In Proceedings of the International Conference on Advanced Mechatronic Systems, Tokyo, Japan, 18–21 September 2012; pp. 626–631.
55. Shiri, A.; Shoulaie, A. Design optimization and analysis of single-sided linear induction motor, considering all phenomena. *IEEE Trans. Energy Convers.* **2012**, *27*, 516–525. [CrossRef]
56. Pucci, M. Direct field oriented control of linear induction motors. *Electr. Power Syst. Res.* **2012**, *89*, 11–22. [CrossRef]
57. Shiri, A.; Shoulaie, A. End effect braking force reduction in high-speed single-sided linear induction machine. *Energy Convers. Manag.* **2012**, *61*, 43–50. [CrossRef]
58. Kang, G.; Nam, K. Field-oriented control scheme for linear induction motor with the end effect. *IET Electr. Power Appl.* **2005**, *152*, 1565–1572. [CrossRef]
59. Accetta, A.; Cirrincione, M.; Pucci, M.; Vitale, G. MRAS speed observer for high performance linear induction motor drives based on linear neural networks. In Proceedings of the IEEE Energy Conversion Congress and Exposition, Phoenix, AZ, USA, 17–22 September 2011; pp. 1765–1772.
60. Mirsalim, M.; Doroudi, A.; Moghani, J.S. Obtaining the operating characteristics of linear induction motors: A new approach. *IEEE Trans. Magn.* **2002**, *38*, 1365–1370. [CrossRef]
61. Abdelqader, M.; Morelli, J.; Palka, R.; Woronowicz, K. 2-D quasi-static solution of a coil in relative motion to a conducting plate. *Int. J. Comput. Math. Electr. Electron. Eng.* **2017**, *36*, 980–990. [CrossRef]
62. Woronowicz, K.; Abdelqader, M.; Palka, R.; Morelli, J. 2-D quasi-static Fourier series solution for a linear induction motor. *Int. J. Comput. Math. Electr. Electron. Eng.* **2018**, *37*, 1099–1109. [CrossRef]
63. Freeman, E.M.; Papageorgiou, C.D. Spatial Fourier transforms: A new view of end effects in linear induction motors. *Proc. Inst. Electr. Eng.* **1978**, *125*, 747–7753. [CrossRef]
64. Paudel, N.; Bird, J.Z. General 2-D steady-state force and power equations for a traveling time-varying magnetic source above a conductive plate. *IEEE Trans. Magn.* **2012**, *48*, 95–100. [CrossRef]
65. Lee, S.; Lee, H.; Ham, S.; Jin, C.; Park, H.; Lee, J. Influence of the construction of secondary reaction plate on the transverse edge effect in linear induction motor. *IEEE Trans. Magn.* **2009**, *45*, 2815–2818.
66. Singh, M.; Khajuria, S.; Marwaha, S. Thrust analysis and improvement of single sided linear induction motor using finite element technique. *Int. J. Curr. Eng. Technol.* **2013**, *3*, 563–566.
67. Jeong, J.; Lim, J.; Park, D.; Choi, J.; Jang, S. Characteristic analysis of a linear induction motor for 200-km/h maglev. *Int. J. Railw.* **2015**, *8*, 15–20. [CrossRef]
68. Abdollahi, S.E.; Mirzayee, M.; Mirsalim, M. Design and analysis of a double-sided linear induction motor for transportation. *IEEE Trans. Magn.* **2015**, *51*, 1–7. [CrossRef]
69. Amiri, E.; Mendrela, E.A. A novel equivalent circuit model of linear induction motors considering static and dynamic end effects. *IEEE Trans. Magn.* **2014**, *50*, 120–128. [CrossRef]
70. Palka, R.; Woronowicz, K.; Kotwas, J.; Xing, W.; Chen, H. Influence of different supply modes on the performance of linear induction motors. *Arch. Electr. Eng.* **2019**, *68*, 473–483.
71. De Gersem, H.; Vande Sande, H.; Hameyer, K. Motional magnetic finite element method applied to high speed rotating devices. *Int. J. Comput. Math. Electr. Electron. Eng.* **2000**, *19*, 446–451.
72. De Gersem, H.; Hameyer, K. Finite element simulation of a magnetic brake with a soft magnetic solid iron rotor. *Int. J. Comput. Math. Electr. Electron. Eng.* **2002**, *21*, 296–306. [CrossRef]
73. Introduction to COMSOL Multiphysics, Version 5.6, ©1998–2020 COMSOL. Available online: https://www.comsol.com/documentation (accessed on 25 February 2021).
74. Sikora, R.; Lipinski, W.; Gawrylczyk, K.; Gramz, M.; Palka, R.; Ziolkowski, M. Analysis of the magnetic field in the end region of induction motor. *IEEE Trans. Magn.* **1982**, *18*, 674–678. [CrossRef]
75. Sikora, R.; Gawrylczyk, K.; Gramz, M.; Gratkowski, S.; Ziolkowski, M. Magnetic field Computation in the End-Region of Electric Machines Using Various Boundary Conditions on Iron Surfaces. *IEEE Trans. Magn.* **1986**, *22*, 204–207. [CrossRef]
76. Blaschke, F. The Principle of Field Orientation as Applied to the New Transvector Closed-Loop Control System for Rotating Machines. *Siemens Rev.* **1972**, *34*, 217–220.
77. Trzynadlowski, A. *The Field Orientation Principle in Control of Induction Motors*; Kluwer Academic Publishers: Amsterdam, The Netherlands, 1994.

78. Boldea, I.; Nasar, S.A. *Vector Control of AC Drives;* CRC Press: London, UK, 1992.
79. Leonhard, W. *Control of Electric Drives;* Springer: Berlin, Germany, 1985.
80. Kaźmierkowski, M.P.; Tunia, H. *Automatic Control of Converter-Fed Drives;* Elsevier: Amsterdam, The Netherlands, 1994.
81. Kovács, P.K. *Transient Phenomena in Electric Machines;* Elsevier: Amsterdam, The Netherlands, 1984.
82. Yamamura, S. *AC Motors for High Performance Applications: Analysis and Control;* Marcel Dekker: New York, NY, USA, 1986.
83. Lipo, T.; Novotny, D. Principles of Vector Control and Field Estimation. In *Introduction to Field Orientation and High Performance Drives;* IEEE Tutorial Course: Minneapolis, MN, USA, 1985.
84. Depenbrock, M. Direkte Selbstregelung (DSR) für hochdynamische Drehfeldantriebe mit Stromrichterspeisung. *etz-Archiv* **1985**, *7*, 211–218.
85. Depenbrock, M. Direct Self-Control of the Flux and Rotary Moment of a Rotary-Field Machine. U.S. Patent 4,678,248, 7 July 1987.
86. Takahashi, I.; Noguchi, T. A new quick response and high efficiency control strategy of an induction motor. *IEEE Trans. Ind. Appl.* **1986**, *IA-22*, 820–827. [CrossRef]
87. Tiitinen, P.; Pohjalainen, P.; Lalu, J. The next generation motor control method–Direct torque control (DTC). *Eur. Power Electron. Drives* **1995**, *5*, 14–18.
88. Kazmierkowski, M.; Kasprowicz, A. Improved direct torque and flux vector control of PWM inverter-fed induction motor drives. *IEEE Trans. Ind. Electron.* **1995**, *42*, 344–350. [CrossRef]
89. May, H.; Canders, W.-R.; Palka, R. Loss reduction in synchronous machines by appropriate feeding patterns. In Proceedings of the International Conference on Electrical Machines ICEM '98, Istanbul, Turkey, 2–4 September 1998.
90. May, H.; Palka, R.; Canders, W.-R. Optimized control of induction machines for dynamic torque requirements. In *Electromagnetic Fields in Electrical Engineering;* IOS Press: Amsterdam, The Netherlands, 2002; pp. 254–259.
91. Woronowicz, K.; Safaee, A.; Maknouninejad, A. Enhanced Algorithm for Real Time Temperature Rise Prediction of a Traction LIM. In Proceedings of the Transportation Electrification Conference and Expo, ITEC 2018, Long Beach, CA, USA, 13–15 June 2018; pp. 616–620.